When Historiography Met Epistemology

History of Modern Science

VOLUME 2

The titles published in this series are listed at *brill.com/hims*

When Historiography Met Epistemology

Sophisticated Histories and Philosophies of Science in French-speaking Countries in the Second Half of the Nineteenth Century

By

Stefano Bordoni

BRILL

LEIDEN | BOSTON

Library of Congress Cataloging-in-Publication Data

Names: Bordoni, Stefano, 1989-
Title: When historiography met epistemology : sophisticated histories and philosophies of science in French-speaking countries in the second half of the nineteenth century / by Stefano Bordoni.
Description: Leiden ; Boston : Brill, [2017] | Series: History of modern science, ISSN 2352-7145 ; 2 | Includes bibliographical references.
Identifiers: LCCN 2016053243 (print) | LCCN 2016054350 (ebook) | ISBN 9789004315228 (hardback : alk. paper) | ISBN 9789004315235 (E-book)
Subjects: LCSH: Science–France–Philosophy–History–19th century. | Science–France–History–19th century. | Civilization, Modern–French influences.
Classification: LCC Q175 .B72275 2017 (print) | LCC Q175 (ebook) | DDC 501–dc23
LC record available at https://lccn.loc.gov/2016053243

Typeface for the Latin, Greek, and Cyrillic scripts: "Brill". See and download: brill.com/brill-typeface.

ISSN 2352-7145
ISBN 978-90-04-31522-8 (hardback)
ISBN 978-90-04-31523-5 (e-book)

Contents

Preface

This book deals with the emergence of a sophisticated history and philosophy of science in French-speaking countries in the second half of the nineteenth century. This historical process represents a meaningful stage in the history of European culture.

In 2015 Anastasios Brenner published a collection of texts belonging to the tradition of French epistemology. He focused on the turn of the twentieth century, and he found that meaningful debates "on the nature and value of science" emerged in the last decades of the nineteenth century, more specifically in 1891 with Poincaré's paper "Les géométries non Euclidiennes" [Brenner (ed.) 2015, pp. 5-6].[1]

In December 2010 a conference on "Epistemologie und Geschichte" was organized in Berlin by the *Max-Planck-Institut für Wissenschaftsgeschichte* and the *Institut d'Histoire et de Philosophie des Sciences, Université Paris 1*. It was not the first conference devoted to a historiographical and epistemological approach that aimed at "bridging the gap between history of science and philosophy of science while combining analytic with continental traditions." The scholars who took part in the conference explored the historical roots of that approach: it was taken for granted that "the central reference point" for the close link between the history and philosophy of science could be found "in the French context during the 1930s and 1940s." In particular Gaston Bachelard and Georges Canguilhem's researches played an important role in the development of that attitude [MPIWG 2012, p. 7].

In 2008, the *Max-Planck-Institut* had already hosted the international conference "What (Good) Is Historical Epistemology?" The organizers Thomas Sturm and Uljana Fest pointed out the problematic status of an intellectual practice whose name disclosed its twofold nature. Historical epistemology could be looked upon as either "a branch of the history of science" or "a philosophical project, namely by thoroughly historicising epistemology."[2] During the conference, the nature and features of this practice were extensively debated. According to Lorraine Daston, historical epistemology could be qualified as the study of "[t]he emergence and articulation of new epistemological

1 Brenner also remarked that the first occurrences of terms such as "epistemology" and "conventionalism" appeared at the beginning of the twentieth century (Brenner (ed.) 2015, pp. 9 fn. 5, and 17 (and fn. 21)).

2 For the first occurrence of the expression "historical epistemology," see Rey 1907, p. 13, Braunstein 2012, pp. 35-37, and Brenner 2016, p. 159.

categories and problems out of knowledge practices." She found that it was a specific intellectual practice that could be placed between "the *History of Knowledge Practices*" and "the *History of Epistemology*." She also stressed that it should focus on meaningful turning points or "moments of epistemological novelty" in the history of sciences. Because of this specific commitment, historical epistemology was more defined and slightly different from "the *history of epistemology*" and "the *history of scientific practices*": in brief, its particular feature was to inquire into "the history of how new epistemological problems come into being" [Sturm and Feest 2009, p. 3; Daston 2009, pp. 35-6].[3]

Apart from the evident difficulty of defining exactly this intellectual commitment, we might reasonably say that historical epistemology involves both historiographical frameworks and epistemological issues. In the end, Jürgen Renn recommended his "recipe for a successful historical epistemology":

> Don't dress up philosophical programs with episodic elements of history, epistemology, or psychology, but rather systematically study – with an open mind as to the results – the large-scale "mutual interactions of quasi-autonomous cultural elements," using the words of Michael Friedman, in the light of insight, as they have been emphasised by Peter Barker, into the architecture and dynamics of knowledge by cognitive sciences, augmented though by more than a grain of philosophical sophistication [Renn 2009, pp. 144].

In a series of seminars held in Berlin, and first published in 2007, Hans-Jörg Rheinberger had remarked that the late nineteenth century witnessed "a crisis of reflection on scientific knowledge," but neither "an immediate solution" nor "a generally accepted alternative to the century's legacy" appeared. He found that a solution, or "a broadly articulated new reflection on science," slowly emerged during the twentieth century, when scholars "began to historicize epistemology in various ways."

I find that this historical reconstruction shifts forward a cultural process that took place some time earlier. It was in the second half of the nineteenth century that "the idea of science as a process replaced the obligatory view of science as a system." More specifically, "the historicization of epistemology" represented a meaningful stage in that nineteenth-century process rather

3 In this context, it seems to me that Ian Hacking's proposal to replace the label *historical epistemology* with "historical meta-epistemology" or to make historical epistemology fall under "the generalized concept of historical ontology" is irrelevant [Hacking 2002, pp. 9 and 24].

than "a decisive moment in the transformation of twentieth-century philosophy of science." However I agree with Rheinberger on two specific features of that transformation: "a considerable part of the work of reflection... was conducted within the sciences and by scientists themselves," and alongside the historicization of epistemology, a converse process of "epistemologization of the history of science" actually took place [Rheinberger 2010, pp. 1 and 3-4].

In brief, I would like to show that the chronological starting point should be moved backwards. I claim that the emergence of sophisticated integration between historiography and epistemology can be traced back to the last decades of the nineteenth century. In French-speaking countries, a new intellectual practice and a new body of knowledge slowly emerged, and new historical and philosophical researches were systematically pursued. Those researches involved mathematicians, scientists, and philosophers, and were deeply linked to contemporary processes that were transforming the cultural and material landscape of Europe. We know that a process of professionalization of scientific practice, and a process of specialization, took place in those decades. New concepts and new technologies also emerged. Meaningful debates on sensitive issues such as reductionism and determinism accompanied that process. Ultimately, I find that the most essential feature of "French epistemology," namely the awareness that the relationship "between philosophy and the sciences can be fruitful only if it rests on the history of sciences" [Braunstein 2012, p. 38], emerged in the late nineteenth century.

It seems to me that the thesis concerning the existence of a crucial turning point in the twentieth century is still widespread. In 2008 Cristina Chimisso pointed out the existence of "a specific tradition in the philosophy of science that intimately connects history and philosophy" in France. Although she found that the main character in that tradition was Gaston Bachelard, she intended to start with "the first decades of the twentieth century." She mentioned Lucien Lévy-Bruhl, Léon Brunschvicg, Hélène Metzger, and Abel Rey, who created a "space for the philosophical history of science at the Sorbonne... in the early 1930s" [Chimisso 2008, pp. 1-2, and 5-6].[4]

In 2003, Brenner had already pointed out the debates on the foundations of science that "took place in France at the turn of the XX century." Although he claimed that "a critical reflection on science" could be traced back to the origins of natural philosophy, he stressed the role played by scholars like Henri

4 David Knight, in a short *Introduction*, agreed with Chimisso on the fact that the history and philosophy of science "took a very different course in twentieth-century France compared to what happened in the English-speaking world" [Knight 2008, p. vii].

Poincaré, Pierre Duhem, and Gaston Milhaud, and "the institutional lineage" that led from Milhaud to Gaston Bachelard through Abel Rey. He mentioned some events that had occurred in the 1890s: in 1891 "Poincaré had stressed the "conventional nature of geometrical hypotheses," in 1892 a chair of "General History of Sciences" had been established at the *Collège de France*, and in 1894 Duhem had looked upon experimental checks as "global" processes. I agree with Brenner on the fact that "the debates at the turn of the XX century can be looked upon as a pivotal stage" in the emergence of the philosophy of science as "it is practiced nowadays" [Brenner 2003, pp. 1, 2, 4-5, and 7-8].[5] At the same time, I find that this meaningful stage emerged a little earlier: in the 1860s, in French-speaking countries, a sophisticated history of science and more critical meta-theoretical remarks on scientific practice began to compete with naïve historical reconstructions and dogmatic views on science.

Poincaré and Duhem represented the last stage of a cultural process that had started some decades earlier. Duhem gave the last touches to an already existing intellectual stream, and both scholars managed to draw the attention of scientific and philosophical communities to the new historiographies and epistemologies.

5 In 1983 Hacking had pointed out that twentieth-century positivism had assumed a sharp distinction between theory and observation, and two main reactions were triggered off by that thesis in the second half of the twentieth century: a "conservative response (realistic)" and a "radical response (idealistic)." In the former, a blurred borderline between "observable and unobservable entities" was assumed, whereas the latter required that "all observational statements are theory-loaded" [Hacking 1983, p. 171]. Even this historical reconstruction overlooks the role played by philosophers and scientists in the late nineteenth century. In Duhem, for instance, a broad positivist background merged with the awareness of the influence of theories on experimental practices. It is worth stressing that this awareness emerged before the turn of the twentieth century, and it was not tainted by what philosophers call *idealism*.

Acknowledgements

I would like to thank Massimiliano Badino for the early confidence in this publishing project, and for fundamental remarks and recommendations on the content and structure of the book.

I thank Jean-François Stoffel for invaluable suggestions during the last stage of my research, more specifically for having discussed and criticised the theses I have put forward the *Introduction*.

I thank Mario Alai for having helped me clarify the complex network of issues I have addressed in my *Afterword*.

I also thank Enrico Giannetto, Mario Castellana, and Salvo D'Agostino for the constant encouragement and informal intellectual exchanges over time, Vincenzo Fano and Gino Tarozzi for having drawn my attention to recent literature in philosophy of science, and Dalmazio Rossi for having discussed my historiographical theses on the late nineteenth century.

I am very grateful to the *Max-Planck-Institut für Wissenschaftsgeschichte*, and in particular Jürgen Renn, Christoph Lehner, and Roberto Lalli for having allowed me to present and discuss my historical studies and historiographical theses in meaningful stages of my research.

Introduction: The Emergence of an Intellectual Stream

Material and Intellectual Context

The emergence of sophisticated histories and philosophies of science in French-speaking countries took place in a context of profound material and intellectual transformations. I must necessarily hint at a complex background where new scientific theories, new technologies, and new standards of living interacted with new philosophical problems, and new attitudes towards scientific practice and scientific achievements. In this *Introduction*, I would like to outline the material and intellectual context, the philosophical background of harsh and mild scientism, the debates on reductionism and determinism, the essential features of those histories and philosophies of science, and my historiographical thesis on the role of buried memories in the history of science. The following chapters analyse in more detail those debates and the emergence of sophisticated historiographical and epistemological frameworks.

The second half of the nineteenth century was the stage for a meaningful transformation in the field of natural sciences. The professionalization of physics, chemistry, earth sciences, and life sciences was accompanied by the emergence of new theories and new practices. Methods and specific contents of science assumed philosophical relevance, and science took the lead. Meaningful philosophical issues spontaneously emerged from scientific theories and scientific practices, and sometimes philosophers had to protect themselves against the faith that a powerful scientific practice represented the only reliable source of knowledge. At the same time, some scientists realized that actual scientific enterprise involved a complex interplay between rational and empirical practices rather than a simplified series of successful experiments and mathematical explanations.

As the historian Geoffrey Barraclough wrote half a century ago, it was "in the years immediately preceding and succeeding 1890 that most of the developments distinguishing contemporary from modern history" first became visible. The last decades of the nineteenth century saw an "industrial and social revolution," and the spread of new technologies: the extent of the differences between social landscape in 1870 and 1900 can mainly be ascribed to the impact of technical and scientific advance. Some historians have spoken of a *second industrial revolution*: it is unquestionable that the role played by new scientific theories was much more marked than in the so-called first revolu-

 | DOI 10.1163/9789004315235_002

tion. A new age "of steel and electricity, of oil and chemicals" slowly replaced the "age of coal and iron." In the decades between the 1860s and the end of the century, new devices appeared that turned out to have a profound social impact: the bicycle, synthetic plastics, dynamite, the electromagnetic dynamo, typewriters, traffic lights, the telephone, the internal combustion engine, the phonograph, electric light-bulbs, pneumatic tyres, cinematography... All these inventions, and more in general "the introduction of electricity as a new source of light, heat and power, and the transformation of the chemical industry," were the result of fundamental scientific advances [Barraclough 1964, pp. 17, 36-8, and 40].

In brief, scientific, technological, and industrial changes "created urban and industrial society" as we still know it. Remarkable progress was also achieved in medicine, hygiene, and nutrition: pasteurisation and sterilisation made new methods of food preservation available to the growing industrial population. One of the most striking effects of those transformations was the decrease of the death rate, and the dramatic increase of population: Europe's population rose by no less than one hundred million in the last decades of the century. The majority of people lived in towns that were changing their dimensions: it is worth stressing that "the emergence of great metropolitan centres was world-wide." A world market governed by world prices also emerged for the first time in the history of civilisation. The new world was forcibly shaped by what has been labelled *imperialism*: the growing dependence of industrialised European societies on "supplies for food-stuff and raw materials" contributed to the exploitation of many land areas of the globe by "the imperial domains of European powers" [Barraclough 1964, pp. 41-3, 45-6, 48, 53, and 55].

Real and clearly perceived scientific progress gave rise to a new ideology, which might be qualified as *rhetoric of progress*. At the same time, nations and governments periodically set up great scientific exhibitions, where scientific progress conflated into the celebration of national strength and success. The first great scientific exhibition took place in London, at Hyde Park, from 1 May to 11 October 1851. In 1889 Paris hosted another *Exposition Universelle* which was also intended as a celebration of the centenary of the French Revolution, in particular the storming of the Bastille as the symbolic beginning of the uprising.[1] The most impressive material representation of that world fair was the Eiffel Tower, which had been completed in the same year, and was placed at the entrance to the fair itself.

1 The second exhibition was held in New York from 14 July 1853 to 14 November 1854. It was followed by Paris in 1855, London again in 1862, and Paris once more in 1867.

Electromagnetic devices had their share of success in these international exhibitions. Electric energy was endowed with specific features, which made it a really new and better source of energy. First of all, it was easily transferable across long distances, and secondly it emerged as a clean form of energy when compared to the smoke and offensive smells given out by steam engines and oil lamps.[2] If we travel backwards through the history of science, we find that during the seventeenth century, when the so-called Scientific Revolution took place, the new science influenced and transformed both the representations of the physical world and the relationship between theoretical and empirical practices. Nevertheless, the transformation did not affect the material conditions of mankind and the habits of ordinary people. On the contrary, a widespread material transformation was the specific effect of scientific practice in the late nineteenth century. To some extent there was a revolution, namely the occurrence of events that deeply transformed both material and intellectual life.

In the last decades of the nineteenth century there was a dramatic increase in theoretical debates and, at the same time, a dramatic increase in technological applications. In those decades the history of technology became closely linked to the history of theories, and from then onwards the links have never slackened, even though we see that the two histories proceed at their own pace. For the first time in the modern age, science took the lead in cultural debate, and at the same time induced meaningful transformations in everyday life.[3] From the mid-nineteenth century onwards, methodological debates spread throughout scientific practice in physics, and a new philosophical and historical consciousness also emerged. Scientists did not entrust philosophers

2 See Lami 1891, p. 743: "En effet, l'électricité fournissant une lumière pure et fixe, ne chauffant pas et ne viciant pas l'air, constitue non pas un éclairage de luxe, mais un éclairage sain et salubre, et, par conséquent, véritablement de première nécessité. Détrônant le gaz pour cet usage, l'électricité ne le bannira pas de la maison: bien au contraire, elle lui ouvrira tout grand son débouché normal, qu'il n'a jusqu'ici envisagé que timidement et comme pis-aller, le chauffage." On the effects of the widespread telegraphic net, see Galison 2003, pp. 174-80.

3 The physicist Robert D. Purrington stressed the "explosive expansion of technology in the last half of the century": it was "both the result and the cause of rapid progress in science on many fronts, especially in physics." In short, "science fed technology, and technology fed science" [Purrington 1997, pp. 3-4]. The historian Richard G. Olson also stressed that most of "major technological innovations since the mid nineteenth century can be directly traced to new knowledge in physics, chemistry, biology or the earth sciences" [Olson 2008, p. 4]. Words such as *science, physics, chemistry, biology and earth sciences* imposed themselves during the development of scientific knowledge that took place in the second half of the century [Ross 1964, p. 66, Bowler and Morus 2005, pp. 3, 6-7, 20, and 53].

with reflections on the aims and methods of science: that meta-theoretical commitment stemmed from actual scientific practice as an inescapable necessity. Moreover the alliance between mathematics and experiments had to be updated: there was a third component that dealt with principles, models, and patterns of explanation. That conceptual or theoretical component, neither formal nor empirical, came to be looked upon as a fundamental component of scientific practice.

The hallmark of the new theoretical practice that emerged in the late nineteenth century was the awareness that the alliance between mathematical language and experimental practice celebrated by Galileo had to be updated. Besides definite demonstrations and sound experiments there was a third component that could be labelled conceptual or theoretical: it involved both foundations and historic developments. Different theories could share the same mathematical framework and make reference to the same kind of experiments: the difference between them could be found just at that conceptual level. Conversely, a given set of phenomena could be consistently described by different theories.[4] The emergence of *theoretical physics* also corresponded to a new sensitivity towards meta-theoretical issues: we find explicit designs of unification, explicit methodological remarks, explicit debates on the foundations of physics, accurate historical reconstructions, and original historiographical frameworks. At this time, all these cogitations were looked upon as intrinsic aspects of scientific practice. That meta-theoretical commitment emerged from actual scientific practice, and influenced various philosophical streams.[5]

In 1858, in a joint article, Charles Darwin and Alfred Russel Wallace had published their theories concerning biological evolution: Darwin's theory of evo-

---

4 A historical reconstruction of the new theoretical practice can be found in McCormmach and Jungnickel 1986, vol. 2, pp. 33, 41-3, 48, and 55-6, Giannetto 1995, pp. 165-6, Kragh 1996, p. 162, Lacki 2007, p. 248, and Bordoni 2008, pp. 35-45. For a historical reconstruction from the point of view of an early twentieth-century scholar, see Merz 1912, p. 199. On the concept of theoretical physics from a point of view of a physicist actually involved, see Boltzmann 1892, pp. 5-11, and Boltzmann 1899, p. 95.

5 See Cassirer 1950, pp. 83-4: "Now not only does the picture of nature show new features, but the view of what a natural science can and should be and the problems and aims it must set itself undergoes more and more radical transformation. In no earlier period do we meet such extensive argument over the very conception of physics, and in none is the debate so acrimonious. [...] When Mach or Planck, Boltzmann or Ostwald, Poincaré or Duhem are asked what a physical theory is and what it can accomplish we receive not only different but contradictory answers, and it is clear that we are witnessing more than a change in the purpose and intent of investigation."

lution by natural selection affected the philosophical debate, and a plurality of interpretation flourished. Both “a highly cooperative form of evolutionary socialism” and “overtly racist ideologies” emerged, and in both cases “Darwinian arguments lent scientific authority and credibility to social doctrines.” Both in physics and social sciences the theme of dissipation and degeneration was at stake, and it also became “a dominant literary theme.” On the one hand, the concept of dissipation was triggered off by the second law of thermodynamics, and it was interpreted as a relentless trend in nature. On the other hand, the concept of degeneration stemmed from medical and physiological researches: it dealt with social problems that could be addressed “through medical and biological intervention.” In some way, both an optimistic and a pessimistic scientism were at stake: science represented the suitable tool for solving technological problems, and for “slowing down the deterioration of the human species” [Olson 2008, pp. 253, 274, 277, and 293]. In the second half of the century, even in the domain of life sciences some scholars inquired into the foundations of their body of knowledge with philosophical and historical sensitivity.[6]

In the field of physics we find a debate between scholars who relied on the mathematical structures of abstract mechanics, and the upholders of specific mechanical models. More specifically, the former could rely on the very general models of analytical mechanics (the *dynamical* approach) whereas the latter relied on detailed physical models in terms of atomic structures and their interactions (the *mechanical* approach). The conflict between “dynamists and mechanists” echoed a more general conflict between “positivists and realists,” even though the intellectual landscape was quite complex, and we cannot univocally attach these labels to all scientists and their actual practice [Purrington 1997, pp. 20 and 22].

With regard to the institutional aspect of the emergence of *theoretical physics*, some decades ago Russell McCormmach and Christa Jungnickel wrote the history of the first Chairs of Theoretical Physics in the last decades of the nineteenth century. They explored German-speaking countries and other neighbouring countries to a certain extent influenced by German cultural traditions. In some universities (Kiel for instance), “theoretical physics was rec-

6 Some scientists and philosophers took for granted that Darwin's theory of evolution by natural selection intrinsically endorsed the myth of progress. In recent studies, historians have remarked that “Darwin was very careful not to link his theory to the linear model of progress.” The philosophical theory of evolution, which had been put forward by the British philosopher Herbert Spencer, rested on Lamarck's rather than Darwin's theory because the former was in tune with Spencer's “ideology of self-improvement” [Bowler and Morus 2005, pp. 147-8 and 150].

ognized as a necessary speciality," endowed with a specific characteristic, "as a link between, and an enrichment of, mathematics on the one hand and natural sciences on the other." In some way, this last feature actually supports the conception of theoretical physics as the conflation of advanced mathematical physics with the most speculative tradition of natural philosophy.[7]

The German institutional framework described by McCormach and Jungnickel shows how difficult a reliable historical reconstruction of theoretical physics as an actual scientific practice in the late nineteenth century really is. Nevertheless a specific and widespread trend can be clearly singled out, and some scholars can be associated with this trend. An original network of general hypotheses, specific models, mathematical tools, historical reconstructions, and meta-theoretical remarks emerged in the last decades of the nineteenth century, and it found its more sophisticated expression in the texts of some outstanding mathematicians, natural philosophers, and physicists. The fact is that, until the early years of the twentieth century physics was practiced by scholars who belonged to various academic categories: mathematicians, physicists, engineers, and natural philosophers. Physics as a definite field of knowledge, a definite academic training, and a definite profession, was the outcome of a historical process that was accomplished in the second half of the nineteenth century. If the emergence of physics as a definite academic discipline was a meaningful event of the late nineteenth century, the emergence of theoretical physics was the most interesting outcome of that process. As already remarked, theoretical physics can be looked upon as a fruitful alliance between the tradition of applied mathematics and the tradition of speculative natural philosophy.[8]

Although the academic recognition of theoretical physics was first achieved in German-speaking countries, sometimes in an ambiguous way, theoretical physics as an actual new practice in physics also appeared in France, Great Britain and then in Italy. We can mention François Massieu, Pierre Duhem and Henri Poincaré in France, Heinrich Hertz, Hermann von Helmholtz, Lud-

---

7 In Prussian universities, the status of theoretical physics remained undefined for some time: it could be identified with advanced physics or mathematical physics. From the point of view of general history, it is worth mentioning that the institutionalisation of theoretical physics was contemporary with German political unification, and the contribution of physics to the development of German industry [McCormmach and Jungnickel 1986, vol. 2, pp. 2, 33, and 41-3].

8 In 1898, the mathematical physicist Georg Helm classified Clausius, one of the founding fathers of classical thermodynamics, as "an outstanding representative of theoretical physics [ein hervorragender Vertreter theoretischer Physik]" [Helm 1898, p. 343].

wig Boltzmann, and Max Planck in German-speaking countries, William Macquorn Rankine, James Clerk Maxwell, Joseph John Thomson and Joseph Larmor in the British Isles, Josiah Willard Gibbs in the U.S.A., and Vito Volterra in Italy. Some of them had been trained as mathematicians, others were engineers. From the academic point of view, Poincaré and Volterra were mathematicians. J.J. Thomson and Larmor had passed the highly selective Cambridge Mathematical Tripos, even though J.J. Thomson had gained his first degree as an engineer. We should not forget that, among the first scholars who built up theoretical thermodynamics, Rankine and Massieu had been trained as engineers, and held Chairs of Engineering in Scotland and France respectively. Gibbs had also been trained as an engineer in the United States, before undertaking his scientific specialisation in Europe. Duhem considered himself a physicist and a mathematician, and his physics was more appreciated by mathematicians than by physicists.

The Philosophical Background: Harsh and Mild Scientism

The faith in scientific practice and scientific progress grew up throughout the nineteenth century. The intellectual trend that can be qualified as *scientism* rested upon two pillars: the unavoidability of human progress, and the close link between scientific and social progress. It is worth remarking that the firm belief in an indefinite social progress triggered by scientific progress emerged quite some time before the great technological achievements that modified human life in the late nineteenth century. I must also specify that we can find different kinds of scientism, more specifically a range of philosophical attitudes that went from a harsh scientism to a milder progressive attitude towards science and history. The last attitude led some philosophers to reconsider traditional issues that had been underestimated after the so-called scientific revolution, and had re-emerged in Kant's philosophical system.

In the French context, the most radical scientism can be traced back to the six volumes of the *Cours de philosophie positive* that Auguste Comte published between 1830 and 1842. In the Foreword to the first volume, he coined the expression "*philosophie positive*" in order to qualify his intellectual commitment, but he regretted that he had been "forced to make use of the term philosophy." He looked upon the noun *philosophy* as the practice pursued by "the ancients, notably Aristotle," who had put forward "general systems of human conceptions"; the adjective *positive* was necessary to specify that he was only interested in theories that dealt with "the organisation of observed facts." In other words, his philosophical system could be qualified as "*philosophy of sci-*

ences" even though it had a wider scope because it encompassed "every kind of phenomena," social ones included. His philosophical researches consisted in "studying the general features of different sciences" in a unified perspective because all sciences had to be submitted "to a single method." At least three strong metaphysical commitments supported his ambitious design: first, the rejection of any question, problem or issue that did not deal with a scientific approach to reality, second, the search for a methodological unification among the different sciences, which were nothing else but "different parts of a more general research plan," and third, the faith in human progress [Comte 1830, pp. VII-VIII].[9]

In the first lesson of his *Course*, Comte put forward his philosophical system as the last stage in the long-term history of civilisation. His broad historiographical framework needed only three historical stages. The first stage corresponded to the dawn of human civilisation, when mankind relied on magic and religion: it was "the theological stage." The second one was qualified as "the metaphysical stage": it corresponded to the emergence and development of philosophy, logic, mathematics, and the practice of rational thinking in general. The last stage was the *positive* one, and corresponded to the faith in scientific practice rather than philosophy or religion. The chronological series of the three stages corresponded to a noticeable progress: the first stage related to a fictitious knowledge, the second to an abstract knowledge, and the third was the stage of fully attained scientific awareness. Comte ventured to qualify his historiographical framework as a law: although it was nothing more than a simple assumption, he made use of expressions such as "the discovery of this law," and mentioned the possibility of "demonstrating this fundamental law." He hinted at "the direct observation" that proved "the exactness of this law," and rational considerations that suggested the necessity of that law or "positive theory" [Comte 1830, pp. 3-8].

Positive philosophy confined itself to problems that could be solved scientifically: it did not inquire into "unattainable solutions" or "the first causes of phenomena." Astronomy had been the first body of knowledge to assume the structure of "a positive theory," and then "earth physics in its strict sense," chemistry, and physiology had been positively transformed. The crucial step in the transformation had been "the interaction among Bacon's guidelines, Descartes' conceptions, and Galileo's discoveries." With the help of these eminent scholars, positive knowledge had managed to extricate itself from the "superstitious and scholastic" constraints that had disguised the search for truth.

9 For the polysemy of the word *scientism*, and its connection with the equally plural meaning of the word *positivism*, see Paul 1968, p. 299, footnote 2. For the origin of the word in French context, see Schöttler 2012, pp. 253-4.

The last step in the achievement of a complete and satisfactory body of *positive* knowledge consisted in designing a "*social physics* [*physique sociale*]." In the end, we find five great categories or bodies of knowledge that corresponded to "astronomical, physical, chemical, physiological and social phenomena." The *Cours de philosophie positive* dealt with "the specific nature [esprit]" of "every fundamental science," and its "relationships with the whole of the positive system" [Comte 1830, pp. 15, 17-19, and 22-5].[10]

The general system of positive knowledge required a sound classification, and different features of scientific knowledge were at stake in the search for a suitable system of classification. The mismatch between *historical* and *dogmatic* classifications was exemplified by the fact that "astronomy had to be placed before physics" even though one part of physics, namely optics, "was essential for the complete exposition" of astronomy.[11] According to the necessity of a reasonable standard, Comte chose a classification where "the study of the most simple and general phenomena" preceded "the most specific and complex phenomena." For instance, "*organic physics*" had to be preceded by "*inorganic physics*."[12] In the end, the search for a rational, clear, universally valid, and historically sound classification of sciences led Comte to

10 In 1966 the historian of ideas Georges Gusdorf reminded readers that the word *science* had changed its meaning over time even though it had maintained the essential feature of "objective and intelligible knowledge to be communicated through language." For long time, astrology and alchemy had been considered as sciences, philology had become an eminent science in the Renaissance, and theology had been looked upon as "the unquestioned queen of all sciences" during the Middle Ages. It is worth stressing that the first Faculty of Science in the modern sense was established by Napoleon in Paris in 1808. In France, the "Facultés de Lettres" and the "Facultés de Sciences" began to separate what in other countries had remained joined in faculties of philosophy. Only at the end of the nineteenth century, the emergence of human sciences such as psychology and anthropology called for a new reassessment of that separation [Gusdorf 1966, pp. 10, 15-6, 20-1, 26-7, 40-1]. Gusdorf also remarked that France had expressed some kind of intellectual dichotomy since the emergence of the first Academies. Differently from other national Academies, the *Académie Française*, which had been founded in 1635, coexisted with the *Académie Royale des Sciences*, which had been founded in 1666. At the end of the eighteenth century, the opposition between humanities and sciences explicitly emerged as a new intellectual attitude [Gusdorf 1966, pp. 26 and 29].

11 It is worth remarking that, in the mid-seventeenth century, Pascal had stressed "the important distinction" between "the purely historical" and "purely dogmatic" (namely rational) methods [Pascal 1872, p. 159].

12 In reality, the history of science was not disregarded by Comte. He thought that the history of every specific science involved "in direct and general way, the whole history of mankind," and the deep comprehension of a science could not be attained without the knowledge of its history. However, he found that the history of sciences should be sepa-

an "encyclopaedic hierarchy." It was only at the end of the two introductory lessons that Comte took mathematics into account, since mathematics played a twofold role. It was both "a constitutive part of natural philosophy," and "the true fundamental basis of that philosophy." More specifically, there was an abstract component of mathematics, which was nothing else but "an admirable, wide expansion of natural logic," and corresponded to the language of all sciences, and a "concrete mathematics," which corresponded to geometry and rational mechanics, and could be regarded as "a true natural science." In the end, Comte found that his classification required six subject matters, namely "mathematics, astronomy, physics, chemistry, physiology, and social physics": the classification was nothing less than "the natural and invariable hierarchy of phenomena" [Comte 1830, pp. 76-9, 83-4, 87, 97-101, and 112-5].[13]

Comte's philosophical commitment might be qualified as a naïve philosophy of science because he offered a simplified and idealised account of scientific practice. Two main issues were at stake in this idealisation: first, the possibility of a sharp separation between science and metaphysics, and second, the historiographical and epistemological thesis of a structural difference between the third stage of science, and the previous stages of religion and metaphysics. With regard to the first issue, Comte was convinced that such a sharp separation was necessary, and could actually be pursued; nevertheless, according to this view, the whole history of science, and the emergence of modern science in the seventeenth century, remained quite a mysterious process. The founding fathers of modern science were strongly committed to putting forward explicit metaphysical agendas. With regard to the second issues, Comte underestimated the fact that every stage in the development of human knowledge relied on more or less explicit beliefs. What Comte termed religious bias was a specific kind of wide-scope belief; in the same way, the various metaphysical theses put forward during the history of philosophy represented another kind of wide-scope beliefs. Modern science itself was based on explicit and implicit meta-theoretical beliefs: confidence in the rational comprehensibility of the natural world, confidence in the active role of mathematics in that comprehension, confidence in the reliability of experiments,

---

rated from "the dogmatic study of science": without the latter, even "the history would not be understandable." Moreover, the historical exposition suited recently established sciences rather than classical sciences like geometry: to follow the series of subsequent developments, and the corresponding original texts, would have been too expensive [Comte 1830, pp. 78-9 and 81-2].

13 The role of mathematics at the top of the encyclopaedic hierarchy was also consistent with the remaining part of the hierarchy: as the classification descended, the importance of mathematics decreased.

confidence in the fruitful alliance between experiments and theories, and so forth. The contents of basic beliefs that made reference to Comte's three stages were obviously different from each other: more specifically, the *positive* or scientific stage was based on its own specific beliefs, which were different from the basic belief of the first and second stages. Nevertheless, explicit and implicit beliefs were also at stake in scientific practice. Comte's naïve scientism prevented him from accepting the existence of meta-theoretical or metaphysical assumptions in each and every body of knowledge.

However it is worth remarking that positivism was both a specific philosophical school, which was put forward by Comte in his *Course de philosophie positive*, and at the same time, "an atmosphere" or a broader intellectual attitude that branched out in different directions. This is consistent with the fact that the philosophical attitude of the founding father of Positivism went through different stages. In the last part of his life he envisioned a secular, *positive* religion or religion of Humanity, which would replace traditional religions. As an acute observer noticed many decades ago, he crossed in reverse order the three stages that would describe the development of Humanity: Comte started from what he called the positive stage but then he "advanced or retrograded to the metaphysical and religious stages" [Benrubi 1926, pp. 16-7]. Although Comte's philosophy underwent meaningful transformations during his life, it represented one of the last instances of a well-identifiable philosophical system. In the following decades, philosophical researches became less systematic, and a less structured style of research spread throughout Europe.[14]

Comte's *Cours de philosophie positive* had an enduring influence in French-speaking countries and abroad during the nineteenth century. Even when his positivistic philosophy was judged too radical and dogmatic, the more general commitment to include scientific methods and contents into philosophical practice could rely on a wide consensus. When we compare Comte's *Cours* with the detailed philosophy of science the British philosopher William Whewell published in 1840, *The Philosophy of Inductive Sciences founded upon their History*, we find similar concerns but a different philosophical attitude. In the second edition of the treatise, Whewell insisted on the recent progress "in Astronomy, in Physics, in Chemistry, in Natural History, in Physiology." The reference to the founding fathers of modern science also appeared from the outset, and Bacon's *Novum Organon* was looked upon as a milestone in the emer-

14 In 1930, the historian of science Hélène Metzger remarked that *Positivism* was something more than a mere philosophical school: it was rather one of the essential components of "the intellectual mood" in the nineteenth century [Metzger 1930 (1987), p. 113].

gence of the new science in the seventeenth century. We also find the confidence in the paradigmatic role of physical sciences: they represented the most "instructive examples of the nature and progress of truth in general." Moreover, "the progress of moral, and political, and philological, and other knowledge" was governed by "the same laws as that of physical science." He claimed that "Astronomy and Geology, Mechanics and Chemistry, Optics and Acoustics, Botany and Physiology" could boast of being "each recognized as large and substantial collections of undoubted truths." They deserved to be qualified as "eminently certain, clear, and definite" bodies of knowledge [Whewell 1847a, pp. v-ix, 7, and 14].

Nevertheless, in Whewell we find more attention to history, and the inescapable necessity of metaphysics in scientific practice. Unlike Comte, who had banned metaphysics, Whewell separated "*good* metaphysics" from bad. He found that Comte's reduction of "all science to the mere expression of the laws of phenomena, expressed in formulae of space, time, and number" was "historically false." It was "a pedantic and capricious limitation of our knowledge, to which the intellect of man neither can nor could submit." To exclude any inquiry into the nature of scientific phenomena would have led us "to secure ourselves from the poison of errour by abstaining from the banquet of truth." The exclusion of the fundamental idea of cause and other ideas of the same kind deprived scientists of important components of "the foundations of science." He found that "the genuine office of science" was "to inquire into the causes as well as the laws of phenomena"; he stressed that such an inquiry could not be avoided. He claimed that "discussions concerning ideas," on the one hand, and "real discoveries," on the other, had always gone "hand in hand in every science." Entities and ideas that "we cannot define," and metaphysical discussions had to be considered as "essential steps in the progress of each science" [Whewell 1847a, pp. x and 1; Whewell 1847b, pp. 321-2, 324, 326, and 329].

Comte had insisted on facts, whereas Whewell insisted on the usefulness and the role of ideas in scientific practice. The complex connection between "Facts and Ideas" was the *leitmotiv* of Whewell's philosophy of science: it was discussed and recast in many ways, and the dichotomy or "*Fundamental Antithesis*" was translated into various names and concepts such as "Thoughts and Things," "*necessary* and *contingent* truths," "Deduction and Induction," "*Ideas* and *Sensations*," and "subjective or ideal, and the objective or observed." He claimed that "[w]ithout Thoughts, there could be no connexion," and "without Things, there could be no reality." According to Whewell, the separation between the two elements of every couple was not required by scientific practice in itself, but by the philosophical analysis of that practice. The words *idea* and *sensation* expressed "more exactly than any of the pairs before mentioned" the fundamental antithesis "in the union of which... all knowledge consists." He found that the dichotomy he had pointed out was a recent implementation of the ancient dichotomy between "*Matter* and *Form*" [Whewell 1847a, pp. x-xi, 18-20, 23, 25, 27, 30, and 33].[15]

The philosophical problems that emerge from this brief outline, and more specifically from Comte and Whewell's different attitudes towards scientific practice, are important steps in my historical reconstruction for at least two reasons. On the one hand, we find in Comte and Whewell the philosophical roots of a harsh scientism and a milder progressive attitude respectively. On the other, we see the development of philosophical issues that had been underestimated after the so-called scientific revolution, and had been reconsidered by Kant. From the point of view of long-term cultural processes, the essential tension between the formal structures of thought and sensorial experiences, which Whewell passionately explored, could even be traced back to Aristotle's *Analytica Posteriora*. That tension had been overshadowed by the naïve and optimistic philosophical trend that had accompanied the intellectual achievements of modern science. The re-appearance of old philosophical problems in the second half of the nineteenth century, and the emergence of sophisticated histories and philosophies of science, can be better appreciated in the light of this wide historical perspective.

Debates on Reductionism and Determinism in French-Speaking Countries

In the debates on science that took place in France from the early 1860s to the early 1890s two main issues were at stake: determinism and reductionism. More specifically, the debate pivoted on two different although intertwined questions, namely how to combine the determinism of physical laws with human free will, and whether physiology and even psychology and sociology could be completely derived from the laws of physics. The most radical reductionism assumed that the recently emerged human and social sciences had to be based on natural sciences, and in turn, natural sciences had to be reduced to mechanics. The supposed determinism of mechanics reverberated in other

15 A further metaphor was offered by the complementarity between man and nature: in Whewell's words, "Nature is the Book, and Man is the *Interpreter*." Sometimes the borderline between facts and interpretations was vague, and scientists were committed to "*interpreting* the phenomena" rather than merely reporting them [Whewell 1847a, pp. 37, 39-41, 44, 48, and 50].

bodies of knowledge. Radical reductionism and radical determinism were in tune with what I have labelled *scientism*, namely the claims that natural sciences represented the model for every reliable body of knowledge, and that social progress depended on, and stemmed from, scientific and technological progress.

The intellectual landscape of *fin de siècle* France was crowded with different characters and different attitudes, and I can only confine myself to outlining a rich and many-branched debate. On the one hand we find scientists, historians, and philosophers who relied on simplified epistemological and historiographical frameworks, and put forward an optimistic cult of scientific progress and human progress in general. On the other, we find other philosophers and scientists who pointed out how problematic some scientific concepts were, and how complex scientific practice was. A sophisticated point of view on science was put forward by scientists and philosophers who did not deny the effectiveness of scientific progress but were able to go beyond the simplified conception of scientific practice as an unproblematic alliance between mathematical and empirical procedures.

In 1861, the French mathematician, economist, and philosopher Antoine Augustin Cournot published a book on the relationship among the specific methods and practices in mathematics, physical sciences ("dealing with inorganic matter"), natural sciences ("dealing with organized and living beings"), and human sciences. Both the essential features of each body of knowledge and the boundaries among them were at stake. In the book title, *Traité de l'enchainement des idées fondamentales dans les sciences et dans l'histoire*, the French word "enchaînement" can be looked upon as carrying a twofold meaning: first, the plain meaning of a list, namely the series of disciplines from logic to sociology on the track of Comte's theory of knowledge, and second, the meaning of a close connection, namely the existence of meaningful and unexpected links among methods and foundations of those disciplines. Cournot himself insisted on the "enchaînement des idées fondamentales" as opposed to their "detailed enumeration" [Cournot 1861, pp. II-III].

In 1851, he had already published a two-volume book on "the foundations of our knowledge." He was aware of a new trend in European culture, namely the increasing professional separation between scientists [savants] and philosophers. The book started from the awareness of that irreversible detachment, and aimed to bridge the gap in a new way, looking for structural analogies between irretrievably autonomous bodies of knowledge. In 1861, he was interested in outlining a theory of knowledge that pursued two different targets and corresponded to two different levels of philosophical investigation. At the first level he took care to include the most important scientific results in philosophical research, and at the second level he pursued a more ambitious target, which he labelled "the philosophy of science." The second level was placed far beyond the old encyclopaedic commitment, and dealt with "principles, methods, and theories" of *positive* sciences. He specified that the present time did not allow him to put forward "theses of *omni scibili*." The time of Aristotle and Saint Thomas had permanently elapsed, and even "the time of Ampère and von Humboldt could not return" [Cournot 1861, pp. VI-VII]. The new stage of philosophy of science led Cournot to overstep the Enlightenment's intellectual horizon and then Comte's philosophical horizon. Philosophy of science required a detailed analysis of the conceptual structure of *positive* science, in order to make its hidden philosophical foundations emerge.

In 1865, the French physician and physiologist Claude Bernard published a demanding book that was intended as an introduction to "experimental medicine." From the outset he expressed the optimistic claim that medicine had "permanently undertaken a scientific pathway." He found that scientific method was nothing else but "the experimental method," but the adjective experimental did not have a purely empirical meaning. Experiments involved a rational practice: some kind of intellectual practice linked ideas and *facts* to each other. The experimental scientist should be "both a theorist and a practitioner." This rational practice in life sciences was not so different from the corresponding practice in sciences dealing with "inanimate matter," even though different phenomena and specific "difficulties of investigation" and application were involved in different sciences [Bernard 1865, pp. 6-8].

In 1867, Bernard remarked that physiology involved "the most complex among natural phenomena"; it was for this reason that physiology had only recently managed to gain "its scientific independence." The simplest sciences, namely the sciences that rested upon "few and very general notions," had already managed to flourish, and life sciences could rely on those achievements. In other words, physics and chemistry offered "both support and instruments" for the advancement of physiology, which was struggling to "get rid of empiricism" but also loose speculations. In brief, Bernard looked upon life as a complex process, where "the physic-chemical features of the environment, and the specific vital features of the organism" interacted with each other. What he labelled "general physiology" rested upon these two pillars: it was an intrinsically *twofold* body of knowledge that resisted any drift towards naïve reductionism and naïve determinism [Bernard 1867, pp. 1-2 e 5].[16]

16 The second half of the nineteenth century saw the emergence of physiology as a science, in particular the partial detachment of physiology from medical practice as an

In 1874, the philosopher Émile Boutroux commented on a scientific landscape where different bodies of knowledge and different methods were at stake. He did not trust in a unified scientific method: contents and patterns of explanations could not be inherited in the passage from one science to another. Physiology and psychology could not be reduced to physics [Boutroux 1874, pp. 80-93 and 143-8]. However, Boutroux's sophisticated antireductionism was not the hegemonic attitude in the scientific community. We find a remarkable trust in radical reductionism in scholars who contributed to the emergence of human sciences, a field of research where old intellectual practices like history and philosophy merged with new disciplines like anthropology, psychology, and sociology. In the late 1870s physiologists, anthropologists, and philosophers debated on the reduction of mind to body, and the reduction of philosophy and theology to brain physiology. For instance, from 1878 onwards, Jules Soury, archivist and palaeographer with interests and serious studies in neurology and psychiatry, attempted to base social processes and the history of civilization on brain physiology [Soury 1878, pp. 9-11 and 34-5]. Other scholars challenged both *scientism* and *reductionism*, and called for a plurality of languages and scientific methods. Some sources and the corresponding debates are discussed in the first three chapters of the present book.

A specific debate on determinism and free will took place around 1880 in French-speaking countries. Some mathematicians, physicists, physicians, and philosophers were involved: the boundaries and the bonds among different disciplines were fiercely shaken. Unexpected meaningful links among mathematical technicalities, the equations of mechanics, sudden release of energy, the emergence of life, and free will were under scrutiny. Unexpected structural analogies among different processes also emerged. The French mathematician Joseph Boussinesq put forward an original research programme, where different traditions of research really converged. We find the integration among mathematical researches, where singular solutions of differential equations were involved, researches in physiology, where concepts like "*Auslösung*" and

autonomous field of research. It is worth remarking that, still in the eighteenth century, both medicine and physiology had been looked upon as part of physical sciences in general, in the same way as the study of heat and magnetism. On the other hand, for almost two centuries, what in the late nineteenth century was labelled *physics* had been practised by two different communities that corresponded to two traditional bodies of knowledge endowed with their own methodological standards: natural philosophy and "mixed mathematics." The former included both theoretical speculation and experimental practices whereas the latter was subsequently labelled mathematical physics [Hankins 1985, pp. 10-11].

"*principe directeur*" had recently emerged, and physical sciences, where transformations of energy in general, and concepts like "*trigger-work*" and "*travail décrochant*" were at stake [Boussinesq 1878, pp. 43-9 and 133-40].[17]

In 1886 the authoritative chemist and politician Marcelin Berthelot published a collection of some papers he had already published in various journals in the 1860s and 1870s. The first essay of the book focused exactly on what we might label the history and philosophy of science. In reality the qualification might be too ambitious because the prestigious chemist confined himself to an apologetic and simplified history of science, and was committed to an equally simplified philosophical analysis of the science of his time.[18] Scientific practice consisted in discovering an ever-wider domain of *facts*. The empirical nature of science was "one of the principles of positive science": in particular no real knowledge could be "established by means of reasoning." Centuries of philosophy were swept away by that flood of facts and by a new, naïve philosophy. He started from the assumption that "the conclusions we draw from… our conceptions" could only be "probable and never certain," whereas certainty was really attained by "a direct observation, which complies with reality" [Berthelot 1886, pp. 10-11].

The intellectual landscape of French-speaking countries also included philosophers who took a keen interest in the latest developments of physics and were interested in putting forward a sophisticated philosophical approach to science. In 1883 a Parisian publisher sent the book of a Swiss philosopher, Ernest Naville, to the printing press. The book dealt explicitly with physics and its history: the title *La physique moderne. Études historiques et philosophiques*, indicated that the analysis would have been both historical and critical.[19] Both

---

17 The debates on determinism stemmed from an essential tension at the foundations of modern natural philosophy. As the historian of science Thomas L. Hankins pointed out some decades ago, philosophers and scientists (in a broad sense, when referring to the seventeenth and eighteen centuries) had perceived "a contradiction in laws of nature." On the one hand, those laws completely "determined events in the physical world," and on the other hand, they "set man free": the contradiction appeared even sharper when human mind was assumed to be part of nature. Obviously the contradiction appeared as such only if human freedom was looked upon as a matter of fact from the outset, in the sense that it was assumed that God or nature had created man free [Hankins 1985, p. 7].

18 From the formal point of view, the essay was a public letter he had addressed to another intellectual father of the Third republic, the historian Ernest Renan.

19 In the same year, Ernst Mach published his historical-critical reconstruction of mechanics, and the following year, a book with a similar target was translated into French: it had been published in 1882 in the United States by the German-born scholar Johann Bernhard Stallo.

scientists and philosophers took part in the debates on the foundations of science, and faced specific issues dealing with physics and life sciences. We find meaningful similarities between the specific contents of those contributions even though sometimes languages and aims were different for scientists and philosophers. The debates around determinism are analysed in the fourth, fifth, and sixth chapters: once more, the conflict between naïve scientism and more sophisticated historical and philosophical reconstructions was at stake.

Combining Histories and Philosophies of Science

The new historical consciousness of scientists gave birth to a series of books on the history of sciences. Alongside histories that stemmed from simplified historiographical frameworks, we find the emergence of a sophisticated historiography wherein both the philosophy of history and the foundations of science were widely reviewed.

In 1885, Berthelot published a history of alchemy where that ancient body of knowledge was regarded as the prehistoric stage of the progressive process that had led mankind to rationality and science. In accordance with Comte's historiographical framework, the contemporary stage, which he qualified as "absolutely positive," had emerged from a previous stage, which he labelled "half-rationalist and half-mystic." In its turn, the latter had emerged from the ancient attitude that leant towards magical practices. In brief, chemistry, "the most positive among sciences," had stemmed from "quirky phantasies." In this progressive historiographical framework, the fast progress of chemistry contrasted "the dark history of alchemy." In spite of his simplified historiography, Berthelot's research aimed at clarifying a complex network of interactions among "ancient Egyptian handicraft, Greek philosophy, and Alexandrine and gnostic mystical dreams" [Berthelot 1885, pp. vii, ix, and 1-2]

In the late 1880s, the engineer Paul Tannery put forward a more sophisticated approach to the history of science and the history of mathematics. He could rely on a remarkable philological competence, and the awareness that the past should not be judged with hindsight. In 1887 he qualified history of science as a practice that focused on processes and problems rather than reports of scattered events. He considered the historiographical thesis of indefinite progress as highly questionable: science itself was the outcome of a partially discontinuous historical process. More specifically, the history of ancient mathematics offered a meaningful instance of "a sharp decline after a stage of remarkable apex." In the same year he published a book on early Greek science, *Recherches sur l'histoire de l'astronomie ancienne*. His historical reconstruction rested on a series of four subsequent stages: the Hellenic, the Alexandrine, and the Greco-Roman stages, and eventually "the age of reviewers." With regard to the first stage, he stressed the need for a careful philological approach, and the necessity of disentangling the specific history of ancient science from the more general history of ancient philosophy. The history of ideas and the history of philosophy could accept some anachronisms, and in general a certain degree of continuity among ancient and modern ideas, whereas the history of ancient science required a detailed analysis of deep discontinuities [Tannery 1887a, pp. VI, 1-4, 7, 11, and 14; Tannery 1887b, pp. 1-2, 4, and 8-9].

Six years later Gaston Milhaud, a teacher in Montpellier, published a book on the origins of Greek science which was explicitly dedicated to "Monsieur Paul Tannery... with esteem and gratefulness." From the outset he pointed out three essential features which a historian of science should be endowed with, and which made the history of science an extremely demanding task: first, "a complete scientific competence," second, "the knowledge of general history and a deep philosophical insight," and third, a specific sensitivity to "philological questions." The history of science appeared as a manyfold activity that required a plurality of skills and points of view. He would have focused on the history of science in the sense of a history "of ideas, methods, and theories," rather than scattered histories of subsequent scientific achievements. He imagined the history of science as a bridge between science and the humanities, where the cultural context was at stake in its "different expressions" [Milhaud 1893, pp. 4-6, and 8].

Science represented a specific achievement of the human mind among many other specific achievements, a "specific language" endowed with a specific, powerful grip on human experience: it was something more complex than "a definite set of objective truths." Milhaud insisted on the regulative power of scientific practice, where the multiplicity of facts was reduced to the uniformity of laws. In Milhaud we do not find Tannery's philological competence, and the huge number of quotations from primary and secondary sources. Nevertheless his more synthetic account is definitely more readable, and epistemological remarks are always in prominence [Milhaud 1893, pp. 9, 11-13, and 21, 23, and 25-6].[20] My seventh chapter is devoted to Berthelot, Tannery, and Milhaud's histories of science.

20 The following year he published another book, *Essai sur les conditions et les limites de la certitude logique*, which was dedicated to Émile Boutroux, with the same words he had addressed to Tannery.

Theoretical physics, the history of physics, and meta-theoretical remarks were mutually interconnected in Pierre Duhem's actual scientific practice. In particular, he managed to keep together the history and philosophy of science in a systematic way. His theoretical design of unification between mechanics, thermodynamics, and chemistry, as well as his re-interpretation of the Aristotelian tradition of natural philosophy, could be pursued only by a scientist mastering science, history, and philosophy.[21]

In 1892, when he was lecturing in physics at Lille University, he published the first paper explicitly devoted to meta-theoretical issues or, to make use of an expression already used by Comte and Cournot, the philosophy of science. At that time Duhem had already published a book on thermodynamic potentials and their applications to different fields of physical sciences, and a demanding essay where he had put forward an original mathematical approach to thermodynamics on the track of analytical mechanics. In 1893, he published a paper that was specifically devoted to the problematic link between physics and metaphysics and was intended as an answer to his critics. He stressed that a theory had nothing to do with truth: it could not be qualified as true or false, but "suitable or unsuitable, good or bad." The plurality of theoretical frameworks corresponding to a set of laws was consistent with this essential feature of theories. He rejected any necessary link between scientific practice and philosophical commitment, and he claimed that his thesis was "neither sceptic nor positivist" [Duhem 1893d, pp. 97-100].

In 1894 Duhem faced some questions emerging from experimental physics. The paper pivoted on three fundamental theses: first, a physical experiment was not a purely empirical process; second, it could not be so powerful as to lead to the refutation of a single hypothesis; third, it was less reliable, even though more precise, than ordinary experience [Duhem 1894c, pp. 147, 151, 155, and 179]. In the same year he published a paper on the history of optics. His history was something more than a mere collection of meaningful facts: from the outset he put forward an original historiographical framework: a long-lasting and persistent stream of progress flowed underneath the short-term theoretical fluctuations that affected the history of science [Duhem 1894d, pp. 94 and 125]. In 1896 Duhem accomplished his scientific design of

21 The most important papers he published on the history and philosophy of science were hosted by the Belgian journal *Revue des questions scientifiques*. The journal was published by the *Société scientifique de Bruxelles*, which was an association of Catholic scientists. The historical and epistemological remarks he had begun to publish systematically in the 1890s were subsequently collected in the book he published in 1906, *La théorie physique, son objet, et sa structure*.

a wide-scope thermodynamics, which was a mathematical generalisation of analytical mechanics. Duhem's meta-theoretical commitment was both anti-mechanical and anti-reductionist [Duhem 1896b, p. 198]. He found that his mathematical physics had realised a new alliance among the tradition of Aristotle's natural philosophy, the more recent tradition of analytical mechanics, and the even more recent developments of thermodynamics. Duhem's complex network of physical researches, historical reconstructions, and philosophical remarks will be dealt with the eighth and ninth chapters.[22]

From the institutional point of view, the first Chair of History of Science was established in Paris in 1870, even though it was in reality a Chair of History of Medicine. In 1892 a Chair of *Histoire générale des sciences* was created at the *Collège de France*. This Chair was held by an orthodox Comtian, Pierre Lafitte,[23] between 1892 and 1903, and by the crystallographer, philosopher, and editor of the journal *La philosophie positive* Grégoire Wyrouboff from 1903 until his death in 1913. After the First World War, the Chair was recreated for Pierre Boutroux (the son of Émile Boutroux), who held it from 1920 until his death in 1922, and afterwards the Chair disappeared. Both in 1892 and 1903, Tannery and his supporters in the academic system were confident in his appointment to that Chair. His hope and subsequent disappointment were also based on the awareness of the objective value of his own researches. With regard to the Sorbonne, only in 1909 was a Chair of *Histoire de la philosophie dans ses rapports avec les sciences* created for Milhaud. Abel Rey in 1919, and then Gaston Bachelard succeeded Milhaud [Chimisso 2008, p. 85, fn. 1; Locher 2007, p. 217; Brenner 2003, pp. 5 and 101; Canguilhem 1979, p. 63].[24]

Comte had been the first scholar to promote the history of science, and the historian of science George Sarton ventured to write that "August Comte must be considered as the founder of the history of science," but in reality Comte did not practice the discipline seriously. For him, the history of science

22 For his general theory, Duhem resorted to the name *Energetics*, but it should be stressed that the label had assumed different meanings in the last decades of the nineteenth century. His energetics should be intended in the sense of Rankine, as a unified mathematical structure for all kinds of physical and chemical actions, rather than in the sense of Wilhelm Ostwald, as an idealisation of the concept of energy [Rankine 1855, pp. 210-8 and 222, and Ostwald 1896, pp. 159-60].

23 Pierre Lafitte was a positivist who followed Comte even in the establishment of the religion of humanity. He held the first Chair of *Histoire générale des sciences* until his death [Benrubi 1933, pp. 22-3].

24 For the institutional activity in the field of history and philosophy of science in the 1920s in the French context, see Chimisso 2008, pp. 85-93.

offered a useful support for "the positivistic ideology" and the law of three stages. Comte's conception of history was not so different from the eighteen-century attitude of looking upon history as "the projection in retrospect of the idea of progress." From the institutional point of view, he began to recommend the settlement of a Chair of History of Science in 1832. One of his followers, the physician and historian of medicine Émile Littré put forward a public petition in favour of that kind of Chair in 1848, and finally in 1857, at the funeral oration in honour of Comte [Sarton 1948, p. 30; Gusdorf 1966, pp. 43-4, 62, and 98-101]. Historians have remarked that Comte's intellectual attitude was not in tune with "the patient and detailed historical researches." The suggestion of a Chair of History of Science at the *Collège de France* had probably stemmed from Comte's drive to expand the influence of positivism: in Comte's perspective, the history of science was instrumental in supporting his philosophical framework and his historiographical sketch based on three stages.[25]

A few years after the appointment of Wyrouboff to the Chair at the *Collège de France*, and the unexpected ousting of Tannery, Gaston Milhaud reminded readers that even Berthelot had endorsed Tannery's appointment to the Chair of History of Science, and the totality of the members of the *Académie* had agreed with him.[26] Other historians made use of sharp expressions such as "tragic history" or "a true scandal" or "tragic and ridiculous injustice" in the description of the *affair*. Probably the historian Georges Gusdorf managed to grasp the sense of the events from 1892 to 1922 when he stated that "the Chair was looked upon as intrinsically positivist" from the outset, and after Wyrouboff's death and the decline of positivism, "the positivist fiefdom" had no reason to exist any more [Gusdorf 1966, pp. 104-6].

---

25 Hélène Metzger found that, in general, the history of science had been looked upon as either an opportunity to endorse a specific philosophical trend or a pleasant exhibition of scholarship "for old scholars in the weekend." She attempted to follow a third way, wherein the reconstruction of ancient intellectual environments could help historians understand our cultural tradition, and the different ways of building up knowledge [Metzger 1935 (1987), pp. 27 and 29]. According to Jean-François Stoffel, the appointment of Wyrouboff to the Chair of the *History of science* was "a sad case" in itself, and it also had long-lasting consequences, namely the mistrust between historians and philosophers of sciences [Stoffel 1996, p. 416].

26 Milhaud's brief report deserves to be quoted: "Un mois plus tard, l'Académie des Sciences, à l'unanimité, ratifiait ce choix. Il ne manquait plus que la signature du Ministre, ce qui, en pareil cas, au moins dans une démocratie soucieuse des intérêts de la science, semblait ne devrait être qu'une simple formalité, et Tannery, qui ne songeait plus qu'à ses nouvelles fonctions, travaillait à sa leçon d'ouverture, quand *l'Officiel* annonça... qu'un autre était nommé. Il n'a survécu qu'un an à cet événement douloureux" [Milhaud 1906, p. 14].

Scientific Progress, Revolutions, and Buried Memories

In 2006 Jürgen Renn pointed out the competition between two historical processes that could also be regarded as two essential features of scientific enterprise: the progressive accumulation of knowledge on the one hand, and the deep transformations or revolutions in the scientific body of knowledge on the other. He discussed which kind of relationship could be found between the two apparently opposite trends, and asked himself whether "a scientific revolution could really change the structure of knowledge." In other words, he wondered to what extent transformations or revolutions affected contents and methods of science. At the same time, the history of science had shown that both the existence of scientific progress and the existence of scientific revolutions were matters of fact. Renn saw progress as "an unstoppable Golem" that had determined social development "for good or for evil" in the last centuries. Nevertheless, that progress was also frail: in brief, under the opposite effects of "chance and necessity," it was an unpredictable but still effective process [Renn 2006, pp. 10-12]. Renn managed to effectively highlight the essential tension that makes scientific practice so specific, and so different from other human activities. Nevertheless, it seems to me that the description is not complete: besides cumulative and revolutionary processes, the history of science shows us the existence of buried memories.

Besides progressive accumulations of knowledge, and revolutionary processes in the architecture and content of knowledge, we find a third historical process, where systematic research programmes or broader intellectual streams disappear despite their intrinsic heuristic power and fruitfulness, but subsequently and unexpectedly re-emerge. Theories or intellectual attitudes can show an actual fruitfulness, and can find a new audience. Sometimes, the original texts and the problems from which the original research programmes or intellectual streams have stemmed are partially or totally forgotten, and frequently the new versions do not perfectly match with the original ones. Nevertheless, the essential features of those theories manage to survive. This is true for Cournot's mathematisation of economics, Duhem's mathematisation of thermodynamics, and the refined historiographical and epistemological theses that were put forward by some scientists and philosophers from Cournot to Duhem. In the twentieth century their meta-theoretical theses, and the critical attitude out of which they had grown, underwent the fate of buried memories: they disappeared or at least faded away, but then re-emerged in a new historical context. Those theses suggested that the wide and clumsily explored land that lay between scientific dogmatism and distrust in scientific practice deserved to be explored. That meta-theoretical attitude echoed Pascal's awareness that "the impossibility of proving, which overcomes any dogmatism [l'impuissance à prouver, invincible à tout dogmatisme]," and "the concept of truth, which overcomes any scepticism [l'idée de la vérité, invincible à tout le pyrrhonisme]," could found a fruitful synthesis [Pascal 1897 (1976), p. 158].

Epistemological and historiographical theses put forward in the late nineteenth century allowed the buried memory of a more ancient and neglected scientific tradition to re-emerge, a tradition that could be traced back to Pascal. That tradition and its search for a suitable balance between a naïve, dogmatic scientism and ineffectual scepticism could offer an appropriate intellectual toolbox for the comprehension of the scientific trends and scientific achievements that had been developing in the last decades of the nineteenth century. The development of a sophisticated history of science definitely required the idea of science as a dynamic body of knowledge that changes over time. Nevertheless, in itself this idea was not in opposition to the conception of science as a progressive accumulation of truths. The transition from a naïve to a sophisticated history of science required a further step: the idea that not only past science but also recent science was a fallible and provisional body of knowledge, and that the relationship between experiences and explanations was highly problematic [Gusdorf 1966, pp. 47-8].

The actual revival of Pascal at the end of the nineteenth century was preceded by the discovery of the original manuscripts of his *Pensées* before mid-century. Some historians have interpreted the revival as one of the effects of the reaction against the naïve scientism that had spread throughout the French cultural environment in the second half of the century. It seems to me that the rediscovery of Pascal also rested upon an objective and more specific analogy between the content of his reflections and the debates on determinism and reductionism that took place in the last decades of the nineteenth century. Determinism and reductionism had emerged together with modern science: they were among the most sensitive issues at the foundation of the new science. In the seventeenth century an explicit debate on determinism and reductionism would have been untimely and potentially explosive for the recently developed body of knowledge. It remained in the background as a silent issue that could not be conjured up. Pascal, who ventured to point out the essential elements of those sensitive issues with sharp irony, was philosophically isolated. Only at the end of the nineteenth century, after a stage of successful and triumphant science, an explicit debate could finally see the light. The time was ripe for the rediscovery of Pascal: it is certainly true that his reflections were in tune with the necessity of overcoming "a dogmatic framework" that positivism had built up [Eastwood 1936, pp. 2, 5, 9, 13, and 26].

It is not strange that Pierre Duhem, who undertook the demanding task of revisiting both the history and the methodological foundations of modern science, came across Pascal and found an intimate philosophical consonance with him.[27]

The French-speaking scholars who put forward a sophisticated history and philosophy of science lived in a time that was subsequently shaken by the fall of the Second Empire, the war against Prussia, the defeat, the insurgency of the Commune, the ideological struggles on the laicism of the state, and the Dreyfus affair. Like the more elderly Tannery, Duhem was a firm believer and, at the same time, an independent thinker: he disliked transforming scientific contents into apologetic arguments. Both Tannery and Duhem thought that the subtle connections among scientific practice, philosophical commitments, and religious faiths could be understood only from the point of view of a clear separation of the three domains. In particular Duhem acknowledged the fruitfulness of some aspects of the Aristotelian tradition; at the same time, he refused to get uncritically involved in the revival of neo-Thomism. This is consistent with his marked preference for Pascal: his historiography and his epistemology were definitely more in tune with Pascal than Saint Thomas [Picard 1922, pp. cxxx et cxxxv-cxxxvii; Paul 1979, pp. 3 and 159; Maiocchi 1985, p. 13; Martin 1991, p. 68, 90 et 115; Stoffel 2002, p. 196 et 345; Deltete 2011, pp. 19-21].

My historical reconstruction covers the time span from 1861 to 1896, more specifically from the publication of Cournot's two-volume *Traité de l'enchainement des idées fondamentales dans les sciences et dans l'histoire* and Duhem's paper "L'évolution des théories physiques du XVIIe siècle jusqu'à nos jours." It seems to me that Cournot's book represents the first attempt to go beyond a naive philosophy of science, and the first attempt to put forward a critical history of science. Duhem's paper represents the fulfilment of that historical-critical design, where a critical historical reconstruction was consciously combined with a new epistemological perspective. However I would like to stress that this trend should not be considered as a definite and unified research programme: we are dealing with a common *esprit* but different aims and different

27 Bas van Fraassen pointed out the silent presence of Pascal's "underground epistemology" in the history of science. It is true that Pascal's apparently unsystematic writings gave rise to "a stream that in the succeeding three centuries has become a powerful river" [van Fraassen 1989, p. 151] but I find that *the river* explicitly emerged only at the end of the nineteenth century. More recently, Jean François Stoffel stressed the deep influence of Pascal on Duhem: more specifically Stoffel pointed out the role played by intuition "besides the certainty of deductive reasoning," in the building up of a third way between scientific dogmatism and scepticism [Stoffel 2007, pp. 299 and 301].

agendas. This is consistent with the fact that the scholars under scrutiny in the present book were not philosophers or historians from the institutional point of view, but mathematicians or scientists involved in different research fields. I am going to focus on texts, more specifically published texts, and I will let the texts speak for themselves, after having sketched some essential features of the context, in order to show the emergence of an intellectual trend from a complex intellectual environment.[28]

These scholars experienced various fates, from public appreciation to substantial isolation. Bernard and Boussinesq arrived at holding Chairs in Paris, whereas Cournot and Duhem did not manage to attain the academic acknowledgement they deserved, and Tannery never held an academic position. None of them was interested in pursuing some kind of worship in the wake of Comte: all of them were open-minded and endowed with a strong intellectual independence. Some of them like Cournot, Boussinesq and Tannery worked almost in isolation. Cournot, Tannery, Duhem, and Milhaud stressed the cultural value of scientific enterprise, more specifically the influence of science on the material and the intellectual environment and vice versa. They also acknowledged the intrinsic historicity of scientific enterprise: entities, concepts, and practices tacitly accepted for a long time could become incomprehensible or meaningless for subsequent generations of scholars. The dissemination of their ideas contributed to the establishment of the French philosophy of science as a specific tradition of research, but also to the establishment of a wider research tradition that is sometimes qualified as historical epistemology. At the same time those ideas played an important role in the emergence of a different research tradition such as logical positivism.

In the late 1950s and the early 1960s, scholars with different agendas such as Alexandre Koyré, Norwood Russell Hanson, and Thomas Kuhn, turned to a then neglected subject, namely the philosophical analysis of the emergence of new scientific practices. When we put that intellectual process in perspective, we find that they revived and reinterpreted a critical attitude that had emerged in the last decades of the nineteenth century, and had survived in France in the first half of the twentieth century. It is worth stressing the role played by Hélène Metzger in the 1930s, and her awareness of the intrinsic historicity of scientific enterprise [Metzger 1933 (1987), pp. 9-13 and 16-9].[29] Koyré, Hanson,

28 It is worth stressing that, in the nineteenth century, important French philosophers, such as Comte and Charles Renouvier, received a scientific training, and never taught philosophy [Chimisso 2008, p. 13].

29 For the important role played by Helène Metzger in the establishment of a *philosophical* history of science in the 1920s and 1930s, see Chimisso 2008, pp. 109-23.

and Kuhn focused on metaphysical foundations, on the logic of research and the logic of discoveries, whereas the previous generation of philosophers of science had paid attention mainly to the logic of accomplished scientific systems. New words and new concepts emerged alongside the words and concept that Cournot and Duhem had already put forward. Koyré, Hanson, and Kuhn revived and transformed an intellectual trend that had faded away, and could be traced back to the historical and philosophical researches that had been undertaken by the main protagonists of the present book.

CHAPTER 1

Critical Analyses of Scientific Method

1 A Historical-Critical Reconstruction of Physics

In France, an explicit attempt to go beyond Comte's philosophy of science was put forward by the mathematician Antoine Augustin Cournot. When he published the book *Traité de l'enchainement des idées fondamentales dans les sciences et dans l'histoire* in 1861, he had already spent his career in French universities, academies, and educational institutions, and had already published a number of books on mathematics, economics and the philosophical foundations of scientific knowledge. In 1838, he had published a short book on a new *science* which he qualified as "political economics": he had ventured to put forward a mathematical inquiry in spite of the economists' distrust of "the use of mathematical formulae." The book, *Recherches sur les principes mathématiques de la théorie des richesses*, dealt with "applications of mathematical analysis to the theory of wealth": he specified that the aim of his mathematical approach was not mere computation, but the establishment of functional relations between quantitative entities. He relied on an authoritative analogy: his economic theory was a mathematical generalization of practical procedures in the same way as "Rational mechanics profitably contributed to practical mechanics by means of general theorems" [Cournot 1838, pp. V and VII-VIII]. In 1843, he had published a longer book on probability theory, *Exposition de la théorie des chances et des probabilités*, wherein two different approaches to probability were at stake: a "*subjective*" approach, where "a certain degree of knowledge" or confidence was involved, and an "*objective*" one, where "the measure of possibilities" was "independent of our knowledge." Another analogy with physical sciences was put forward: probability theory could compute a reasonable outcome of stochastic processes in the same way as "the mathematical theory of heat" allowed scientists to determine "*the final state*" of a physical system independently of "the variability of the initial state" [Cournot 1843, pp. III-V].[1]

1 Cournot entered the *École Normale Superieure* in 1821, but the following year the *École* was closed by the government for political reasons; he continued his studies in mathematics, and attended various lectures at the *Sorbonne* and the *Académie des Scieces* together with his close friend Lejeune Dirichlet (who was to became an authoritative mathematician, and

 | DOI 10.1163/9789004315235_003

In the book he had published in 1851 on the foundations of knowledge, *Essai sur les fondements de nos connaissances et sur les caractères de la critique philosophique*, Cournot had oscillated between two poles: an old-fashioned encyclopaedic design on the one hand, and a new and sophisticated analysis of scientific practice on the other. He had perceived that philosophical tradition could be preserved and revived only through a process of cross-fertilisation between that tradition and the recent "positive knowledge," namely the whole body of scientific achievements. This involved a detailed inquiry into "the contemporary state of affairs in science." The dialogue between the old and illustrious body of philosophical knowledge, and the more recent body of scientific knowledge required an exploration of their foundations. The analysis ranged from mathematical probability to psychology through "the ideas of space and time," vital actions, continuous and discontinuous processes, "analytical and synthetic judgments," philosophy of history, and "philosophy of science" [Cournot 1851, pp. I and II].[2]

In 1861 Cournot pointed out how much the recent developments in physical and natural sciences had highlighted the differences among "contents, principles, and methods" of the various sciences. At the same time, he stressed the usefulness of "principles and methods of natural sciences" in the study of human "languages, habits, ideas, institutions, and history" in general. In his view, the process of specialization and professionalization was widening the gap between physical sciences and natural sciences but it was narrowing the gap between natural sciences and human sciences. Recent developments in life sciences had transformed them into the intellectual knot where the network of "our ideas and our scientific knowledge" converged. In some way, life sciences represented a centre of symmetry for the classification of disciplines, and even the book's table of contents mirrored that symmetry. A series of bodies of knowledge was sorted into a hierarchical order: "logical and mathematical sciences," physical sciences and natural sciences were to be analysed in

succeeded Carl Friedrich Gauss at the University of Göttingen). He then worked as a tutor while continuing with his researches, and in 1829 received a doctorate in mathematics. He was appreciated by Poisson, who helped him during the first stages of his career: on Poisson's recommendation, he was appointed to a chair of mathematical analysis in Lyon in 1834, and in Grenoble the following year. After becoming Dean in Grenoble, in 1838 he became General Inspector of Public Education; in the same year he published his book on the mathematical theory of wealth [Moore 1905a, pp. 528-35].

2 He had also published *Traité élémentaire de la théorie des fonctions et du calcul infinitésimal* in 1841, and *De l'origine et des limites de la correspondance entre l'algèbre et la géométrie* in 1847. For Cournot's biography, see Moore 1905, pp. 521-43.

the first volume, and the study of "human societies" together with "history and civilisation" in the second volume [Cournot 1861, pp. III-V].

Cournot undertook his analysis in ostensible accordance with Comte's hierarchy of sciences, but his methods and aims were different. He was interested in understanding how the emergence of modern science had changed our patterns of explanation. The conceptual transition between the natural philosophy of the Middle Ages and Kepler, Galileo and Newton's modern science had required "a new key," namely new conceptual pathways. That new attitude had transformed Copernicus and Tycho's "purely mathematical ideas," or purely geometrical models, into physical models. From a different point of view, that transition could be looked upon as the passage from a timeless science to a science involving time and history. Periodical eternal motions only needed logic and geometry. Modern science, and modern physics in particular, had emerged when time and history had come into play, when the time evolution of natural processes had been considered worth studying. The introduction of time and history into the tradition of natural philosophy had led to the introduction of new concepts such as "*initial conditions*," and forces or "physical causes" that triggered off the passage "from a previous state to the present one" of any material system [Cournot 1861, pp. 118-22]

Cournot was aware of the problematic status of the concept of force, and of the relationship between force and matter. Physics was based on a sort of "*duality*" and complementarity between matter and force, and the concept of inertia, could be looked upon as an ingenious implementation of that duality. Matter was supposed to be unable to exert force: the passive nature of matter and the relational nature of force were looked upon as "independent from each other." That duality represented an interpretative tool which overcame "the conclusions that could be drawn from experience": it was "a metaphysical hypothesis," which could be opposed to the metaphysical hypothesis of "essentially active monads." An essential feature of inertia deserved to be pointed out: it was "insensitive to the difference between rest and motion," and it did not change when a body at rest was put in motion. This was another fundamental hypothesis indeed, and it marked the "crucial difference" between the ancient "theory of motion, which was geometrical, and mechanical physics" [Cournot 1861, pp. 162-4].

Cournot stressed that the "nature and aim" of his book were different from those of "an elementary and didactical treatise": he was interested in a critical and historical analysis of science, and different attitudes and alternative pathways had to be explored. According to Cournot, mechanics was not a monolithic body of knowledge: at least two different approaches were at stake. The transition from "the abstract truths of geometry to the fundamental principles of physics" could be undertaken in two ways. Mechanics could be put forward as a generalisation of kinematics, "namely the geometrical theory of motion," or as a generalisation of statics, where "the composition and equivalence of forces" were involved. The generalisation of mechanics from kinematics, and the generalisation from statics rested upon different principles: "*the law of independence of motions*" in the first case, and "*the principle of proportionality between forces and velocities*"[3] in the latter. However, the different principles and the different conceptual frameworks converged towards the same empirical outcome. In brief, different interpretations at the theoretical level could correspond to the same result at the empirical level [Cournot 1861, pp. 165-6, 174, and 176]

When Cournot focused further on inertia, he found that the assumed duality did not bear the weight of a deeper analysis. A mass of lead fastened to a mass of iron opposed its inertia to the magnetic force coming from the iron. As a consequence, the inertia of lead could decrease the acceleration of iron. In other words, the inertia of lead resisted the magnetic force in the same way as a force resisted another force. In this sense he could "compare inertia with a force," and he could put forward a cautious correspondence between two terms which involved "opposite concepts." He found that the complex concept of inertia involved both physics and metaphysics, or rather, "a shared land" where the principles of physics had their natural seat.[4] Together with other principles, the principle of inertia appeared in physics textbooks as the consequence of experience, but in reality general principles could not be submitted to experience in the same way as specific physical laws could. They were interpretations of actual and possible experiences rather than necessary consequences of experiences. Experiments on the Boyle-Mariotte law for elastic fluids, the Snell-Descartes law for the refraction of light, or the Coulomb law for electrified bodies, had actually been performed in order to test the accuracy of specific mathematical laws dealing with specific effects. On the

3 Here Cournot makes use of the term "velocity" in the sense of the velocity variation or incremental velocity.

4 In 1858, in the first lines of the book *La métaphysique et la science ou principes de métaphysique positive*, the French philosopher Étienne Vacherot had remarked that history had not spoken the last word on metaphysics. Metaphysics dealt with the free practice of "analysis and critics," and allowed philosophers to protect themselves against "unreasonable dogmatism and regrettable scepticism." He found that metaphysics needed to be updated, and philosophers should restart from the point "where Kant had left it." Although he was aware that his intellectual task would not have been endorsed by the majority of scholars, he ventured to "reconcile metaphysics with science" [Vacherot 1858, pp. V-VI, XV, XXXV, 52, and 94].

contrary, the law of inertia could be looked upon as "a necessary truth" or a fundamental definition in the context of modern science. Speculating on such a principle dealt with "philosophy of science rather than science in itself or positive science." *Philosophy of science* was the name of the borderland between philosophy and science [Cournot 1861, pp. 179 and 181-3]

With regard to the received view on inertia Cournot mentioned Laplace's *Mécanique céleste*, where the great mathematician had expressed his firm belief in the close link between the principle of inertia and experience. More specifically, Laplace had stated that "when we see a variation in the motion of a body, we assume that it stems from an external cause." Cournot advanced a methodological question: why had Laplace automatically linked the variation in the motion of a body to an unknown cause rather than to the failure of the inertia principle? He answered that Laplace could not have given up the inertia principle because the principle had great explanatory power. Physical principles and other general laws involved "both facts and ideas, or better, the rational interpretation of facts." More in general, he stressed that when scientists observed an unexpected variation in a natural phenomenon, they preferred to assume the existence of a perturbation rather than the failure of a principle or law [Cournot 1861, pp. 186-8; Laplace 1798-1824, tome 1, p. 14]. According to Cournot, the separation between causes and perturbations was a rational choice that could not be directly deduced from experience.

> Let us suppose that pressure is roughly proportional to the reciprocal volume for gases submitted to usual conditions, but the results are at variance with this law when pressures exceed a given threshold. The simple relationship originally observed and the smallness of variations might suggest that two different causes are involved, the latter being subordinate to the former. The main cause gives rise to the original simple law, whereas the variations can be associated with incidental and disruptive causes. Therefore we can decide to theoretically separate two effects that are not experimentally separated: to make or not to make this choice leads to different interpretations of the same fact [Cournot 1861, p. 188].

Cournot also stressed that extra-empirical and extra-mathematical elements came necessarily into play in the building up of a scientific explanation. Logical and metaphysical elements allowed scientists "to grasp a set of phenomena in its totality," in order to make them rationally understandable in terms of "order, unity, and simplicity." In other words, along with "perceptible facts, on which experience can irrevocably decide," and "mathematical truth, which involves a formal, thorough demonstration," there was a third element, namely "a philosophical conception, on which scholars can reach agreement," but which was not submitted to "an experimental proof or mathematical demonstration." Facts had to be "interpreted by an idea," and that interpretation affected "experiences involving both incidental and consequential facts." The more general a physical principle was, the less empirically sensitive it was, because many concurrent facts were involved, and their influence could only be interpreted by means of the principle itself. Principles and very general laws relied mainly on rational evidence, which was based on "the order and regularity" they introduced "in the explanation and interpretation of observable facts." Once more Cournot remarked that the analysis of those principles pertained to "what should be labelled the philosophy of sciences" [Cournot 1861, pp. 189-90].

Essential features of scientific theories such as order, unity, simplicity and regularity did not imply that all phenomena were subject to the same laws, or that different research fields could be reduced to the same set of principles. The passage from one field to another appeared as a sensitive issue indeed. The transition from geometry to mechanics, or rather the passage from "the geometrical theory of motion to physical mechanics," had required a dramatic conceptual transition, and the emergence of modern science in the seventeenth century testified to that dramatic nature. The subsequent transition from "physical mechanics to physics itself" could be looked upon as less dramatic but nonetheless problematic. Some questions were at stake. Should the whole body of physical sciences be submitted to mechanics? Why had neither chemistry nor optics nor "any other field of physical sciences" aspired to such a "supremacy or universality"? Was mechanics a field of physics among other fields, or was it different in its nature? Cournot's first answer made reference to history and the supposed stability of mechanics over time: neither geometry nor mechanics had experienced as much "progress and revolutions" as other bodies of knowledge had done in the last centuries. The second answer dealt with a specific feature of mechanics, namely its *ideal* nature. In other words, mechanics dealt with simplified mathematical models. The difference between mechanics and physics was exemplified by the difference between the theory of ideal fluids and the theories of real fluids, where viscosity, friction, and other phenomena were involved. Some scholars hoped that a satisfactory comprehension of natural processes from the mechanical point of view would be achieved when the microscopic landscape of matter and motion had been clarified, but Cournot found that, for the time being, it was a mere "philosophical hypothesis" [Cournot 1861, pp. 191-2]. Ultimately, Cournot acknowledged that mechanics had played, and continued to play, a specific and important role in science as an ideal model, but he found that any attempt to reduce other sciences to mechanics was chimeric.

Moreover he pointed out that two kinds of mechanics had been at stake in the history of science: on the one hand, "the mechanics of geometers, especially applied to the motion of heavenly bodies," and on the other hand, "the mechanics of mechanics, engineers,…." The separation between the two mechanics echoed the traditional mediaeval separation between the theoretical body of knowledge of mathematics in a wide sense, astronomy included (the *quadrivium*), and the empirical body of knowledge of the mechanical arts. According to Cournot, the separation had survived after the emergence of modern science, and had found a new implementation in the separation between the mechanics of actions at a distance, and "the mechanics of machines," where "*living force* and *work*" were involved. That opposition represented a feature of continuity in the history of science, namely the persistence of ancient classifications throughout the profound transformations that had overturned natural philosophy in the seventeenth century. In some way, the difference between "the laws pertaining to heavenly motions" and "those ruling sub-lunar phenomena" had survived. He interpreted the lively debate between the *Cartesian* and *Newtonian* schools after the scientific revolution in accordance with this historiographical framework. The former had promoted a universal pattern of explanation in terms of "pressures and corpuscular collisions" of the *sub-lunar kind*, whereas the latter had claimed that all interactions could be reduced to actions at a distance of the *heavenly kind* "on the track of Newtonian attraction" [Cournot 1861, pp. 195-7 and 199].

2 A Plurality of Scientific Methods

Cournot pointed out the existence of "two sections, or leading categories" in physical sciences: on the one hand, "the physics of *perceptible bodies*," and on the other, "the *infinitesimal, or corpuscular, or molecular physics*." For instance, crystallography could be looked upon as a purely geometrical science, but at the microscopic level involved "the field of molecular physics." In the same way, the equilibrium of liquids in connected vessels pertained to mechanics or "perceptible physics," but the "dynamical explanation" of viscosity, friction, and capillarity required "the mysteries of molecular physics." The duality between perceptible and molecular, or between macroscopic and microscopic, also affected the role played by chemistry in the classification of sciences. Should chemistry be looked upon as a part of molecular physics or as an autonomous body of knowledge? Chemistry dealt with "the *hetereogeneity* of bodies," and the composition and decomposition, or "analysis and synthesis" of "*homogeneous* elements." At the molecular level, the heterogeneity could be interpreted as a consequence of "the essential and primitive homogeneity of the elements of ponderable matter." Nevertheless this reduction to molecular physics was not in tune with the interpretation of chemical actions in terms of *affinity*. Chemical affinity could not be reduced to a specific instance of physical force because it was not endowed with intensity and direction in the physical sense. The word 'affinity' stood for a complex network of "imponderable entities" and hidden forces that resisted a naïve reduction to mechanics [Cournot 1861, pp. 208, 210-12, and 214].

Another sensitive issue prevented chemistry from undergoing a naïve reduction: the existence of sudden transitions between different states, namely sudden combinations and decompositions between qualitatively different substances. Chemistry was a science of transformations, and chemical transformations could be violent or marked by a sharp discontinuity. Cournot insisted on the "sharp and sudden transformations from one state to another" as opposed to "the law of continuity" that ruled "mechanical phenomena." The fact is that continuity in itself was not an essential feature of mechanics: unlike the continuity of the equations of motion, Cartesian mechanics, or the mechanics of collisions, was based on discontinuous processes. Nevertheless, he found that different kinds of discontinuity emerged from chemistry: discontinuity in the sense of abrupt and energetic transformations, discontinuity in the sense of qualitative transformations of chemical compounds, and discontinuity in the sense of rearrangements of chemical substances in accordance with specific, integer ratios between their weights. Cournot omitted a detailed analysis of these different kinds of discontinuity, and confined himself to synthetically remarking that "chemical actions give place to sudden combinations and decompositions with definite proportions." However he also mentioned the difference between mechanical mass, which was linked to "weight and inertia," and "what we could label chemical mass," which was linked to "the capacity of saturation." In the end he found that chemistry could not be reduced to "the notions of mechanics" [Cournot 1861, pp. 214-5].

After having guarded against a naïve reductionism, namely a continuous transition between contiguous bodies of knowledge, Cournot focused on the opposite aspect. The transition was neither smoothly continuous nor sharply discontinuous: the boundaries between physics and chemistry were blurry rather than clearly defined. In the case of clusters of molecules, the law of definite proportions among elementary components "maintained its theoretical validity" but the contingent addition or subtraction of a single atom "could not be appreciated." The discontinuous nature of chemical phenomena disappeared when the difference in weight was smaller than experimental accuracy. Blurred transitions between different fields were at stake even in the case of "the physics of imponderables" where phenomena like "light and radiant heat" were involved. Waves could be approached by means of "pure kinematics," where the mere superposition of wave motions was involved, or "hypothetical dynamics," where "hypothetical forces between atom and atom" were assumed. Anyway the reduction of the *imponderables* of physics to mechanics remained highly questionable. In the end, neither scientific method nor actual scientific practice could help decide between the two meta-theoretical interpretations, namely mechanics as the foundation of the whole of physical sciences, and mechanics as "nothing else but a field of physics" [Cournot 1861, pp. 216, 220-2].

The patterns of explanation offered by mechanics and physics could be useful in life sciences: some "organs and functions of animals" could rely on the analogy with "mechanical machines and engines" whereas others could rely on the analogy with "physical and especially chemical effects." Nevertheless life sciences could not rely on two very general principles that all physical and chemical sciences had in common: a principle of linearity or superposition, and a principle of invariance over time. With regard to the first principle, two material systems, or parts of the same system, could act on each other as if the other systems did not exist. The interaction between two planets took place independently of the presence of other planets. The presence of a molecule C did not affect the intensity of the interaction between two molecules A and B. On the contrary, in a living structure, the action of a part of the structure on another was affected by the systematic link with "the structure and the functions of the system" as a whole. With regard to the second principle, physical processes were ruled by "unchangeable laws over time": there was a mutual independence between "the flow of time and processes taking place through time." On the contrary, living beings experienced an evolution, more specifically "essential changes in the intensity of the active principle" that ruled their evolution. Living species had appeared and then disappeared over time: *Nature* was not compelled "to act always in the same way in the same situations." Different ages could involve different laws, and therefore "time was involved in an intrinsic way in the laws ruling Nature" [Cournot 1861, pp. 223, 272-3, 277, and 284].

According to Cournot, the dichotomy between physical sciences in a wide sense and life sciences stood beside another dichotomy that involved the whole domain of natural sciences, and corresponded to two different approaches to the natural world. He synthesised the dichotomy by means of two couples of words, *Nature* and *world* [*Monde*], or "*physical* sciences" and "*cosmological* sciences." The former dealt with "what the ancients had qualified as science in general, which abstracted from individual objects." Physics and chemistry belonged to this field. Astronomy and geology belonged to the second field, because they dealt with "specific or individual objects such as the sun, the Milky Way, Saturn's rings, the moon or the earth." Cosmological sciences involved the specific history of every object, where the word "*history*" had to be intended "in its widest philosophical sense." In some way, cosmological sciences bridged the gap between physical and life sciences, and at the same time, opened another meaningful gap. They introduced history into some of the physical sciences, and this fact stressed their distance from other physical sciences. Cournot outlined a hierarchy of historicity that led from cosmological sciences to anthropology through life sciences. The "triple character of simplicity, constancy, and regularity" of physical laws gave way to less definite laws. In subject matters like meteorology or volcanology, he found that order and disorder, and predictability and unpredictability were mixed with each other. Sometimes order and "regular, permanent or periodical phenomena" emerged from provisional stages of disorder and irregularity [Cournot 1861, pp. 279-81 and 305].

With regard to the link between life and inorganic processes, Cournot leant towards a mild vitalism. He preferred looking upon complex organised structures "as a consequence of life rather than life as the consequence of a material system somehow stirred up by purely physical forces." He assumed "a creative and pliable power, a vital energy" that oversaw "the development of the organism," and at the same time was influenced by the transformations that took place in the organism itself. At that stage he outlined a meta-theoretical framework where "*physical* and *cosmological* sciences" stood on the one hand, and "*natural* sciences and *natural history*" on the other. The process of specialisation inside the scientific community had given birth to new branches stemming from the tree of *natural* sciences. Therefore he put forward a slightly different framework which satisfied him because it seemed endowed with a more symmetric structure: in the first place logic and geometry, and then mechanical forces, molecular and chemical forces, vegetative life, animal life, and human intellectual life. The borderline between the third and fourth level represented the most meaningful transitions between bodies of knowledge. It also corresponded to the ideal line that Cournot interpreted as an axis of symmetry. He found that the label "*biological* sciences" suited the field of life sciences in general [Cournot 1861, pp. 319, 321, 323, and 329].

He knew that, in the context of biological sciences, a network of new research fields and corresponding labels was emerging. As a consequence, a more refined classification could be envisaged, wherein labels such as *morphology*, *physiology*, and *psychology* were involved. Medicine was closely linked to physiology but their relationship was quite complex. Medicine could be looked upon as "mother and daughter of physiology at the same time": it was "the mother from the historical point of view, and the daughter from the theoretical point of view." Once more, both dichotomies and symmetries were at stake in Cournot's classifications, and he arranged the bodies of knowledge in couples whose relationship was of the structure-function kind: logic could be associated to psychology in the same way as anatomy could be associated to physiology. In the end, some kind of circular taxonomy emerged. Logic, at the top of the classification, was closely linked to psychology, which was placed at the bottom. Human psychology in a wide sense could be considered as the living source of logic, and logic as the intimate structure and formal support of psychology [Cournot 1861, pp. 334-6 and 339].

A complex relationship among bodies of knowledge slowly emerged from Cournot's historical and meta-theoretical reconstruction. For instance, he insisted on the close relationship between biology and history. Not only did he stress that "in the domain of life, everything depends on history" but he also remarked that the transition from physics to biology involved an increasing rate of history and philosophy. More specifically, history dealt with the origin and development of beings whereas philosophy dealt with *harmony* and *finality* in the structures and functions of living beings. In the context of life sciences, the interplay between science and philosophy was quite complex: on the one hand, "scientific speculations" on the origins of life required a suitable "philosophy of Nature," and on the other, "philosophical speculations on life processes" required that the outcomes of "positive science" be taken into account. According to Cournot, physiologists and physicians could not shy away from that symmetric relationship, even though they should suitably abstain from "undertaking ontological debates" on the nature of life and the soul [Cournot 1861, pp. 339 and 344-5].

In reality, medicine and physiology had experienced a dramatic experimental turn in France. An important professional process had taken place during and after the Revolution: the unification between medical and surgical practices. The previous separation had its roots in a long–lasting and more general separation between theoretical and practical bodies of knowledge. The conflation of surgical expertise with medical theoretical practice can help understand the greater success that vivisection had in the French environment when compared to the British and German ones. Differently from the German and British context, in France the most effective experimental practice consisted of experiments on animals. Vivisection obviously required that "the aversion to manual procedures" be overcome, and at the same time that any moral objection be dismissed. It also required a positive bias in favour of a cruel practice that disrupted the biological balance and the necessary integration among the parts and functions of every living system. Finally, it relied on the assumption that "the mode of action of medicines was the same for men and animals" [Lesch 1984, pp. 5-6, 10, and 157; Mendelsohn 1965, p. 217].[5]

In 1836 the influential physician and physiologist François Magendie had regretted that medicine and physiology were not yet "true sciences": chemists in different countries of the world agreed with each other on the interpretation of a chemical reaction, whereas such an agreement had not been attained in medical sciences. He stressed that "the most authoritative scholars" defended "the most heterogeneous systems" when they faced the interpretation of "physiological or pathological events" [Magendie 1836, p. 4].

> Many people have arrived at the discomforting conclusion that medical science and physiology are not real sciences yet. Although in different corners of the world, both in Paris and New York, both in London and Calicut, chemists agree with each other on what happens at the bottom of a crucible, a similar agreement has not been achieved for medical theories. Instead of a fortunate harmony in the interpretation of physiological and pathological phenomena we find that the most authoritative scholars of every time, our time included, endorse the most different theories [Magendie 1836, p. 4]. [6]

Nevertheless, some progress had been made at the borderline between chemistry and medical sciences. In the first decades of the century new substances were studied and chemically isolated: among them, the active principles of *opium, nux vomica, St. Ignatius bean,* and prussic acid. Around mid-nineteenth

---

5 In 1820, in the report of his visit to the influential French physician and surgeon Xavier Bichat, the English physiologist John Cross expressed his vivid impressions of "the mania of vivisection," and the "unlimited confidence in this manner of studying physiology" [Lesch 1984, p. 80].

6 Magendie developed his physiological interests coming from surgery and anatomy, and in 1816-17 published the influential *Précis élémentaire de Physiologie* in two volumes. In 1821 he founded the *Journal de Physiologie expérimentale et pathologique*, which he edited for ten years, and established an experimentalist tradition in medical sciences. His scientific approach had a marked surgical character, and relied on animal vivisection, a surgical practice that merged with clinical practice, veterinary, and pharmacy. He collaborated extensively with pharmacists and chemists: the surgical practice on animals required the injection of various drugs and poisons, and the observation of the corresponding effects. In 1821 he published a *Formulaire pour la préparation et l'emploi de plusieurs nouveaux médicaments* that went through eight editions between 1821 and 1834, and grew rapidly from 84 to 438 pages in this time span [Lesch 1984, pp. 90-1, 100, 109, 127, 137-9, 155-7, and 162-3].

century, "physics and chemistry had become indispensable for the study of organic functions and the construction of physiological concepts," but the role of physics in physiology was widely debated. Both in France and in German-speaking countries, a distrust in "earlier vitalistic and/or teleological explanations of Nature" was generally professed, but in German-speaking countries the reaction was definitely sharper because of the previous hegemony of *Naturphilosophie*. The problematic link between life sciences and the physico-chemical body of knowledge had already been debated in the time-span between Lavoisier and Laplace's first experiments on "respiration, animal heat and transpiration," and Friedrich Wöhler's 1828 synthesis of urea. In brief, "the major tendency in German physiological theories was reductionist" whereas French scholars, and Claude Bernard in particular, "insisted none the less that there was need for special biological laws" even though "the importance of physical-chemical techniques and laws for understanding biological phenomena" was not underestimated. As I will show in a following section, the French physician Claude Bernard was one of the main protagonists in the debate on scientific method in the context of life sciences [Mendelsohn 1965, pp. 203-4, and 215, and 217].[7]

Bernard's life and scientific enterprise in the 1840s and 1850s can be placed at the crossroads between the reduction of life sciences to physics and chemistry, and the claim to specific principles and specific experimental practices. In the early 1860s, he had already brought many important contribution to the emerging science of physiology: he had discovered the liver's glycogenic function, the pancreas' digestive function, and "the existence of the vasomotor nerves." From the mid-1840s onwards, he had progressively moved away from medical practice, and had become a full-time physiologist. In 1848 Bernard contributed to the foundation of the *Société de Biologie* together with the physician and naturalist Charles-Philippe Robin, and the physician Pierre Rayer, who was its first president. This society and its *Comptes rendus et mémoires* marked the passage from medicine to *science*: physiology in the sense of experimental and general physiology, took the lead, and medical practice followed. In other words, Bernard "and other scientifically-minded members of the Paris medical community" made the move from "a medically oriented

7 The professionalization of pharmacy and veterinary medicine had been accomplished before the professionalization of physiology. However in 1795 we find the first occurrence of the term *physiology* in the title of a scientific journal: it was the *Archive für Physiologie* that had been founded by the German-speaking physician Johann Christian Reil. In 1800 Bichat published the book *Recherches physiologiques sur la vie et la mort* [Lesch 1984, pp. 15-6 and 28].

physiology to the ideal of a general science of life" [Lesch 1984, pp. 1-2, 196-8, and 218-23; Mendelsohn 1965, p. 217].[8]

3 Physical and Biological Actions

The most sensitive issue in life sciences was "the concept of vital force," and Cournot took for granted that it was "intrinsically different from the concept of mechanical force." First of all, vital forces could not rely on a definite intensity and direction like ordinary vector forces, a negative feature that was shared by the above-mentioned chemical affinity. Secondly, vital forces did not have "a definite seat": they did not stem from, nor were experienced by, nor acted between definite particles or living tissues.[9] However the different features of the two forces did not prevent them from acting simultaneously, and achieving an effective collaboration. More specifically, vital forces could set into action physical forces in the same way as a negligible friction could trigger the explosion of a chemical mixture. In other words, "the principle of life" could not be added to physical forces as something of the same kind. Neither could it balance a physical force by acting in the opposite direction, but it could "give a suitable direction" to chemical and physical forces. According to Cournot, this distinctive feature of vital forces could help scholars overcome the opposition between the two main "physiological schools," namely *vitalists* and "the adversaries of vitalists" or reductionists. He found that the directive power of vital forces could rely on a meaningful historical analogy: in the seventeenth and eighteenth centuries, some scholars had mocked Descartes' followers because their physical *world* was put in motion by an initial divine impulsion [chiquenaude]. Cournot found that even scholars who poked fun at that supposedly divine flick of the fingers could find reasonable "a directive power" whose consequences were not so different from chemical and physical forces [Cournot 1861, pp. 339 and 367-72].

8 Everett Mendelsohn synthesised the difference between French and German traditions of research. See Mendelsohn 1965, p. 219: "Bernard in France, Schwann in Germany, took their theories of the organism in different directions. Schwann reacting against the past utilized physics and chemistry to *repudiate* the earlier German biological traditions. From his efforts a whole generation of physiologists emerged embracing physicalism as the acceptable guide for the formation of concepts in physiology. In France, Bernard absorbed the past of biology and welded to it new theories and techniques of physiological chemistry."

9 It is worth mentioning that, in the book Bichat had published in 1801, *Anatomie générale*, a certain number of organic tissues were looked upon as "the ultimate, irreducible constituents of all bodily structure" [Lesch 1984, p. 67].

In the end, Cournot was not able to solve the problem he had raised on the exact nature of vital actions and their interaction with physical forces. Vital actions were different, but their way of action was not so different from the way of action of the latter. They could stimulate the full deployment of physical and chemical forces in the same way as a little expenditure of physical work could stir up an avalanche, or trigger off an explosion, or activate an electric circuit. This interpretation of life and vital actions was interesting even though not completely convincing. He should have explained in which sense vital actions had a "directive power." Obviously the adjective *directive* in the context of life sciences had a different meaning from *directive* in the spatial sense or in the sense of vector forces: he himself had already specified that vital forces lacked a direction in a physical sense. The analogy between vital forces, and what Cournot labelled "moral commitment [excitation morale]" did not help clarify the concept. The analogy in itself was reasonable: a moral decision could "set in motion" a muscular force that had not found "the opportunity to be deployed" even though in no way could it "produce a muscular force that did not previously exist." The permanence of life required that "the directive power of the vital principle" could "bring into play (electric) currents or other chemical forces" which could provisionally counterbalance both "the affinities that led to the decomposition of organic matter," and molecular forces that led to "the blood clotting." In the end, in Cournot's sketch, the attempt to explain the difference between physical and vital forces gave way to a mild symmetry: not only could the former be triggered by the latter, but also the latter could be triggered by the former [Cournot 1861, pp. 376-9].

It is worth stressing that questions about the relationship between triggering actions and their physical or chemical effects had already emerged in the context of medical sciences before the turn of the mid-century. Those questions involved the quantitative and qualitative relationship between causes and effects, and more specifically the quantitative discrepancy between a slight triggering cause and possibly huge effects, as well as the different nature of a given cause and the corresponding effect. The German-speaking physician and physiologist Robert Mayer had outlined the problem in a letter to the physician and psychiatrist Wilhelm Griesinger in 1844, where he had stressed how questionable the meaning of the words "cause, effect, and transformation" really was. In the field of mental processes, might we say that "cerebral activity" is the cause of the book a scholar is writing? The sentence could be accepted in a very general sense, but it would be definitely pointless to say "the cause, namely the cerebral activity, transforms itself into the effect, namely the book." In the field of physiology, it was known that physical activity could "improve breathing, heartbeat, and warmth," and could "accelerate metabolism."

In this case, which was exactly the quantitative link "by pounds and ounces" between causes and effects? In the field of physical and chemical processes, the transformation of a cause into an effect was not less problematic. If a spark triggered off an explosion, might we say that the former was the cause of the latter? [Mayer 1844, pp. 98 and 100-102].[10]

Mayer had attempted to analyse the various steps in a process that involved both physical and chemical actions. The spark set fire to the gunpowder, the blaze released a certain amount of heat, and finally heat was "in part transformed into the mechanical effect" of explosion. The series of two transformations showed an ostensible symmetry. We could say that the spark (a) was "the cause of the gunpowder explosion (b)," and in its turn, the latter was "the cause of the earth blowing up (c)." Nevertheless, there was a "definite proportion" between (b) and (c) but neither (b) nor (c) could be put into a definite ratio to (a). Comparable explosions might be "triggered off by a spark or by a torch," and in this case the first cause involved two different sources of *force* or energy. According to Mayer, from a logical point of view, we were "not allowed to label *causal relation* two relationships that are so different" as (a) to (b) and (b) to (c) really were. He found that two alternatives were on hand: we could give up looking upon one of the two connections as a causal connection, or we could give up any demand for "a logically consistent language." In other words, a conceptual and linguistic revision of basic physical entities was in order [Mayer 1844, pp. 98-99 and 101].

According to Cournot, the transition from inanimate matter to living beings involved both the temporal and spatial scale: more specifically, living structures had appeared long after the appearance of molecular structures, and the spatial dimensions of the former were far greater. Which was the structure that could be looked upon as the transition from inorganic to organic? Perhaps eggs, or other "organic machines" such as spermatozoon, which required an act of fertilisation, or spores, bulbs, buds and cuttings, which could give rise to a living being without any fertilisation. However every search for the first step

10 In 1842 Robert Mayer had stressed the two essential features of forces [Kräfte] or causes [Ursachen]. First, they could not be destroyed, and second, they could be transformed into each other. The former was a quantitative feature, and the second a qualitative one. Forces shared the two features with matter, but differently from matter, they were imponderable: in brief they were "indestructible, transformable, and imponderable entities." Every cause produced a corresponding effect [Wirkung], and the effect had to equal the cause, as in the case of a falling body, where "the distance between the weight and the ground" corresponded to a specific "quantum of motion" gained by the weight [Mayer 1842, pp. 4-6 and 9].

towards life was unsuccessful because all "problems of origin" were pointless: living structures could only emerge from other living structures [*Omne vivum ex vivo*]. Moreover, the questionable reduction of life to inert matter called into play determinism in the context of life sciences. In spite of the computational complexity, the future of a physical system could be theoretically determined provided that "the invariable laws that rule the inert matter" and "the current situation of all parts" were suitably specified. On the contrary, "phenomena submitted to the influence of life" could not enjoy that determinism, in the sense of "determination or prevision *in futurum*." The reason rested upon an essential feature of living beings that he had already pointed out: vital forces might "change over time independently of external influences" [Cournot 1861, pp. 385, 401-3, 406, and 409-11].

The necessity of "a directive power," or "a principle of finality and coordination," also dealt with "the lack of intermediate shapes in paleontological series." That principle, which he also labelled "principle of finality and harmony," "harmonic unity," and "creative synergy," had to be looked upon as a "natural mode of operation." At the same time, the necessity of such a principle in biological processes highlighted an intrinsic weakness of "the system of our conceptions," or "a discontinuity in its theoretical connections." The discontinuity between the two kinds of forces rested upon the impossibility of a definite correspondence between biological forces and "a physical *substratum*." In the recent history of electricity, some scholars had envisaged the origin of vital actions in "galvanic actions," but the enthusiasm had slowly faded away. At a higher level, namely the level of human feelings and consciousness, it had appeared that "memory and sensitivity" could be associated with an actual "individual subject," but even this possibility had vanished as soon as scholars had realised that sensitivity was nothing else but "one of the ways in which Nature allows living beings to get in touch with external objects." Even memory was a relationship between a subject and some external events rather than a feature or activity of the subject. In general, phenomena like sensitivity and imagination did not require any "substantial support" or "substantial subject for the *self*." In the end, Cournot took shelter in "Leibnitz's conception," wherein force was not "a feature of a substance" but the origin of "substantiality or identity." This meta-theoretical option allowed him to retain the concept of vital force as something endowed with "an intrinsic and necessary role" [Cournot 1861, pp. 466-8, 471-2, 478, and 480-1].[11]

11 At the end of the XVIII century, the term and concept of "*principe vital*" or "vital force" or "special principle of life" was already in use in order to identify neither mental nor

After having pointed out the difference between "*vitalism*" and "*animism*," Cournot focused on the relationship between vitalism and *finalism*. It was a complex problem indeed, because two other issues were involved: the role played by chance, and the existence of different levels of finality. On the one hand, some "animals and plants" survived because their parental germs were preferred to other germs, but that preference was due to "mere chance." On the other hand, biological finalism involved a competition among different kinds of purposiveness. The individual organism aimed at preserving itself, the species aimed at the same target, and even the whole environment aimed at preserving "the general harmony of the living universe, and the balance among species." In some way, biological finalism was a blind finalism: frost could be a danger for a population of insects but an advantage for plants that were infested by those insects. Massive extinctions over time were obviously bad for the species and individuals involved but were good for other species and individuals that had the opportunity to flourish. From the point of view of the natural environment, periodical extinctions represented "a source of further progress through variety." The adjectives *good* and *evil* had "a relative meaning for species and individuals," but acquired a more definite meaning when "the wealth, the harmony, and the beauty" of the world were involved. Pleasure and pain of "sensitive beings," and their provisional lives were embedded in a sort of *absolute* or higher-level *good* that involved the natural world as a whole [Cournot 1861, pp. 483-6 and 492-3].

Differently from the concepts of "*race*" and "*variety*," the concept of biological "*species*" appeared to Cournot as the clearest and most reliable because it made reference to a set of features that were definitely "the most essential and fundamental." In brief, a species was a set of individuals that could mate, and give birth to fertile beings. The definition could arbitrarily be extended to asexual reproductions, but in this case it lost its empirical meaning. In reality, even in the case of sexual reproduction, how could mutual fertility be checked in all specific cases? Neither had anybody ever performed such an empirical inspection, nor had anybody seriously considered doing so. Moreover, how could the inspection be performed for extinct species that had been living in ancient ages, in conditions very different from the current ones?[12] In brief, a

mechanical processes taking place in living beings. It can be found in the texts of the physician Paul Joseph Barthez, then in Bichat's texts, and eventually in Magendie as a principle that was "ontologically distinct from the conscious mind or soul" [Lesch 1984, pp. 25, 65, and 93].

12 Cournot also stressed the impossibility of watching the emergence and transmutation of a species.

thorough definition of species was "an illusion," but this did not prevent scientists from attaining a widespread agreement on the classification of species. In the end, the species classifications should not be "arbitrary conventions" but rather "natural classifications" or suitable representations of something intrinsic to Nature. Classifications involved something that belonged to neither the logical domain [la forme d'un jugement logique] nor the empirical domain [une mesure précise de la proximité]. There was a third domain, which dealt with a pragmatic, "instinctive judgement [appréciation instinctive]." However pragmatic and provisional it could be, a theoretical option or judgement could be effective and reliable. In particular, the concept of species was a fruitful theoretical entity [Cournot 1861, pp. 412-3, 416-8, 424-6].[13]

4 The Emergence of Physiology as a Science

In reality Darwin's theory did not attract many French physicians and physiologists because it was far from their actual practice and interests. More specifically, some physicians and physiologists were interested in fostering the acknowledgement of physiology as an experimental science, and Darwin's theory could not help attain this target. Bernard's purpose was exactly to establish experimental physiology as an autonomous discipline, namely a discipline endowed with "intellectual, institutional, and pedagogical independence." He had begun his activity as *préparateur* for Magendie's course in physiology at the Collège de France in 1842, and he had made use of the "inadequate laboratory provided by that institution." Only in 1854 did he obtain the Chair of General Physiology at the Faculty of Science in Paris, but even then he did not rely on a laboratory "for research or teaching purposes." This date could be looked upon as the first institutionalisation of experimental physiology in France, even though Bernard continued to fight for the complete acknowledgment of physiology as a science in the following years. The physiological laboratory became the symbol of the emancipation of physiology as a science: it should have been the seat of research, "discovery and training for discovery," wherein both "a body of knowledge and a corps of investigators" could emerge. Physiology had to go beyond the observation of organic processes, beyond

13 The inquiry into the concept of species and its origin appeared to Cournot as a philosophical rather than a scientific issue. Although some questions could not be "scientifically solved," they were not meaningless: they belonged to the field of research he had already labelled "philosophie de la science" [Cournot 1861, pp. 449-50].

organic chemistry, and even beyond medical practice, pathological anatomy included. The comprehension of physiological and pathological processes required an active practice: chemistry and histology had to be accompanied by an invasive and cruel practice, namely vivisection. [Coleman 1985, pp. 50-1, 55, and 57; Lesch 1984, p. 121].

In the book Bernard published on "experimental medicine" in 1865, he stressed the peculiarity of biological processes, and at the same time the necessity of a scientific explanation: both physical determinism and biological guiding principles were at stake. Although Bernard had been trained in medicine and physiology by Magendie in accordance with the reverence for facts and empirical practice, his book, *Introduction à l'étude de la médicine expérimentale*, was intended to go far beyond. It dealt with the foundation of the experimental method, the principle of scientific determinism in the context of life sciences, and a critical analysis of experimental practice. The experimental method called into play determinism, because determinism was nothing else but the possibility of reproducing experiments. More specifically, the experimental method required that, in every specific science, "*the conditions of existence of every phenomenon*" were "*defined in an absolute way*." It was an axiom of science that "in identical conditions, every phenomenon identically happens." Life sciences could not represent an exception. Determinism corresponded to the universality of scientific laws [Bernard 1865, pp. 116 and 119-20; Virtanen 1960, pp. 7, 13, and 22].[14]

According to Bernard, along with the experimental method there was "the *experimental analysis*," which decomposed "every complex phenomenon into a series of ever more simple phenomena" till their reduction to "elementary conditions." A reductionist strategy required that physiologists and physicians should represent every vital process in complex living beings as "a game among vital organs," and the latter as "properties of vital tissues or specific organic elements." At the same time, a too rough reductionism did not manage to grasp the complexity of living beings: in particular, "the links between a body and its *environment*" could not be overlooked. A problematic and demanding balance was in order. On the one hand, "if we isolate a body in an absolute sense, we definitely annihilate it"; on the other hand, if we accept taking into account "its connections with the external environment, we multiply its features." The

14 Bernard insisted on the antiscientific nature of the word *exception* [Bernard 1865, p. 120]. It is worth stressing that in the same year the German physiologist Carl Ludwig was appointed to the chair of physiology at Leipzig University, and then began his research activity in the Institute that was to become the most important physiological laboratory in Europe.

study of life required that "the close interactions between organism and environment" were taken into account, but a complete knowledge was impossible because it "would require the knowledge of the whole universe even for the simplest phenomenon" [Bernard 1865, pp. 123-5, 128-9, and 140].

Bernard made reference to the harmony of the universe, and pointed out that both harmony and determinism were involved in life sciences: determinism was not only very complex but at the same time "harmonically ordered." The role of the environment and the intrinsic harmony among all parts were the hallmarks of living beings: "the harmony of the whole" was as important as "the mechanism ruling every part." Nevertheless, the existence of a harmonic finality should not prevent physiologists from "performing experiments in medicine" as some "followers of vitalism" claimed. Once more, a demanding balance between different meta-theoretical attitudes was in order. Bernard was convinced that the banishment of systematic experiments from physiology would have led to a slowdown of scientific progress; at the same time, the underestimation of "the harmonic unity of the organism" would have led scientists to misunderstand the essential features of living structures. Differently from chemistry, the properties of living beings did not depend "only on the proportions between different kinds of matter, but also on the layout," namely their complex architecture [Bernard 1865, pp. 150-1, 153-4, and 156-7].

However Bernard remarked that the insufficiency of reductionism also emerged in chemistry itself. The essential features of water could not be directly derived from the features of hydrogen and oxygen, even though water was a definite and predictable combination of them. In physiology, in the interaction among various "physiological components," new features emerged, and those features did not belong to the components. In brief, the properties of a harmonic set of physiological parts were different from "the sum of the single properties." Scientific method and scientific determinism required a twofold strategy, where both "organisms as a whole" and "their parts" had to be taken into account. According to Bernard, that duality echoed a more general, meta-theoretical dichotomy between the specificity of empirical observations and the generality of rational explanations. He saw two opposite pitfalls: "the excess of specificity," which could become unfruitful and *antiscientific*, and "the excess of generalisation," which could lead to "an ideal science without any connection with reality" [Bernard 1865, pp. 157-8].

Even the relationship between physiology and medical practice involved a similar dichotomy. Physiology required careful experiments and rational generalisations whereas medical practice dealt with "the *single* human being" and his/her morbid conditions. Individuals and individual conditions called into play physicians' competences, whereas structures and functions of "living beings in general" called into play physiology as a science. Bernard was oscillating between the two poles of a scientific practice whose foundations he was attempting to set up. On the one hand he dealt with a process of reduction of life sciences to physics and chemistry, and on the other he had to make reference to the specific features of living beings and "the essence of life." The more demanding task was the clarification of that specific nature or essence. He found that life required a sort of "guiding *idea*" or principle, or "creative idea," which "manifested itself in the organisation" of living beings. In other words, the difference between inanimate and animate matter was self-organisation, and self-organisation required a specific principle or power [Bernard 1865, pp. 159 and 161-2].[15]

The scientific method he was devising was anything but naïve, and the problematic link between hypotheses and experiments played an important role in it. An observation involved only *facts* whereas an experiment required a network of hypotheses. Moreover, hypotheses might lead to the discovery of new facts, and the design of new experiments. At the same time, the combination of facts and theories was based on the greater reliability of facts. Bernard looked upon theories as "partial and provisional" entities, which could be replaced by other, wider-scope theories. That process of replacement and evolution of theories represented the core of *true* scientific progress. In no way could a scientific theory be considered "conclusive and... absolute." The fruitfulness of a theory was more important than its alleged truth: independently from their specific content, "hypotheses and theories... might lead to new discoveries." Great discoveries "might stem from bad theories," in the same way as chemistry had stemmed from alchemy. In brief, theories were nothing else but "intellectual tools, necessary for the evolution of science" [Bernard 1865, pp. 285, 287-90, and 299-300].

Theories must not be confused with principles. Unlike theories, principles could rely on a greater stability. In some way principles were "absolute" because they were like intellectual guidelines: they assisted researchers in the observation and interpretation of natural phenomena. At the top of scientific practice he placed two very general principles: "experimental determinism

15 Here Bernard seems not so far from Cournot. It is worth remarking that, in 1862, in a letter to the Scottish classical scholar Lewis Campbell, James Clerk Maxwell had hinted at the problem of "action and reaction between body and soul," where *soul* stood for mind. He confined himself to remarking that the action was not "of a kind in which energy passes from the one to the other," as some instances could easily show. [Maxwell 1862, p. 712]. In short, Maxwell stressed that the transformation of a given amount of energy should not be confused with the negligible activation energy that triggered off that transformation.

and philosophical doubt." It seems that Bernard hinted here at methodological principles rather than principles in the sense of foundations of specific scientific theories. Scientific progress could and should modify theories but should not modify the fundamental principles, since these principles allowed progress to take place. He claimed that "scientific principles and methods" were of greater importance than theories: a sound scientific practice, which he called "experimental criticism" could "cast doubt on everything, apart from scientific determinism" and other methodological principles [Bernard 1865, pp. 302-3].

In any case facts and interpretations had to be carefully separated. Facts required a strict agreement among scientists whereas a plurality of interpretations could be accepted. In reality that plurality corresponded to the vitality of scientific practice: only the debate on "interpretations and ideas" could lead to "new researches and new experiments." The most suitable interpretation had to be chosen among a bundle of possible interpretations: unfortunately Bernard did not specify which criterion could lead to the choice of the best explanation. In some way, Cournot had made a similar move when he had hinted at the possibility of "a natural classification" [Cournot 1861, pp. 416-8 and 424-6]. However, Bernard guarded against any radical empiricism: physiology should not be looked upon as a purely empirical science. Empiricism corresponded to the first stage of every science, and had to be overcome by more rational practices. There were sciences such as astronomy that pivoted on observations. Astronomy was a meaningful instance of science that could predict phenomena but could not "modify or master them." On the contrary, physics and chemistry had already gone beyond the stage of the pure description of specific facts: they had attained an actual comprehension, and the power of transformation of reality. This was exactly Bernard's purpose: physiology could go beyond the stage of passive observation and mere description. It could become an active or experimental science [Bernard 1865, pp. 314, 332 and 334].

The emergence of life sciences represented a very sensitive stage in the history of science, and the problematic link between determinism and reductionism was more in prominence than in other fields of research. According to Bernard, science was "nothing else but the determinism of the conditions of phenomena," and scientific practice aimed at "explaining the unclear and the most complex by means of something clearer and simpler." Life was definitely the most complex among the natural processes, and therefore it had to be studied by means of the knowledge already developed in the physical and chemical domains. Life could not be the explanation of anything: in this sense *vitalism* had to be rejected. At the same time, a specific and natural design allowed living beings to sprout and grow. In some way, there was a peculiar force or impulsion that "fed and organised," even though in no way could that force determine "the features of living matter." These passages are not completely clear and convincing: they show how narrow and uneven an intermediate way between *mechanism* and *vitalism* really was. His conclusion might be qualified as a sort of mild reductionism: it was wiser "to reduce the features of living beings to physical-chemical features" rather than to reduce the latter to the former [Bernard 1865, pp. 352-4].[16]

In the last pages of the book, Bernard outlined an epistemological framework, and defined the meaning of words such as hypotheses, theories, systems, and doctrines. A theory emerged from the alliance between hypotheses and experimental practice: a theory had a logical, conceptual, and empirical content. A *system* corresponded to a theory without any sound connection with a consistent body of empirical knowledge. A doctrine corresponded to a theory that was looked upon as permanent and unchangeable. In some way, systems and doctrines represented a sort of barren, dead knowledge. Only theories were fruitful entities, and they were so because they were provisional and changeable. They were frail but also pliable and resilient at the same time. A good theory should "always be modified," and should be submitted "to the criticism of continuously emergent facts." The difference between unfruitful systems and doctrines on the one hand, and scientific theories on the other corresponded to the difference between stiff "philosophical systems" and a critical and fruitful "*philosophical attitude* [*esprit*]." Scientific theories could be looked upon as an implementation of a very general intellectual commitment to attaining "the knowledge of the unknown" [Bernard 1865, pp. 385 and 387].

In Bernard's meta-theoretical framework, the relationship between science and philosophy was not a contingent one. The fruitfulness of the close bond between the two bodies of knowledge depended on their open and dynamical structure: *change* was the key concept. He claimed that "neither science nor philosophy should be systematic," and neither the former nor the latter

16 It is worth stressing that Bernard was determinist without being mechanist. He attempted to find a frail balance between the reduction of life sciences to physics, and the sharp separation between them that had been claimed by "French vitalists and German natural philosophers." Probably Bernard was not "the Newton of living beings" as Canguilhem emphatically asserted, but he built up a fruitful integration between opposite attitudes. If Bichat had maintained that life sciences required a different method and a different language, Bernard softened the opposition: life sciences could rely on the same method as physical sciences, but a different language and specific concepts were in order [Canguilhem 1979, pp. 139 and 148-9; Canguilhem 1965, pp. 157-8 and 161].

"should prevail over the other." Without a fruitful interaction, science would become a collection of scattered researches, and philosophy would begin "to wander through the clouds." Science and philosophy represented two different but intertwined ideals of knowledge: he looked upon the English natural philosopher Joseph Priestley and Blaise Pascal as instances of this twofold commitment. He praised their pursuit of scientific research as an endless and fascinating enterprise. Only freedom and an open mind could fuel scientific progress, and in its turn scientific progress fuelled "human progress." The last words of the book guarded against "the excess of scholarship, and the overgrowth of systems," and once more promoted "independence of mind" [Bernard 1865, pp. 388-9, 391-2, and 396].

Bernard's opposition to a naïve reductionism in life sciences was in tune with Comte's attitude, and even his opposition to making use of mathematical methods was in agreement with the philosopher's leanings. However, he did not discuss explicitly Comte's theses in his published works but he remarked that positivism, which had opposed philosophical systems in the name of science and scientific progress, had become another unfruitful philosophical system. He insisted on his dissatisfaction with philosophical systems and their attitude to take advantage of specific scientific contents in order to endorse very general views. On the other hand, he stressed the positive role of philosophy in raising continuously new questions and problems. More specifically, he could not share Comte's lack of confidence in experimental practice in the context of life sciences, and Comte's sharp dichotomy between life and inorganic matter. Bernard's scientific practice and philosophical commitment resists any attempt to link him to a specific philosophical school, even though, in his works, we find some references to Leibniz and Pascal, and the implicit presence of a philosophical tradition that could be traced back to Aristotle. We do not find any appreciable influence of Darwinism, and this is in tune with Bernard's search for an experimental foundation of life sciences. Probably Darwin's theory appeared to him as too "remote and speculative" since it was looked upon as "neither susceptible of experimental verification nor applicable in medicine" [Bernard 1865, pp. 387, 390, and 393-4; Virtanen 1960, pp. 41-4; Canguilhem 1943, pp. 21-2; Canguilhem 1958, pp. 67-8; Benrubi 1926, p. 17].[17]

17 Canguilhem stressed the scientific lineage going from Bichat to Bernard through Magendie: Comte and Bernard's mistrust in the mathematisation of life sciences could be traced back to Bichat, the founding father of histology. Canguilhem also remarked that Comte, Magendie, and Bernard had been looked upon as representatives of the same tra-

When Bernard mentioned the subsequent stages of human development, he qualified them as the ages of "*feeling, reason,* and *experience,*" and this seems in tune with the Comtian historiographical framework. Nevertheless, what he called "experimental method" rested upon "the three legs of this fundamental [immutable] tripod," namely feeling, reason, and experience. In other words, sound scientific practice needed not only the intervention of rational and empirical practices, but also the intervention of a more ancient practice that could not be superseded by reason and experience. Feeling, intuition, and the faculty of personal interpretation stood beside reason and experience; they could not be disregarded but had to be included in a more advanced synthesis. Moreover, feeling and intuition took the lead in "the search for truth" in the scientific context, and then reason followed. Intuition had to be clarified by reason, and reason had to be led by experience. It was the combination of ideas, intuitions, and hypotheses that triggered scientific research: then logical analysis and experience allowed scientists to put forward a reasonable interpretation of the phenomena under scrutiny. It is worth stressing that, according to Bernard, the *fundamental tripod* was at stake in every kind of scientific practice: naturalists, physiologists, physicians, chemists, and physicists shared the same method for the interpretation of reality [Bernard 1865, pp. 50-1 and 57].[18]

The following year Bernard published the book *Leçons sur les propriétés des tissus vivants,* which was a list of academic lectures on specific physiological organs and functions. The first chapter had been composed in 1864, and contained historical and critical remarks: the first passages dealt specifically with the methodological foundations of physiology as a science. Physiology, "or *life science* or *biology,*" involved phenomena that were "far more complex than

dition by "physicians and biologists of the Second Empire." Although Canguilhem found that Bernard owed much to Comte [Canguilhem 1943, pp. 29-30 and 32; Canguilhem 1958, p. 73], I must stress that Bernard did not trust in philosophical systems in general, and in a normative philosophy of science in particular [Bernard 1865, pp. 387-94]. For the references to Leibniz and Pascal in Bernard's writings, the imprint of Pascal in the last writings, and "the paucity of references to the great figures of the Enlightenment," see Virtanen 1960, pp. 32-42.

18 In 1926, in his synthetic reconstruction of French philosophical trends, the historian of philosophy Isaac Benrubi highlighted the importance of Bernard's "immovable tripod on which the experimental method rests." He also stressed that Bernard stood "under the influence of Positivism" but "in one essential point" he broke away from Positivism: he had not admitted "the empirico-imperialistic conception of exact science" [Benrubi 1926, pp. 88-9].

those of the inorganic domain." At the same time, there was one and only one scientific method that encompassed both life sciences and physical-chemical sciences. Once more he was looking for a balance between the application of the scientific method of physics, and the fact that "living beings follow their own laws." What Bernard labelled "general physiology" corresponded to the search for "*the elementary conditions for the existence of life*," or in other words, the determination of basic functions in living beings. The insistence on the adjective *elementary* corresponded to the fact that the basic functions were "identical in all animals." In the domain of general physiology, classifications in terms of "class, genus, and species" were not important, and the concept itself of "comparative physiology" made no sense [Bernard 1864, pp. 4-6 and 8-9].[19]

Bernard considered the physician, surgeon, and anatomist Xavier Bichat as the founding father of general anatomy: he had singled out some fundamental living tissues as the seat of "elementary vital processes." Living beings had been represented as the result of a complex interaction among those "elementary parts," in analogy with the finite number of *simple* chemical elements that could give birth to the most complex chemical compounds. Afterwards the focus on cells and their nucleus had opened new perspectives in anatomy or *histology*: all organs could be considered as "derived from cells," and egg cells could be looked upon as the common origin of all cells. The "elementary parts" of living beings were however different from the elementary parts of chemistry: the transition from the biological to the chemical level was a dramatic transition indeed since it involved the disappearance of "all vital features." The correspondence between elementary parts, and specific "vital features" was more evident in the higher forms of life because those features were better "separated from each other." In this sense, the lower forms of life had to be considered as more complex than the higher ones. As a consequence, the results of physiological experiments were clearer when performed on higher forms of life [animaux élevés]. In the domain of life sciences, the word *experiments* literally meant experiments on "living animals." According to Bernard, the cruel practice was necessary because no "secret of life phenomena" could be drawn from a dead body [Bernard 1864, pp. 15-19 and 22-3]. His sophisticated and pliable scientific method for life sciences stood beside his harsh experimental practice on real living beings.

19 Bernard pointed out that attempts at "simplifying and generalising" could be traced back to the most ancient times, but the establishment of "general anatomy and general physiology" as definite sciences were recent achievements. They were born in the nineteenth century, and therefore their history was "easy to tell" [*Ibidem*, pp. 11-2].

In Cournot and Bernard we find the awareness of the complex and branched structure of the scientific body of knowledge and scientific practice. The systematization of life sciences challenged the traditional view of scientific method. The actual scientific practice required an active commitment to putting forward and clarifying conceptual and theoretical issues alongside the traditional commitment to logical accuracy and experimental soundness. Scientific practices in astronomy, physics, chemistry, and life sciences called for logical coherence, experimental reliability, and a network of mutually consistent assumptions. Nevertheless astronomy, mechanics, and other physical sciences also required a sophisticated mathematical toolbox and detailed mathematical models, whereas physiology involved partially predictable systems ruled by a specific kind of natural finalism. In Cournot and Bernard we find the explicit acknowledgement of a plurality of scientific methods that had in common the fruitful interaction among logical-mathematical language, experimental practices, and the above-mentioned theoretical commitment. Specific methods of specific sciences resorted to a specific combination of the three main ingredients. The specific features of different sciences prevented Cournot and Bernard from endorsing an unproblematic reductionism: life sciences could rely on physics and chemistry, but could not be reduced to the latter. Every transition from one science to another was highly problematic also because all scientific concepts, from inertia to vital actions, were intrinsically problematic. Moreover the two scholars acknowledged the existence of meta-theoretical commitments or very general principles that profoundly influenced scientific practice: well-defined principles like determinism or broader beliefs like the principle of harmony actually oriented scientific research. Ultimately, both Cournot and Bernard insisted on the plurality of scientific methods and on the necessity of a wide network of concepts and assumptions which allowed scientists to combine logical-mathematical structures with the set of available experiences and experiments. A critical analysis of that network was one of the specific tasks of a philosophy of science dealing with actual scientific practice rather than abstract philosophical prescriptions.

CHAPTER 2

Between Experimentalism and Mild Naturalism

1 The Experimental Hallmark of Physiology

In 1867 Bernard published *Rapport sur le progrès et la marche de la physiologie générale en France*, and important methodological and historiographical issues were at stake. It was one of the periodical, official reports on the advancement of different sciences in France. According to Bernard, physiology was achieving "its scientific independence": it was fighting "to get rid of empiricism" together with speculations and hypotheses that had hampered its development. In other words, physiology had to guard against two complementary risks: excess of empiricism, and excess of speculation. Physiology was a young science, and its progress would require "new discoveries" together with "new methods and ideas." The landscape of sciences offered two kinds of bodies of knowledge: on the one hand, the sciences that had already developed "their principles and methods" long ago, and on the other, sciences that were taking their first steps. Physiology belonged to the second group. He credited Lavoisier and Laplace with having clearly stressed that physical and chemical processes in living beings followed "the laws of general physics and chemistry" which ruled inorganic matter. At the same time, in the domain of physiological processes, those laws could find their specific expression or their "specific form" in a completely different way from inorganic matter. Physical and chemical laws stood beside the specific "vital features of the organism": this was the twofold foundation of "general physiology" or "experimental physiology" [Bernard 1867, pp. 1-2 and 4-5].

In Bernard's historical reconstruction, the key adjective was indeed "experimental," and after Lavoisier and Laplace's scientific foundations, and after Bichat's "general anatomy," the third momentum in the development of physiology had been the assumption of "the method of experimental sciences," namely the "experiments on living beings." It had been Magendie who had strengthened the experimental character of physiology, even though he had confined himself to "*empirical experiments*," and distrusted "every interpretation or deduction." In the physiological domain, the expression "physiological experiment" was synonymous with vivisection. The questionable reliability of experiments performed in extreme conditions of unbearable pain, and the ethic aspects of vivisection, were not taken into account by Bernard. He acknowledged that this practice had raised many objections, but he stressed its

 | DOI 10.1163/9789004315235_004

necessity. Physiology had to be based on three pillars: "physical-chemical sciences, anatomical sciences, and experiments on living beings." Physiology as a science had emerged from a plurality of practices and traditions: alongside Magendie's empirical pathway, a different pathway had been undertaken by the Scottish anatomist Charles Bell. Differently from Magendie, who had been "a passionate experimentalist," Bell had taken pleasure "in a network of speculations," had put forward "a systematic interpretation" of brain and cerebellum, and had "a great repulsion for experiments on animals" [Bernard 1867, pp. 5-7 and 11-14].[1]

The practice of vivisection was consistent with Bernard's confidence in the unification between physiology and pathology, and "the actual continuity between pathological and physiological processes" [Canguilhem 1943, pp. 33-4].[2] It was assumed that pathological conditions, such as the extreme conditions of physical and psychological agony during the procedure of vivisection, did not modify normal physiological functions. This assumption allowed Bernard to rely on vivisection as a suitable practice for the comprehension of physiological processes. But vivisection had its limits, and other experimental practices were useful "in the study of elementary physiological features." Powerful poisons could be transferred to living tissues by blood, and could then "act directly on the histological elements." Curare, for instance, had been shown to be useful in order to separate "the contractive feature of muscles from the motive power of nerves." General physiology had managed "to separate the physiological elements," then "to determine their conditions of vital activity," and finally "to establish their mutual physiological relationships in the network of vital mechanisms."[3] According to Bernard, physiology must not remain in the "contemplative domain of natural sciences" but had to become an experimental science which acted on living beings. In several passages he insisted on the power and effectiveness of physiology, which aimed at systematically modifying complex "organic mechanisms" [Bernard 1867, pp. 16, 18, 30, 36, and 39].

1 Bernard specified that Magendie had also held "private lessons of experimental physiology based on vivisection." According to Bernard, Bell had confined himself to dissecting "brains, cerebellums, spinal marrows, and the emerging nerves" [Bernard 1867, pp. 7 and 11].

2 Canguilhem claimed that the assumed continuity between pathology and physiology was in tune with "the idea of continuity between life and death, and between organic and inorganic matter" [Canguilhem 1943, p. 37]. It seems to me that the hypothesis of continuity between inanimate matter and living beings would not have been endorsed by Bernard without careful specifications.

3 According to Bernard, life was nothing more than an extremely delicate mechanism that depended on "the well-balanced functional activity of all histological elements" [Bernard 1867, pp. 39-40].

The specific features of living structures were discussed in different sections of Bernard's report. He remarked that, in the continuous exchange of matter and energy "between the mineral and the organic kingdom, *nothing is created and nothing is lost*," but that invariance was implemented in different ways. More specifically, chemical basic elements were invariable and could not perish; on the contrary, biological basic elements underwent many transformations, and could die. Moreover living cells experienced an evolution that could be approximately forecast in its main stages, but the evolution could not be triggered off by the material "*substratum*" of cells. Matter could only offer "the conditions for the manifestation" of those processes [Bernard 1867, pp. 93 and 110]. There was a principle of auto-organisation that was hosted by matter but did not emerge from matter.

> Not only are living structures able to regenerate themselves by means of a specific, organic creation, but they can also preserve and fix themselves. Therefore we find a difference between living beings and inanimate bodies in the fact that the former are born, live, and die out, but also in that they can be ill and then recover [Bernard 1867, p. 214].

Bernard imagined that some modifications of nutritive and developmental conditions in the early stages of living structures would have allowed the physiologist "to change the evolutionary direction" of those structures, and therefore "their final organic expression." In other words, he was confident that scientists would be able "to scientifically produce new organised species" in the same way as new chemical compounds could be synthesised. He specified that he did not aim "to create organised matter, and directly build up living beings" in the same way as physical machines. He rather confined himself to modifying "the duration, intensity, and even the nature of vital features." It was quite an ambitious research programme that could only be roughly sketched. For the time being, the actual procedures remained quite mysterious but the target was clear: the mastery of biological processes. Physiology had already experienced the transformation from "natural science" into "experimental science." Bernard had already insisted on the difference between a sophisticated experimentalism and a naïve empiricism devoted to a mere accumulation of facts. The key point was the distinction between observation and experiment: the former corresponded to the *contemplation* of nature whereas the latter involved the forecast of natural phenomena. Natural sciences were contemplative sciences whereas experimental sciences were *explicative* and active. Experimental practice entailed three actions: to explain, forecast, and conquer Nature. The conquest of "living nature" was the role of physiology as well as the conquest of "mineral nature" had been pursued by physics and chemistry [Bernard 1867, pp. 113, 128-9, and 131-2].

Instability and organisation became other key words for qualifying the nature of living beings: instability consisted in the continuous disorganisation of a living structure under the influence of external actions. What Bernard labelled "the cause of life" was nothing else but "the organising power that establishes life, and continuously mends its deterioration." However his attempts to clearly define this generative power were not successful: it seemed that, in some way, the concept had to be accepted as a primitive one. He confined himself to assuming the identity between life and organisation, or the existence of "phenomena of organisation that cannot be found in inanimate bodies." He was also aware that the word *force*, and the associated concept were problematic even in the domain of physical sciences. The concept of force was an idealisation, and the concept of *vital force* required a further idealisation. Those idealisations made the concept unclear, and he attempted to replace the expression vital force with "*organ-trophic* [*organotrophiques*] or *nutritive* phenomena," which was definitely less clear. Further specifications made things even worse. The fact is that, in Bernard's meta-theoretical framework, the cogency or reliability of a definition was not so important. The actual practice, and the effectiveness of physiology as a science, was far more important. He insisted on the pragmatic side of scientific practice: once more the final aim of physiology was "to conquer living nature" [Bernard 1867, pp. 133 and 137-9].

The pragmatic turn, pivoted on the effectiveness of scientific practice, was enlivened by aggressive expressions towards Nature. Scientists' mission was "action and domination": it aimed at "explaining life phenomena, acting on them, and submitting them to its will." The practice of experimental science involved "an actual power" over Nature, which was the intrinsic target of every "truly scientific action." What he called "the modern concept of science" was nothing else but "to conquer Nature." Physiology had to dominate living Nature in the same way as physics and chemistry had already learnt to dominate inanimate Nature [Bernard 1867, p. 142]. The concept was so important that he devoted a long note to it, and one passage deserves to be quoted in order to appreciate the conflation of his meta-theoretical commitment with a coarse historiographical framework.

> Experimental physiology is the science that proceeds at the conquest of living Nature. Ancient science had not managed to put forward this new scientific idea because observational or *contemplative* sciences had emerged before *executive* and experimental sciences. Nevertheless

> mankind has realised that passive contemplation is not its aim but rather active handling and progress [Bernard 1867, p. 233].[4]

As Canguilhem remarked half a century ago, the concept itself of an experimental medical science involved some kind of "scientific domination of living Nature" that reversed the Hippocratic reference frame. Both vivisection and the more active treatment of disease were in contrast with a tradition that required cautious observation, patient waiting, and treatments that fostered the natural trend of living beings: disease had to follow its natural development. What Bernard put forward was a new philosophy, a different attitude towards living beings and disease, even though he preserved the final aim, namely human health. The aim could be pursued at the expense of other living beings' health. Although the expression "experimental medical science" had already been put forward by Magendie, it was Bernard who drew the extreme consequences [Canguilhem 1979, pp. 131-4].[5]

In 1869 Bernard became member of the *Académie Française*, and in his inaugural dissertation, he presented physiology as a body of knowledge with "its roots in physical sciences," and "its branches climbing up to the philosophical sciences of the mind [esprit]." He claimed that philosophy could receive "necessary support" from the new science of life: the practice of "free thinking" required the harmonious collaboration of organic, chemical, and physical conditions in the brain. He acknowledged that the problematic link between mind and brain required "not to confuse the features of matter with the functions they fulfil." However, no contradiction could be found "between physiological and metaphysical sciences": they approached the same problem from opposite sides. Physiology linked "the study of mental skills to their organic and physical conditions" whereas metaphysics dealt with "the expressions of soul [âme]," and disregarded those conditions. The relationship between physiology and psychology was quite different: according to Bernard, it was a relationship of closeness rather than complementarity. More specifically, he remarked

4 Aggressive expressions towards Nature can be found in Bacon, and in Kant in the *Foreword* to the second edition of his *Kritik*, where he compared scientific practice to a violent inquisition [Kant 1787 (1853), pp. 17-8].

5 Canguilhem saw the expression and the corresponding concept as an outcome of "the demiurgic dream" that involved all industrialised societies around the mid-nineteenth century, when science became "a social power." It is true that this power, and the linguistic translation of this power, convinced people that disease could and should be repressed rather than mildly accompanied to its natural end [Canguilhem 1979, p. 140].

that "no actual borderline" could be found between the two disciplines because physiology was "the immediate support for psychology" [Bernard 1869, pp. 6 and 20-23].

2 History of Science and Philosophy of History

In 1872 Bernard reprinted the report he had published in 1867 on the progress of "general physiology." The title, *De la physiologie générale*, was different but the content was definitely the same: he had simply changed some words, and transformed some sentences from interrogative into affirmative. After five years, "nothing important" had to be added: scientific practice in life sciences was following the same pathways as "five years ago" even though recent progress had became definitely faster [Bernard 1872, p. VI]. In other words, no meaningful novelty had appeared on the experimental side of physiology. This publication and its content allow me to point out two issues: the heroic stage of the emergence of physiology as a science had already elapsed, and had taken place in the 1860s. However, Bernard probably found that some kind of cultural pressure was in order because the institutional weight of physiology as a science was still weak.

In the meantime, the mathematician-philosopher Cournot had continued to explore the role played by life sciences in the landscape of knowledge. At the same time, he continued to explore the scientific tradition with historical and philosophical sensitivity. In 1872 he published another book that was devoted to the history of science in the context of the history of European culture, *Considérations sur la marche des idées et des événements dans les temps modernes*. More specifically, the emergence of modern science was looked upon as part of a more general development of ideas and events that had happened in the sixteenth and seventeenth centuries. Only after this original historical-philosophical inquiry into the foundations of science, in another book he published in 1875, did he return to life sciences. With his characteristic discreetness, he showed some shortcomings of the marked experimentalism in physiology. Cournot's whole intellectual pathway deserves to be explored in this chapter.[6]

6 In the meantime Cournot had published *Principes de la théorie des richesses* in 1863, and *Les institutions d'instruction publiques en France* in 1864. These books, and the others he had published before 1861, show the plurality of his interests and professional activities, which included mathematics, economy, theory of probability, history and philosophy of science, not to mention his duty as inspector of public instruction.

In 1872 he discussed various systems of classifications for the heritage of human knowledge, and in particular systematically developed bodies of knowledge. In the *Foreword* to the book, he showed that historical, logical, and pragmatic hierarchies could reasonably be put forward. The intrinsically plural, provisional, and historical nature of every classification was at stake when he focused on the role played by science in the history of civilisation. His historiographical sketch went far beyond Comte's horizon since he acknowledged the possibility of different historical perspectives. He remarked that his contemporaries were more interested in Galileo, Descartes, Pascal, Newton, and Leibniz achievements than in religious and political debates that had taken place at that time. Nevertheless, when the founding fathers of modern science were studied in the context of the seventeenth century, it could easily be shown that those debates had attracted seventeenth-century scholars at least as much as the first scientific achievements. According to Cournot, every historical reconstruction depended on specific philosophical insights: the choice of the book's title, *Considérations sur la marche des idées et des événements dans les temps modernes*, had stemmed from his focus on "philosophy of history" rather than "some kind of historical composition." However, he specified that his reference to philosophy of history did not mean that he presumed to have discovered "any law about history." He would have striven to pursue "a strict balance between induction and hypothesis," namely between the cogency of facts and the rational necessity of putting forward interpretations [Cournot 1872, pp. II-V].[7]

The concepts of "*chance*" and "*independence*," and adjectives such as *accidental* and *contingent* were at the core of his philosophy of history. With regard to this specific subject he could certainly rely on his reputation as a mathematician and serious researcher in the fields of economy and the theory of probability. He stressed that chance did not mean ignorance or unreliability: it was rather a matter of fact.[8] Chance was nothing else but "the mutual inde-

7 With regard to "the classification of human knowledge in Comte and Cournot," see Audierne 1905, pp. 509-19. Comte had assumed "the homogeneity of the objects of science, and the correspondence between things and thoughts, with an optimistic attitude," whereas Cournot had insisted on "differences and difficulties" [Audierne 1905, p. 519].

8 In 1812 the mathematician Pierre Simon de Laplace had published *Théorie analytique des probabilités*, and two years later a less demanding *Essai philosophique sur les probabilités*. In the latter he had claimed that "the most important problems of life" dealt with probability. In other words, probability was an essential feature of human knowledge. This meta-theoretical attitude does not appear as a contingent one, because he insisted on the probabilistic nature of scientific knowledge, and specifically mathematical knowledge. See Laplace 1825, pp. 1-2:

pendence of several series of causes and effects that took part *accidentally* in a given phenomenon." Chance had its laws, and those laws were no less reliable than the laws of physics and astronomy. Both statistics and philosophy of history rested on chance, even though Cournot preferred the label "*aetiology of history*" for the latter. That aetiology dealt with the study of causes or the disentanglement of different kinds of causes. More specifically, the historian had to separate causes, even seemingly weak causes which produced essential and long-term effects, from other causes — even the apparently strongest ones — which had given rise to short-term and irregular effects. Here the scientist and the historian converged on the acknowledgement of an intrinsic tension between "law and fact, the essential and the accidental." Both in Nature and history, and both in the history of Nature and the history of civilisation, that separation was necessary, but what was really contingent or accidental depended on the context and the intellectual reference frame. The motion of a comet around the Sun could be predicted and accurately computed. If its orbit had met the Earth, it would not have been an unexpected contingency from the astronomical point of view, but it would have represented an unfortunate accident for life on the Earth [Cournot 1872, pp. 1-5].

The historian of philosophy Isaac Benrubi stressed the central role played by "the idea of Chance" and "the idea of Probability" in Cournot's meta-theoretical researches: the "original stamp" of Cournot's view rested upon his *probabilism*, which could be traced back to the *Essai* he had published in 1851. Cournot had sketched a theory of knowledge that did not aim at dogmatic certainty without drifting towards scepticism. It is also true that *probabilism* kept him away from a naïve reductionism, and allowed him to accept a plurality of scientific languages and practices. The specificity of living beings and organic processes and the specificity of social processes required a plurality of specific standards of rationality. His reflective, critical attitude, which was the hallmark of his meta-theoretical enterprise, eluded any attempt to describe it by means of "a definite summary and a sharp classification." The fact is that Cournot's theory of knowledge stemmed from his actual researches in mathematics and economy, more specifically his mathematisation of economy, where statistics and probability played an important role. It is worth remarking that Comte had looked upon the theory of probability and political economy as "false sci-

---

"On peut même dire, à parler en rigueur, que presque toutes nos connaissances ne sont que probables; et dans le petit nombre des choses que nous pouvons savoir avec certitude, dans les sciences mathématiques elles-mêmes, les principaux moyens de parvenir à la vérité, l'induction et l'analogie se fondent sur les probabilités." In the first edition of Laplace's book (1814), "l'induction et l'analogie" had not been mentioned [Laplace 1814, p. 1].

ences," and in general, the application of mathematics to social sciences appeared as unrealistic to him [Cournot 1851, 1 vol., pp. 171-2; Mentré 1908, p. 11; Benrubi 1926, pp. 89-90].[9]

In 1843, Cournot had remarked that statistics really was "a modern science" because it dealt with the search for precision, which was one of the hallmarks of modern science. Statistics involved "a principle of compensation": the computation of mean values was nothing else but the compensation of irregularities. More specifically, that principle allowed scientists to disentangle "the influence of regular and permanent causes" from the effects "of irregular and chance causes." Therefore statistics was something more than a mere computational tool: it had a philosophical and scientific meaning. From the meta-theoretical point of view, statistics showed a striking power of unification: it involved methods and procedures that could be applied both in the domain of natural sciences and in the domain of social sciences. In 1851 he saw statistics as a specific kind of intellectual practice that was intrinsically different from the traditional scientific one: the latter could be qualified as *rational* whereas the former was a merely *computational* practice. Nevertheless, statistics was a meaningful source of knowledge: not only was it at stake whenever scientists had to deal with actual measurements, but it kept scientists in touch with the complexity of the natural world. There were two kinds of mathematical sciences of Nature: "the geometrical and mechanical" one, which could account for the domain of inorganic phenomena, and the statistical one, which explored "the domain of living beings" in their biological and sociological aspects.[10] Cournot found that sometimes mathematical precision in the context of living beings and social life was even greater, and the events more predictable, than in the inorganic domain, where only "the blind forces of nature" were involved. A wide set of phenomena could be unified by the

9 The acknowledgement of that epistemological plurality, and on the other side, the plurality of his interests in mathematics, economy, history, and philosophy might explain the persistence of an influence that was quite discreet, hardly visible. The extent and discreetness of that influence has been stressed by other scholars over time. According to the French philosopher François Mentré, Cournot had always been epistemologically cautious: he had relied on probability rather than certainty. That *discreetness* was both a personal leaning and an epistemological commitment. On the contrary, Comte appeared as quite dogmatic, and inclined to willingly preach and pontificate [Mentré 1905, p. 483; Mentré 1908, pp. 644 and 646].

10 In 1990 Hacking pointed out both the emergence of probability and statistics as a respectable and useful body of knowledge, and the application of this knowledge to social matters in the first decades of the nineteenth century. He insisted on the metaphor of "avalanche of printed numbers" of statistical origin as an effective illustration of a new commitment by many European institutions [Hacking 1990, pp. ix and 3].

mild determinism of statistics, which might be qualified as probabilistic determinism [Cournot 1843, pp. 84, 181-4, and 205-6; Cournot 1851, pp. 418-9; Faure 1905, pp. 409-10].[11]

With regard to predictability in science, Laplace had already stressed the "ignorance of the links" that connected every event "to the whole system of the universe." At the same time the intelligibility of the physical world required the well-known "*principe de la raison suffisante*." According to the principle, present events were enchained to past events: in other words, a thing could not occur without a triggering cause. He represented "the present state of the universe as the effect of a previous state, and as the cause of the following." He hinted at a hypothetical "intelligence" or mighty mind, who "should know all forces acting in nature at every time," and would be able to submit that information to mathematical analysis.[12] A scientist could not attain that kind of cleverness, even though "the perfection" of astronomy could be looked upon as "a weak outline [faible esquisse]" of it. Although mathematical physics aimed at approaching the power of the superior mind Laplace had envisioned, the human mind would always be "infinitely distant" from that kind of intelligence. He had remarked that the mathematical laws ruling planetary motions also ruled "the path described by a simple molecule of air," but it was practically impossible to know the huge number of microscopic motions. Probability was a sort of bridge that filled the gap between our body of knowledge and our ignorance [Laplace 1825, pp. 2-4 and 6-7]. It seems that Laplace's text does not support a widespread received view, namely Laplace as a champion of strict determinism. It seems reasonable to think that the mythology of Laplacian determinism was a late reconstruction, and the renowned German physiologist Emile Du Bois-Reymond played an important role in the emergence of that mythology.[13]

---

11 Cournot found that statistics discouraged scientists from pursuing the Pythagorean dreams of mystic numerology or mysterious "ideas of harmony" [Cournot 1851, p. 418]. In 1905, Fernand Faure, politician, professor of law, and then professor of statistics in Paris, remarked that Cournot's researches on statistics passed almost unnoticed because they were "too philosophical for statisticians and too statistical for philosophers" [Faure 1905, p. 396].

12 It is worth stressing that the concept of an almighty mind that could follow the chain of all causes and all events might be traced back to the Stoics, as the Latin scholar Cicero pointed out in *De divinatione* [Russo 2013, p. 318].

13 The hypothetical mind Laplace described was actually hypothetical: the verbs he employed to describe that power were conditional verbs, and the sentences had the typical conditional structure *if it were…, then it would…*. On the emergence of the so called *Laplacian determinism*, see Cassirer 1937, pp. 7-9, 11, 16, and 32, Hacking 1983a, pp. 455-60, and van Strien 2014b, p. 25.

It is worth remarking that, in the same year (1872) of Cournot's *Considérations*, in the first part of a lecture he delivered to the German association of scientists and physicians, the physiologist Du Bois-Reymond took a different pathway. He claimed that scientific knowledge consisted in "reducing all transformations taking place in the material world to atomic motions." Since mechanical laws could be translated into mathematical language, they could rely on "the same apodictic certainty of mathematics." The universe was ruled by mechanical necessity: its present state could be "directly derived from its previous state," and could be looked upon as "the cause of its state in the subsequent infinitesimal time." He mentioned Laplace's Mind [ein Geist], and represented *It* as a powerful entity that would be able to "count the number of hairs on our heads." Although the human mind will always be remotely distant from this perfect scientific knowledge, what he labelled Laplace's Mind represented "the highest conceivable stage of our scientific knowledge" [Du Bois Reymond 1872, pp. 441-4 and 446].[14]

According to Du Bois-Reymond, the best implementation of Laplace's *Geist* was astronomy, where "past and future position and motion" of parts of the universe could "be computed with the same accuracy." Astronomical knowledge represented the most perfect attempt to grasp the behaviour of matter and force by means of our limited ability. The possibility of reducing "mental processes in nerve fibres and ganglion cells to specific motions of specific atoms" would have been "a great triumph." At the same time, that triumph would emphasise "the unsolvable contradiction between the mechanical world view, and free will and ethic." In the end, he found that the actual scientific practice was strongly limited by two impossibilities: on the one hand, the impossibility of grasping the actual nature of matter and force, and on the other, "the inability to reduce mental processes to their material conditions" [Du Bois Reymond 1872, pp. 455-7 and 459-60].

Cournot's probabilistic attitude was not shaken by those impossibilities because he could accept both necessity and contingence. Nevertheless a sharp

14 With regard to the metaphysical nature of Laplace's celebrated determinism [van Strien 2014a, pp. 171], I find that Laplace's approach was deeply rooted in his physics, and it might be qualified as *metaphysical* only in a very broad sense, in the same sense as we label metaphysical the hypotheses that lay at the basis of his physics. In other words, his determinism stemmed from the metaphysical foundations of his physics. I obviously agree that Laplace did not merge his physical determinism with the mathematical issue of the existence and uniqueness of solutions in differential equation. We know that the issue was to be clarified in the late nineteenth century. However, as I have already remarked, I find that Laplace's determinism was milder and more pragmatic than its subsequent idealisations.

separation between essential and long-term processes on the one hand and accidental and short-term processes on the other, appears questionable in the scientific context, and its application to the historical context appears even more problematic. Equally problematic is Cournot's identification between *essential* and *long-term*, as well as the identification between *accidental* and *short-term*. However he confidently remarked that societies changed slowly over centuries because of "intrinsic and general causes whose action can be disentangled from all mishaps of history." At the same time he acknowledged that his simplified historiography could be overturned by the appearance of sudden "revolutions" that could take centre stage "here and there." Although revolutions were triggered off by "local and accidental causes," their effects might be deep and long lasting. According to Cournot, a revolution might influence a whole historical age, and might even spread its effects "over the whole civilised world." He insisted on the difference between "the necessary and fortuitous, the essential and accidental," because it allowed scholars to grasp "the true nature of history" but in the end, his historiographical framework rested upon two different kinds of processes. Ordinary processes, together with the separation between essential causes and accidental processes, stood beside extraordinary processes, which he labelled "colossal accidents" or revolutions. During revolutions, apparently meaningless contingencies could lead to long-term and extraordinary important effects [Cournot 1872, pp. 5-6].

When he focused on the history of science, he stressed that a historical reconstruction should offer "a leading direction." History of science should be neither a disordered heap of discoveries nor a predictable and logically determined series of events. In the first case, "annals rather than history" should be suitable for sciences, and in the second case, the history of science would be nothing but a chronological table of triumphant discoveries. In both cases, which corresponded to two "extreme hypotheses," the history of science lost its intellectual interests. He remarked that, in reality, the development of science and the process of professionalization in the nineteenth century had shifted the history of science towards the second extreme. More specifically, when the number of researchers increased and "the ways of communication among scholars" improved, scientific practice became more intensive, systematic and predictable, and distanced itself from "the *historical stage*" [Cournot 1872, pp. 7-9].[15]

---

15 According to Cournot, the objects of scientific practice persisted over time, and were not affected by "the specific tastes and whims of every researcher." This fact marked the important difference between the history of science and other histories such as "history

3 Continuism in the History of Science

Cournot remarked that science and industry had played an important role in recent history, or "in the civilisation that we qualify as modern": science and technology had introduced an element of "growth, progress, and indefinite improvement." European culture had not always been influenced by this progressive momentum: more specifically the landscape of European science had only experienced a slow progress from Greek civilisation to the Renaissance. For many centuries two bodies of knowledge had been at stake: theoretical sciences such as the four sciences of the "*quadrivium*," and a set of scattered empirical practices, "extraneous to the official lecturing." Only the first field of systematic classical sciences could rely on sound foundations, and Cournot suitably acknowledged that the source of this foundation could be found in Aristotle's *Analytica Posteriora*. If Plato had devised a rational world "ruled by geometry," and a physics based on geometry, Aristotle had "set in order" a wider *corpus* of natural philosophy, and had offered a very general logic or "general grammar" of knowledge. The theoretical knowledge that Cournot labelled in modern terms as "speculative physics and chemistry" had been systematised by Aristotle and his followers, and had given birth to a long-lasting "*natural philosophy*" that had survived through the Middle Ages. The body of empirical practices had survived as well, and could be tracked down across mechanical, medical, and apothecary arts [Cournot 1872, pp. 23-4, 50, 55-6, 60-2, 65 and 73].

The cultural reference frame that had endured for centuries until the sixteenth century was overturned by new astronomical hypotheses. Cournot pointed out both elements of continuity and elements of discontinuity in that transition. He claimed that the sixteenth century was "a revolutionary century" indeed, and the Copernican turning point had overthrown a body of knowledge that had had enduring influence on "philosophical and religious doctrines" involving the place of man in the universe. On the other hand, "from the point of view" of some sensitive issues that "deeply involved the structure of science," in Copernicus and Tycho there was no innovation because their astronomy dealt with "anything else but the geometrical theory of heavenly motions." In other words, according to Cournot, their geometrical approach, namely a new astronomy without a new physics, was on the

---

of languages, religions, arts, and institutions," where human behaviour was involved. In the end, making history was an open enterprise, the history of sciences included. He was unsatisfied with Hegel's perspective, and even Vico's perspective, wherein historical laws and "the repetition of the same series of stages" were at stake [Cournot 1872, pp. 17-8].

track of the old natural philosophy. Moreover he stressed that the accomplishment of the transition between old and new science was slow, and consisted of three main stages: the first stage corresponded to Copernicus and Tycho's astronomical achievements, the second was associated with Kepler, Galileo, and Newton's "physical mechanics," whereas the third could be identified with the century that had elapsed from the publication of Newton's *Mathematical Principles of Natural Philosophy* to the publication of the first volume of Laplace's *Mécanique céleste* [Cournot 1872, pp. 130-1].[16]

When Cournot undertook the analysis of science in the sixteenth century, he specified that he would confine himself to singling out the main ideas that had influenced "the progress of human mind" in a wide sense, and "the progress of science" in particular. With regard to mathematics, he mentioned the Italian Tartaglia, Cardano, Ferrari, and Bombelli, and the French Viète, who had transformed "algebra into a language" that allowed science to advance. With regard to theoretical mechanics he mentioned Leonardo da Vinci and Simon Stevin. Nevertheless the main progress could be found in astronomy, and that progress was looked upon by Cournot as "the victory of reason over experience, fantasy, and bias." The inner engine of the astronomical revolution had not been an improvement in observational practices: it had been an intellectual transformation rather than an empirical one. He also remarked that sometimes empirical and instrumental improvements without any theoretical development might even lead to a stage of regression in science. In that case, only a new theoretical interpretation could solve the conflict between new data and old geometrical models. For instance, only Kepler's new geometrical model was able to solve the problem triggered off by "the improvement of astronomical tables" [Cournot 1872, pp. 138-40, 143, and 146-7].

According to Cournot, the "peculiar and exceptional character of greatness" of the seventeenth century was actually due to "the progress and the revolutions of sciences." That century had shown that scientific progress required a fruitful alliance among "the power of cogitation and reason," "patient observations" and "complex experiences," and great "intuitions." A simple al-

16 Cournot also attempted to sketch some meaningful connections between the most important events in the history of science and the most important events in general history: the first stage, more or less the sixteenth century, corresponded to "the supremacy of Spanish power," the second, around the seventeenth century, corresponded to "the supremacy of France," and the third, namely the following century, corresponded to "English political and commercial influence." More specifically, the last stage corresponded to the time interval "between the English revolution and the French revolution" [Cournot 1872, p. 129].

liance between mathematical laws and purely empirical practices would not have allowed scholars to build up new scientific theories. That "wonderful alliance" had also involved "abstract conjectures" that had transformed "discoveries into revolutions." In a scientific revolution three basic elements were at stake: first, logical deductions and mathematical computations, second, observations and experiments, and third, a network of rational conjectures. Only the combination of the three components, logical-mathematical, empirical, and philosophical, could realise the fruitful synthesis among the general laws of sub-lunar motion, gravitational actions, and "the theory of positions and motions of heavenly bodies." The scientific revolution of the seventeenth century had stemmed from that deep and extensive integration among different bodies of knowledge and different traditions. It had given birth to new bodies of knowledge, new scientific standards, new scientific communities, and new ways of communication, academies and journals included [Cournot 1872, pp. 259-62].[17]

Cournot stressed the role played by Galileo in the emergence of modern science: not only did he see in the Italian scholar the roots of mathematical physics and experimental physics but also the ability to discover important physical laws "in the most trivial phenomena" such as "the fall of a stone or the oscillations of a hanging lamp." Even biases and illusions had played an important role in the development of modern science. He found that Kepler's "Pythagorean illusions" had been fruitful, because they had led him to "the mechanical explanation of planetary motions," and moreover had allowed science to go beyond Pythagorean tradition itself. Other illusions had emerged, and among them the belief that all physics could be based on mechanics. Although some authoritative scholars of the seventeenth century had claimed that it must be so, Cournot thought that the developments of physics in the nineteenth century had called this belief into question. Other illusions dealt with the principle of inertia. Once more he stressed the non-empirical nature of the principle. It was a fundamental law that allowed physicists to interpret "any further experience." If some experiments had put the principle in danger,

17 Cournot warned readers against a possible misunderstanding: the emergence of new academies was a historical process quite different from the process of professionalization of scientific practice that was taking place in the second half of the nineteenth century. The latter, namely the establishment of well-defined scientific trainings, and well-defined scientific careers, had taken place long after the scientific revolution, and was leading to the establishment of academic institutions that could be looked upon as an updated implementation of "mediaeval universities" or the "priestly corporations" of ancient civilisations [Cournot 1872, pp. 263-4].

physicists would not have given up the principle but would have envisaged the existence of new "hurdles or forces." The law or principle had to be looked upon as "a law of Nature on which we can always rely," however complex were the effects that surrounded it. In general, a strong philosophical commitment had been at stake in the emergence of modern science even though that commitment had frequently been silent or implicit. That commitment was still active, and frequently it continued to be implicit [Cournot 1872, pp. 269-70, 278-9, 282-3, 285].[18]

Newton's scientific enterprise appeared to Cournot as the result of a successful balance between "a serious science and a cautious philosophy": Newton had managed to distance himself from both "bold speculations and barren empiricism." Cournot labelled him "scientific legislator," in the sense that he had been able to transform experiments and computations into a theory. Nevertheless in no way was Cournot's historiographical framework based on the concept of scientific genius: he thought that, "if an unfortunate chance had deprived mankind of Newton's achievements," Huygens and Leibniz's followers "would have put forward" something similar to Newton's theories. He thought that, in the scientific context, a counterfactual history would not have led to sharply different events. The Bernoulli family, Euler, Clairaut, and d'Alembert would probably have attained Newton's result "fifty years later," but "the dates of Lagrange and Laplace's great works would have been the same." His key concept was "the *maturity* of discoveries": the scientific landscape was influenced by previous events, and those events could give rise to various chains of effects, but the final results would not be too different. Once more he pointed out that the mere alliance between "the most impressive experiences" and "the most demanding computations" would never have generated meaningful scientific achievements without a strong theoretical commitment [Cournot 1872, pp. 287-9].[19]

In Cournot's historiographical framework, the history of science dealt with two different traditions. On the one hand he saw the history of systematic sciences, which had been systematic since the age of Greek civilisation, and had experienced a striking revolution in the sixteenth and seventeenth cen-

18 Cournot remarked that, in the seventeenth century, "a mathematical theory of chance" had also emerged, and Fermat and Pascal's researches affected the development of mathematics, logic, and philosophy [Cournot 1872, p. 275].

19 However important Newton's achievements may be, Cournot stressed that no stage could be the conclusive one. A subsequent fruitful alliance between the new mechanics and differential calculus would have given rise to "a more or less new science, rational mechanics" [Cournot 1872, pp. 268-9].

turies. On the other hand, he saw the less defined history of scattered bodies of knowledge, such as "theories of heat, magnetism, and electricity," that had preserved their "childish condition" of semi-empirical sciences throughout the seventeenth century and even afterwards. Only "the domain of pure mathematics and physical mechanics" underwent "a revolutionary crisis." There was therefore a strong asymmetry between the tradition of applied mathematics and astronomy, on the one hand, and the tradition of empirical practices or arts, on the other hand. From the point of view of scientific instruments, there was an ostensible symmetry between the domain of systematic and empirical sciences: the discovery of the telescope in the context of astronomy could be compared with the discovery of the microscope in the context of natural sciences. In reality, according to Cournot, that kind of symmetry between the extremely great and the extremely little could not bear close scrutiny. The microscope had not managed to clarify "phenomena that we label chemical or molecular," or the functions and structures of microscopic living beings. The "scientific consequences of the discovery of the telescope" were in synchrony with the development of new astronomical theories, whereas microscopic observations lingered in a condition of "mere curiosity" throughout the seventeenth and eighteenth centuries. Much time was to elapse before the microscope became "a device in tune with actual scientific practice" [Cournot 1872, pp. 292-4].

In the context of life science the main discovery of the seventeenth century appeared to Cournot the circulation of blood. In Harvey's discovery he found an important contribution to "the new philosophy." Harvey had managed to show "the inferiority of ancient doctrines, the weakness of Scholastic lines of reasoning, and the power of observation" in a domain that was closer to human experience than Jupiter and Saturn's motions.[20] In other words, the overturn of old physiological theories affected human life much more than the new astronomical theories. Nevertheless that important discovery had not led to "a revolution in medical science," neither in the theory nor in actual medical practice. Cournot managed to grasp an essential feature of the so-called scientific revolution: despite Bacon's dream, seldom deep intellectual transformations had given birth to useful applications. From a more general point of view, Cournot found that "the scope of a scientific discovery" depended "on the

20 In reality Cournot missed an essential feature of Harvey's theoretical framework, namely his Aristotelian philosophy. Harvey was a modern scientist and a traditional philosopher: blood circulation was closely linked to the mythology of circular motion [Gusdorf 1969, p. 173].

maturity of the scientific environment more than on the intrinsic value of the discovery" itself. Cournot's complex historiography allowed him to separate effectiveness and success from the intrinsic relevance of scientific achievements. He also stressed that the value or the relevance of a theory could be evaluated only *a posteriori* [Cournot 1872, pp. 295-6].

He insisted on the close link between new science and new philosophy in the seventeenth century, and more specifically on "the intrinsic alliance between the inventive genius in science and the reformative genius in philosophy." Descartes, Pascal, Newton, and Leibniz had been "at the same time first-class geometers and great philosophers." Although they had been talented pioneers in science, they represented the crowning achievement of a long-lasting philosophical tradition rather than the emergence of a new philosophical trend. Cournot remarked that, "after the seventeenth century, after the great Leibniz," the most authoritative mathematicians had completely given up philosophy, and "looked upon philosophical cogitations as a pointless intellectual activity," or had confined themselves to scattered meta-theoretical remarks. Conversely, philosophers continued to reject any intrusion into their domain coming from "an extraneous science." In a wider historical perspective, the "divorce between exact sciences and philosophy" appeared to Cournot not so different from the process that had taken place "in Greece at the time of Alexander." In spite of the chronological distance, he found that the two historical processes had had "the same consequences," and those consequences had equally been long-lasting in our civilisation [Cournot 1872, pp. 298-9].

With regard to Bacon, Cournot rightly remarked that the Lord Chancellor belonged chronologically to the seventeenth century but his intellectual attitude was in tune with the previous century. Although he had promoted systematic observations and experiences, "he had never become acquainted with Galileo's experiments, or had not taken them into account." At the same time, he was the founding father of "the future religion of progress" and the "scientific cult of Nature." He could be enlisted "in the family of pioneers or utopians of the sixteenth century," and "his anticipations of the future state of affairs" led Cournot to place him "at the rear-guard of his renowned contemporaries." Bacon's "redundant list of instances and forms" was structurally similar to Scholastic forms of syllogism. He had not been completely aware of the fact that "the knowledge based on a list of well-ordered facts cannot be a science yet." That illusion had fuelled the subsequent circulation of naïve empiricism and naïve scientism till the recent emergence of positivism. The fact that only "a rational principle" could lead scholars "from the fact to the law" had long been underestimated. The human mind could deduce from experience more than the experience itself contained. Moreover Cournot saw an inescapable "vicious circle" in Bacon's philosophy. On the one hand Bacon had warned against mistaken mental attitudes that could transform the comprehension of Nature into "a deceptive and fantastical portrayal." On the other, he had offered a naïve solution that consisted in pursuing "a mindful study of Nature." The remedy to deceptions stemming from the human mind had been entrusted to the mind itself [Cournot 1872, pp. 300-305].[21]

Cournot found another vicious circle in the philosophy of another founding father of modern science, Descartes, who had put forward a more rationalist approach to the knowledge of Nature. Descartes had claimed that the reliability of our ideas rested upon God's perfection, but that idea of perfection stemmed from the human mind. In brief, he had deduced "the existence of a perfect being from the idea of a perfect being." His radical dualism entailed the existence of "entities endowed with extension but unable to think," on the one hand, and "entities endowed with thought but without any extension," on the other. That dualism had led Descartes to envisage "the nonsense of animal-machine," which subsequently would be rejected by scientific progress itself. A strict mechanistic world-view could not account for "the great separation between organic and inorganic domains." From a more general point of view, Cournot interpreted Descartes' rejection of empty space as a consequence of the persistence of Aristotelian views. He saw Descartes under the yoke of "peripatetic ontology, which took into account nothing else but substances and their features." Even "the physical novel" of his swirls could properly be understood in that context, as a necessary consequence of his philosophy. According to Cournot, scientific progress could only be attained by overstepping "the boundaries of Cartesian hypotheses": this was what Newton and Leibniz had attempted to do [Cournot 1872, pp. 309 and 311-14].

Cournot focused on Newton and Leibniz's theories from the point of view of the problematic relationship between force and matter, which he had already explored in 1861. He saw two extreme meta-theoretical options, or "two systems opposed to each other: pure atomism and pure dynamism." Atomism removed "the idea of force as superfluous," whereas dynamism made "rationally superfluous the idea of atom." While Peripatetic and Cartesian scholars

21 The historian Georges Gusdorf found that an empiricist and inductive attitude encouraged the awareness of the intrinsic historicity of scientific enterprise. For this reason, he looked upon Bacon as "the founding father of modern thought" whereas he saw Descartes' rationalism as dogmatic, on the track of Scholasticism, and insensitive to historical processes [Gusdorf 1966, pp. 47-8]. It seems to me that this appraisal was somewhat unbalanced; Bacon was overestimated by Gusdorf, whereas Cournot had managed to grasp the most essential feature of Bacon's view.

had deduced everything from the idea of substance, Leibniz had attempted to deduce everything from the idea of force. Newton had put forward a different dynamics: he had attempted to put together "the *attractions* that bodies exerted on each other," and the inertia of matter. Matter was passive or inert because "no body was able to change its state of rest or motion by itself," and because attraction did not deal with "the essence of bodies." According to Cournot, Newton's dynamics was more suitable for "the astronomical problems from which it had stemmed." The scientific competition between Newton and Leibniz's *systems* had started from "two different interpretations of mechanics" but then it had rather become a competition between "two sects." Leibniz had blamed Newton for having "deified physical space" but he himself had "humanized space": he had transformed physical space into the subjective ability to set things in order, or even "the mere idea of order." In the end, what Descartes, Newton, and Leibniz had in common was "the suppression of nuances, namely the search for a clear-cut intelligibility." Nevertheless their efforts had led to "a displacement of darkness" rather than a true enlightenment: Descartes had transformed "living beings into machinery," Newton had assumed "an action at a distance that remained quite mysterious," and Leibniz an even more striking "pre-established harmony" [Cournot 1872, pp. 316-18, 320, 322-3, and 348-9].

In the end, Cournot's historical and critical analysis acknowledged the existence of a meaningful scientific revolution in the seventeenth century and at the same time, the frailty of its foundations.[22] In general, he pointed out the weakness of every systematic natural philosophy, and every wide-scope worldview. This issue deserves to be clarified because it highlights the specificity of Cournot's historiography and epistemology. He found that the philosophical foundations of modern science, wherein Bacon's empiricism, Descartes' rationalism, Newton's natural philosophy, and Leibniz's metaphysics were as points in case, were not more firmly established than the foundation of ancient sciences, mainly in their Aristotelian version. The impossibility of a firm founda-

22 Gusdorf saw Cournot's historical inquiry as the consequence of "the brilliant career" of the tradition of "the history of reason" or "meta-history" that had emerged with Fontanelle, and had been labelled "philosophie de l'histoire" by Voltaire. Gusdorf also found that, from the structural point of view, Hegel, Comte, Marx, and Spencer had pursued the same target. See Gusdorf 1966, p. 86: "La philosophie de l'histoire apparait comme un exercice de haute école intellectuelle, un peu comme les théologies systématiques de jadis." It seems to me that Cournot's critical attitude allowed him to devise a philosophy of science and a philosophy of history more pliable and sophisticated than Comte's and Spencer's.

tion was what ancient and modern sciences had in common. In this sense, historiographical continuism and discontinuism stay hand in hand in Cournot's historical reconstructions. The epistemological side of this criticism is that the constellation of scientific methods cannot converge onto a single normative epistemology. It was the plurality of epistemological attitudes and philosophical foundations that had allowed modern science to develop. This historiographical and epistemological perspective was not in tune with simpler evolutionary historiographies and simpler epistemologies such as Comte's.

4 Sharp Anti-reductionism and Mild Naturalism

Two years later, the young philosopher Émile Boutroux published his doctoral dissertation, which was a remarkable piece of writing, even though it did not contain any explicit reference to philosophical literature. The frailty of scientific foundations was also at stake. The title of the book, *De la contingence des lois de la nature*, really expressed the author's view on the epistemic status of scientific theories.[23] From the outset he focused on two main issues: the relationship between scientific and philosophical knowledge, and the relationship between scientific statements and experience. The question was: how, and to what extent, can we attain an actual comprehension of the natural world? In a brief historical sketch, he claimed that in the first stage of natural philosophy, when the human mind could only rely on its sensitivity, the natural world appeared as a collection of events. In the following stage, when "a purely descriptive science" had appeared unsatisfactory, the human mind had looked for "an explicative knowledge." Mere experience could not help scholars to attain this target: the human mind entrusted the responsibility of "classifying, interpreting and explaining" empirical data to its more abstract skills. According to Boutroux, this development in the history of natural philosophy had had an unexpected consequence: an increasing distance between philosophy and the real world. The alliance between the order of mind and the plurality of experiences, between the multiplicity of facts and the unity of the natural law that should explain those facts, inevitably led to a paradox: the enlargement of the gulf that must be bridged. In Boutroux's representation, human knowledge continuously swung between the two poles of "variety and unity," or "contingency and necessity," or even "change and permanence." Under the

23 Boutroux had graduated in 1865 from the *École Normale Superieure*, and had then moved to Heidelberg in order to perfect his philosophical education.

pressure of this intrinsic tension, could science account for "the objects of its inquiry"? [Boutroux 1874, pp. 2-4].

According to Boutroux, a rationalist approach to the natural world relegated every contingent event and free will to a sort of "delusive world," or a domain of imperfect knowledge where the reasons of those events and "the causes of our actions" remained substantially unknown. In some way, a rationalist approach led to fatalism, both in the context of natural sciences and in the context of "psychology, history and social sciences." The Introduction of the dissertation ended with a question, or better hinted at a possibility: if the natural world exhibited a certain degree of actual contingency, then rationalism would not be "the final point of view" on the natural world. He insisted on the dichotomy between rational and empirical practices, and more specifically on the unbridgeable gap between the "purely formal" structures of logic, and experience. That gap did not allow us to transform "constant links" among real things into "necessary links." The ideal of the most radical rationalism would be the attainment of a single, very general law which would encompass "all the laws of the universe as specific instances" of it, but any change in the least detail of the universe, would lead to a complete destruction of the rational scaffolding. In the end, he found that the comprehension of the natural world required a milder rationalism [Boutroux 1874, pp. 5-11].[24]

Boutroux insisted on the opposition between the stability of scientific laws and the variability of the actually experienced world. He claimed that "an absolutely permanent law" could not be attained, and "the reality of change" was as evident as "the reality of permanence." No abstract law of causality could account for the fundamental tension between change and permanence that was the essential feature of "life and real existence." This problem also dealt with the reliability and explanatory power of classifications. Was it wise to assume that nature contained a well-defined number of genera? How could the presence or absence of specific features exactly qualify these genera? Could we rely on the meta-theoretical belief that "everything changes, apart from the law of change"? Once more Boutroux suspected that the confidence in

24 He made use of a Kantian linguistic toolbox, where adjectives like *analytic*, *synthetic* and *a priori* were at stake. See specifically pp. 15-6: "On peut donc admettre la possibilité d'une nécessité de *fait* à coté de la nécessité de *droit*. Celle-ci existe lorsque la synthèse que développe l'analyse est posée *à priori* par l'esprit et unit un effet à une cause. Lorsque cette synthèse, sans être connue *à priori*, est impliquée dans un ensemble de faits connus, et qu'elle est constamment confirmée par l'expérience, elle manifeste, sinon la nécessité du tout, du moins la nécessité de chaque partie, à supposer que les autres soient réalisées."

the laws of Nature and strict determinism inevitably led to a sort of fatalism or pessimistic concept of nature. The complexity of empirical data prevented scholars from attaining a "natural classification," namely a perfect correspondence between actual experiences and abstract laws. Differently from Comte, Boutroux did not rely on the possibility of replacing artificial classifications with natural classifications: every classification was artificial, and what could be actually attained was nothing more than genealogical links. He saw Nature as a battlefield, where "a radical contingency," and a tendency towards "a fundamental state of dissemination and chaos," opposed logical order, namely the tendency to converge towards definite species and genera. In the end, the desire for law and order had to give way to "indeterminism and contingency" [Boutroux 1874, pp. 17-19, 30-31, 39, and 45-7].

According to Boutroux, the mathematisation of natural philosophy had introduced a new, heterogeneous, feature in Nature, namely continuity. Continuous space, continuous time, and continuous matter were in tune with the idea of continuous motion. Although the hypothesis of continuity had allowed natural philosophers to geometrise Nature, he challenged the Platonic mythology of a geometric world that had been revived at the dawn of modern science. To the confidence in geometry as the ideal model of an imperfect world Boutroux opposed his interpretation of the geometrisation of Nature as an impoverishment of reality. He put forward the radical thesis that geometrisation represented "a negative feature" in the sense of an oversimplification of the luxuriant variety of Nature [Boutroux 1874, pp. 50-2 and 55]. In brief, mathematisation could only offer a simplified overview of the universe. In the intellectual context of the second half of the nineteenth century, Boutroux' historiographical thesis on the dangerous effect of mathematics might appear irritating and outmoded. The fact is that Boutroux aimed to emphasise the flaws in the foundations of modern science, and at the same time the value of philosophical tradition.[25]

25 It is worth reporting a passage that echoed Aristotle's distrust of the mathematisation of the physical world: "Un tronc d'arbre qui, vu de près, est tortueux, paraît de plus en plus droit, à mesure qu'on le voit de plus loin. Quel besoin avons-nous de notions *à priori*, pour achever ce travail de simplification, et éliminer par la pensée tous les accidents, toutes les irrégularités, c'est-à-dire, d'une manière abstraite et vague, celles que nous voyons et celles que nous ne voyons pas ? Par là, sans doute, nous n'acquérons pas l'idée des choses supérieures à la réalité. C'est, au contraire, la réalité appauvrie, décharnée, réduite à l'état de squelette. Mais est-il donc si évident que les figures géométriques soient supérieures à la réalité; et le monde en serait-il plus beau, s'il ne se composait que de cercles et de polygones parfaitement réguliers ? [Boutroux 1874, p. 56]."

The apparent anti-modern turn did not prevent Boutroux from remarking that the *mechanistic* view, namely the reduction of physics to matter and motion, or "geometry and motion," had attained an unquestionable success. He acknowledged that the mechanical theories of heat and light corresponded to real scientific progress. Nevertheless, he insisted on "the contingency of details" that eluded the absolute determination of general laws, even in the domain of astronomy, where the revolutions of planets did "not reproduce exactly the same periods." In the domain of everyday experience he mentioned "the remarkable effects" produced by minor variations in the position of weights around their equilibrium configurations, both in simple and complex natural systems: a disruptive avalanche could be generated by "the seed fallen from the beak of a bird on a mountain covered by snow." In the domain of the kinetic theory, the reduction of thermal phenomena to motion could not account for the intrinsic difference between macroscopic and "molecular motion": something else, "a new element" of non-mechanical nature, had to be tacitly added in order to catch that difference. The invisible motions of microscopic particles called into play "new and more sophisticated properties" of matter. Boutroux seemed at ease with some aspects of physical sciences. He was also familiar with the argument of *mechanists*, and Laplacian mythology: if scientists were able to know "*all* the mechanical parameters" of a natural phenomenon, they would be able to predict it "with absolute certainty." He objected that the concept of the totality of mechanical conditions was an abstraction, and it was questionable whether something like "a finite number of totally determined mechanical conditions" really existed for a natural phenomenon [Boutroux 1874, pp. 57, 68-70, 73, and 76].

According to Boutroux, the reduction of Nature to matter and motion was undermined by discontinuities and qualitative transitions taking place in some physical and chemical processes. He focused on specific physical effects such as "the intermediate state" of an electric circuit between the establishment of an electric tension between the poles of the battery and the appearance of light in carbon electrodes. Then he pointed out a whole class of phenomena wherein a sudden transition took place, and huge amounts of energy could be released or transformed "by the addition of a little amount of motion." Boutroux knew contemporary physics, and was aware of the debate raised by the emergence of thermodynamics as a new systematic body of knowledge. He stressed that the principle of conservation of energy determined "the *intensity* rather than the *way*" of physical transformations. Both in ordinary life and in cosmological processes, the second principle of thermodynamics determined the direction of events. He saw a restless universe, ruled by the competition between the attractive and creative power of gravitation and the dissipative effects involved in heat transfers. On the one hand he saw "elementary cosmic matter" that condensed into stars "endowed with light and heat"; on the other, a process of dissolution that reduced stars to scattered particles. He envisaged a complex universe, where the existence of living beings defied the mechanical world-view: physical and chemical forces could not account for the emergence of complex living systems, or even the existence of "elementary living matter." The "extreme instability of living systems," and their "hierarchical order" called into play specific features that "could not be reduced to physical properties." The new, essential features could be qualified as "*individualisation*" and "inner *correlation*" [Boutroux 1874, pp. 80-1, 84-5, 87, 89, 91, and 93].

In the transition from a level of complexity to another, Boutroux found it legitimate to study "the physiological conditions of psychic life," as well as "the physical conditions of organic life," and "the mechanical conditions of physical transformations," but he did not see any possibility of a satisfactory reduction of one level to another. Some kind of qualitative discontinuities were at stake in every transition: physics cold not be reduced to mechanics, physiology could not be reduced to physics, and psychology could not be reduced to physiology. Even though the activity of the human mind could be quantified in terms of energy, the relative value of every thought did not depend on the amount of physical energy corresponding to it. There was a sort of incommensurability between the two levels: the differences in the physical activity of the brain could not account for the difference between "genius and madness." Moreover, in man, actions like heroism and sacrifice could overwhelm "the strongest resistances" offered by biological laws. Free will allowed people to challenge those laws: man's actions did not depend on the mere preservation of the physical body. He questioned the dependence of mental processes on physiological ones: he dared to claim that it was more probable that "the latter depend on the former." It was precisely this specific behaviour of human beings that made so difficult a unified science for inorganic and organic matter [Boutroux 1874, pp. 113, 115, 132-3, 143 and 147-8].

In the very long *Conclusion* of his dissertation, Boutroux raised an even more dramatic issue: modern science, and specifically "deductive sciences," had assumed the invariability of the laws of Nature. This meta-theoretical belief qualified modern science as an *abstract* science, which excluded both progress and *decadence*. Nevertheless, in the real world, creation and dissipation were both involved, and qualitative transformations stood beside the quantitative ones. In the context of quantity, homogeneity and permanence could be looked upon as "*essential* and *absolute*," but they became "*accidental* and *relative*" in the context of qualities. According to Boutroux, the human mind was under pressure from two sides: the formal structures of mathematical laws and logical procedures on the one hand, and the wide set of natural events and experiences on the other. The stability of mathematical laws was balanced by the contingent flux of becoming. The bodies of knowledge that Comte had qualified as "positive sciences" rested on "the conservation of being": they dealt with *change* as far as it could be brought back to *permanence*. In the wider field of human experience a further principle, namely "a principle of creation," balanced the recently established principles of conservation. The natural world was in a state of dynamic equilibrium between mere preservation and creative transformation [Boutroux 1874, pp. 151, 153, 155, and 158].[26]

Boutroux hinted at a peculiar and radical epistemology: some kind of "*dynamical* sciences" had to be put beside and above what he labelled as static or positive sciences. Not only did dynamical sciences have to inquire into the *nature* of things, but also into their *history*. This perspective would have allowed philosophers to merge liberty with necessity, and the "possibility of change" with the "possibility of permanence." The comprehension of a plural and dynamical universe required a plural and dynamical intellectual practice. Moreover the natural laws of scientific tradition had to be consistent with the "moral and aesthetical" structure of the world. Natural laws did "not possess an absolute existence" because they represented only a historically determined stage in the history of knowledge. The apparent immutability of scientific laws corresponded to "the intrinsic stability" of the ideal model from which they had been generated [Boutroux 1874, pp. 164-5, 167, 173, and 192-3].

In the end, Boutroux' radical anti-reductionism conflated into a broad naturalism, where variability and instability replaced invariance and stability of mathematical laws. In some way, this confused and enticing natural philosophy represented a new mythology, and it was as unbalanced as positivistic mythology. Contrary to what has sometimes been claimed, Boutroux' philosophy was not antiscientific but was based on the assumption that sharp reductionism and determinism were not necessary foundations for scientific practice. Boutroux's 1874 text was rich in claims and assertions rather than cogent argumentations: for this reason, it is not easy to put forward a detailed evaluation of its effects on the development of the philosophy of science in French-speaking countries. Sometimes his naturalism echoed the representations of the universe as a living entity on the track of sixteenth-century natural

26 See also p. 159: "… dans les mondes inférieures, la loi tient une si large place qu'elle se substitue presque à l'être; dans les mondes supérieures, au contraire, l'être fait presque oublier la loi."

philosophers. He did not despise science, and kept abreast of recent scientific achievements: he firmly opposed the positivistic trend, and was in search of a new natural philosophy where the wisdom of philosophical tradition and those recent achievements could interact with each other.[27]

The following year, Cournot published another book on the relationship between science and philosophy: both the role of life sciences and the debate on reductionism were in prominence. Cournot's anti-reductionist attitude towards science was not so different from Boutroux's, even though the philosophy of science of the former was more cautious and pliable. His naturalism was milder and more inclusive.

With regard to life sciences, he thought that physics could offer "the material framework for the domain of organisation and life," but the specific features of living beings could not be understood in physical terms. The transition from physics to life sciences could not be considered as "a smooth development or a continuous progress": it involved a discontinuity. The most recent developments in life sciences had shown their proximity to chemistry, and at the same time their heterogeneity. The specific feature of a living being was "the convergence of all functions towards a common aim," which he labelled as *instinct* and "*creative energy*." Finality or purposiveness could not be excluded from the domain of natural sciences, and even some founding fathers of modern science had acknowledge such a necessity. He sketched a broad historical genealogy that started from Aristotle's *entelechy* and led to Leibniz's *monads* through Van Helmont's *arké*. Cournot interpreted Leibniz's *dynamism* as an attempt to unify "the phenomena of the two worlds, organic and inorganic": when compared to Newton's theory, Leibniz's monadology had led to "a more complex approach to physics" because it was more suitable for, and in tune with, biology. He saw Leibniz's natural philosophy as a theoretical framework that had stemmed from biological models, and had then been extended to physical models [Cournot 1875, pp. 87-9, 103, and 106-7].

27 How difficult it is to qualify Boutroux' philosophy is shown by Benrubi's manifold qualification of Boutroux as "an adherent of the critical, idealistic, and rationalistic schools." He stressed that Boutroux' contingency should not be confused with chance: it was close to the idea of natural freedom, in the sense of free and unpredictable unfolding of natural laws. In other words, contingency occupied the intermediate place between chance and necessity [Benrubi 1926, pp. 154-7]. In the same decade, the French philosopher Léon Brunschvicg stressed the role played by Boutroux in the emergence of a mature philosophy of science. According to Brunschvicg, Boutroux' 1874 dissertation represented the starting point of a new awareness: more specifically, "the critique of scientific knowledge became aware of itself" [Brunschvicg 1922, p. 271]. For the influence of Boutroux on Poincaré, see Crocco 2016, pp. 212-215.

In his overview of recent developments and the problematic foundations of life sciences, Cournot also focused on the experimental side of physiology. He discussed physiological practices that submitted animals to "cruel experiments." He thought that animals should not be looked upon as passive objects of experimental practices: they were endowed with "sensitivity, memory, and knowledge." In this sense, they were not so different from human beings, and deserved sympathetic treatment. At the same time, the position of human beings in the animal kingdom was intrinsically ambiguous, and it was doubtful whether experiments on animals could cast light on human physiology and psychology. On the one hand, humans appeared as the fulfilment of "organic creation," but on the other, they could be interpreted as "a singularity or an anomaly" among other animals. Beside the outstanding development of nervous system and sensitivity, human beings showed the hallmark of a peculiar *regression*. Children entered the world "too early," and in a condition of relative imperfection: training and education had to complete what purely physiological processes had not managed to accomplish [Cournot 1875, pp. 170, 172-4, and 177].[28]

Cournot's mild naturalism and his view on life sciences as an intermediate body of knowledge between physical sciences and human sciences were not in tune with the hard experimentalism that had emerged in physiology. Cournot's critical and cautious attitude allowed him to grasp the complexity of scientific practice, and the limits of methods and research programmes that had emerged around mid-century. He could not share Bernard's full confidence in vivisection as a means of testing normal physiological functions. But he certainly shared Bernard's concerns about the position of life sciences in the classification of sciences: physiology deserved to be considered as a science, and physiological functions depended on chemical and physical processes. Nevertheless, unlike chemical compounds, whose basic chemical elements could not be created or destroyed but could be rearranged from one combination to another, complex physiological functions did not experience a law of conservation: when dismantled in any individual, they could not be restored. Cournot understood that the emergence of life sciences required both new scientific classifications and new epistemologies: statistics and probability appeared as new intellectual and computational toolboxes that allowed

28 According to Cournot, another specific feature of human beings was their extreme variability, both inside their race, and among different races. However he thought that the most interesting differences among human beings were the cultural ones: he stressed the presence of mythological poems "at the roots of the Aryan civilisations," and their absence in the first stages of other civilisations [Cournot 1875, pp. 178-80].

scientists to replace certainty with reliability as the main aim of their scientific practices. Beyond the mythology of mechanical determinism, Cournot's mild naturalism could encompass both determinism and contingency in the domain of natural events. Boutroux emphasised the role played by contingency in natural processes and focused on the gap between the stability and rationality of scientific laws, and the contingency and variability of human experiences. Cournot's *probabilism* was a philosophical attitude that could be based on reliable mathematical procedures: it offered a new equilibrium between the stability of laws and the contingency of facts.

CHAPTER 3

Different Attitudes Towards Reductionism

1 New Classifications

In the second half of the nineteenth century, the emergence of physiology as a science and the development of life sciences in general played an important role in reshaping philosophical reflections, political commitments, ethical codes, and social order. In spite of the increasing specialisation and professionalization that took place in those decades, the boundaries between scientific practice and philosophical commitment remained quite loose until the end of the century, and they could be crossed in both directions.[1] Some philosophical issues such as necessity and predictability of physical laws, the role played by chance and contingency, the reduction of physics to mechanics, and the reduction of all sciences to physics were widely debated. Other questions involved the nature of biological processes and their relationship with physical forces and chemical actions. Boutroux' radical anti-reductionism was not the hegemonic attitude in the scientific community: in the late 1870s we find many different attitudes, and different degrees of reductionism. Some physiologists and physicians were less cautious than Cournot in outlining the complex connections between life sciences and the recently emerged human sciences, and relied on a radical reductionism.

The title of the book Cournot published in 1875 consisted of two parts: the first, "*Matérialisme, vitalisme, rationalisme*," made reference to a rather traditional classification that covered a wide range of disciplines from physics to the philosophy of science through chemical sciences and life sciences, psychology and anthropology included.[2] The second part, *Études sur l'emploi des*

1 As already remarked in the *Introduction* to the present book, the establishment of definite boundaries between science and philosophy was one of the achievements of scientific practice in the late nineteenth century. Even the word *scientist* does not seem suitable for some geographical contexts. On the process of specialization and professionalization taking place at the end of the nineteenth century, see for instance Ross 1964, p. 66, and Morus 2005, pp. 3, 6-7, 20, and 53.

2 The choice of words such as *matérialisme, vitalisme*, and *rationalisme* was definitely unsuitable, because they suggested philosophical attitudes rather than disciplines. Cournot confined himself to pointing out that the title of the volume would have repelled readers "in search of easier readings."

 | DOI 10.1163/9789004315235_005

données de la science en philosophie, made reference to a more specific commitment to inquiring into the philosophical meaning of scientific concepts. He found that his book looked like "a talk on Natural philosophy in the sense of English scholars" rather than "Philosophy of nature in the more ambitious sense of German philosophy" [Cournot 1875, pp. I-II]. In brief, the book consisted of a critical analysis of the foundations and open problems of the different sciences.

Cournot started from physics, and specifically from the concept of matter-mass: he immediately pointed out the contingency of physical data, and the ostensible arbitrariness of chemical theoretical frameworks. The Earth contained some chemical compounds rather than others by "pure chance": other planets and other stars could contain different elements and compounds. Moreover the search for basic, simple elements had led to the awareness that the number of simple elements could not be defined *a priori*. However, other problematic issues involved matter: for instance, the relationship between matter and force he had dealt with in 1861. To consider matter as the seat or "substantial support" of force was nothing more than a convenient linguistic choice. According to Cournot, both physics and chemistry relied on useful conventions: the function of atoms in chemistry was not structurally different from the role of matter and mass in physics. Scientists had no direct experience of atoms: they had only access to "chemical equivalence or the law of definite proportions." The fact was that science needed "both facts and theories," and the human mind was in search of general ideas and general models for encompassing and explaining a wide set of facts. The human mind needed extensive frameworks where both facts and laws could find room [Cournot 1875, pp. 7-9, 11, and 27].

Cournot stressed that chemistry had become an autonomous body of knowledge endowed with its own foundations; chemistry could not be reduced to mechanics. Chemistry dealt with discontinuous processes: this fact suggested to him a structural analogy between chemistry and very different disciplines such as "the theory of numbers and abstract syntax." He did not see qualitative differences between inorganic and organic chemistry, whereas a real qualitative discontinuity could be found between organic chemistry and life sciences. The specific feature of organic chemistry was "the smaller stability… and the greater complexity" of its compounds: the binary compounds of inorganic chemistry gave way to "ternary or quaternary combinations among the four radicals" involved in the composition of living beings (hydrogen, carbon, nitrogen, and oxygen). According to Cournot, the emergence of chemistry as a science corresponded to a meaningful stage in the history of civilisation, which could be compared with the emergence of the Bronze Age or Iron Age.

Progress in both inorganic and organic chemistry, and the huge number of useful applications, represented one of the hallmarks of scientific progress in the nineteenth century [Cournot 1875, pp. 28, 30, and 33].

With regard to life sciences, Cournot now differed from the position he had taken in 1861 [Cournot 1861, pp. 412-3]. He now found that the concept of *species* in biology was nothing else but "a relic of the mediaeval scholastic world." It had to be banned from the linguistic toolbox of life sciences and replaced by the concepts of "race and genus." He discussed the problems raised by the concept of species in the same way as in 1861, but the conclusion was different: he claimed that the concept had stemmed from "coarse empiricism" rather than "scientific accuracy." His recent distrust was based on the fact that more detailed observations had shown that "the fecundity of hybrids in the first and following generations" was not impossible. Although he had cast doubt on the suitability of species as a fundamental entity for biological evolution, he extensively discussed a pillar of Darwin's theory, namely the thesis of "competition for life," from the point of view of species. He found that the existence of competition could suitably explain the fact that we do not observe unbounded proliferations of species. He also acknowledged that the analogy between natural and artificial selection, as put forward by Darwin, was a fruitful analogy indeed. Furthermore he had no objection towards the supposed common origin of all living beings. Both vegetal and animal beings might have stemmed from "the same archaic cell" or from "identical cells in different places at different times" [Cournot 1875, pp. 144, 147, 150, 157-60].

Cournot managed to grasp the core of Darwin's theory, and the differences between Darwin's theory and other evolutionary theories. What he found unsatisfactory was the implicit continuity in the process of evolution by natural selection. With regard to the emergence of the elephant's trunk over a long time interval, he asked himself what evolutionary advantage for the struggle for life could have been given by the intermediate stage of a longer nose which was not yet a trunk. In general, even when a given final state of an evolutionary process was considered suitable for successful selection, it was questionable whether the intermediate states would have been really successful. He saw a serious inconsistency "between the assumption of slow transformations, and the principle of natural selection." Moreover there was no paleontological evidence of "the huge number of intermediate forms required by Darwin's theory." In reality, this empirical objection was not so important for Cournot, because he was aware of the partial and provisional nature of every empirical evidence in the domain of natural sciences. However, he found that the first objection was insuperable in the context of Darwin's original theory. In the end, he saw two possibilities: either "a radical shortage of paleontological observations" prevented naturalists from recording a series of slight transformations, or "the chain of slow actions" had to be replaced by "stages of crisis under the effect of extraordinary causes" [Cournot 1875, pp. 162-6].

However, according to Cournot, the replacement of a slow transformation with "a sudden transition" did not solve the main problem, namely "the necessity of an internal principle, which could explain the final harmony" of deep transformations such as the transition "from reptiles to birds." He found that "a mechanical process of selection" could account for the increase or decrease of already existing "organic features," but it was not as powerful as would have been required by the radical transformations "from one zoological class to another." In other words, natural selection could account for limited biological transformations rather than extraordinary events such as the emergence of new classes of living beings. Alongside natural selection, he envisaged the existence of stronger processes that acted "at certain critical times, and for a suitably short time interval." Processes of this kind could obviously explain the lack of evidence for intermediate biological structures. In the end, in spite of the mysterious nature of those extraordinary causes, he found that his theoretical outline exhibited two advantages. On the one hand, from the specifically scientific point of view, it could offer a more sophisticated scientific explanation for complex evolutionary processes. On the other hand, from the philosophical point of view, his hypothesis was compatible with "the metaphysical idea of a highest cause," or "the religious idea of Providence" [Cournot 1875, pp. 166-9]. Once more, we meet Cournot's preference for a plurality of possible explanations, the awareness of the shortcomings of every explanation, the recognition of tacit metaphysical assumptions, and the search for possible conciliations between different points of view.[3]

He stressed the existence of "a sharp borderline between physical laws and the laws of life," and at the same time the existence of a fuzzy borderland between the study of human societies and "the conditions of organic life." In other words, the emergence of creativity and rationality from the domain of biological processes was less astonishing than the emergence of life from the domain of physical-chemical processes. He saw a meaningful link, or better a structural analogy, between *culture* and *Nature*, or better, organic Nature. There also was an intermediate domain between Nature and culture: it was the anthropological domain, which dealt with the emergence and establishment

3 Generally speaking, he hoped that, in an indefinite future, "human reason" and scientific practice would not have banished "any religious emotion, and poetical inspiration" [*Ibidem*, p. 169].

of human civilisation. According to Cournot, the structure of human communities was not so different from the structure of living beings: the life of communities was structurally much more akin to the life of animals or plants than to "the life of individuals" in the sense of their psychological activity. Anthropology and sociology dealt with biology rather than psychology or philosophy: moreover, it was psychology that was based on sociology and anthropology. He specified that his view had nothing to do with reductionism of psychology to human biology. It was rather an anti-reductionist move: "the highest activities of individuals" were much more affected by the features of the communities, nations, and races to which they belonged than by "the biological conditions of the individuals" themselves. It was reasonable to assume that Homer had been influenced by "his contemporary Hellenic society" much more than by "the combination of some anatomic elements." In general, the individual received from society "more than society received from the individual" [Cournot 1875, pp. 188-91].

The relationship between different bodies of knowledge as outlined by Cournot was in contrast with Comte's classification of disciplines, which appeared as a natural hierarchy wherein "a given science was based on the preceding one, which was less complex." According to Cournot, a given body of knowledge dealing with complex entities could even evolve before the complete systematisation of the body of knowledge dealing with the elementary components of those entities. Meaningful researches on plants and animals, which were assumed to be "collections of cells," were performed before the discovery of cells. The overturn of Comte's classification also involved the role of language, which he looked upon as "an organic structure." Social organisation and language appeared as an immediate consequence of vital momentum; more specifically, life presided "over the development and maintenance" of language. Life was "a principle of instinctive organisation," and the emergence of language depended on that *instinct*. In many passages he insisted on the concept of language "as the outcome of instinctive work, a vital energy" rather than "a creation of the higher skills" of mankind. In other words, language did not deal with the desire for intellectual elaboration but had stemmed from the vital necessity of expression and communication typical of highly-developed biological communities. He found that recent researches in linguistics were consistent with the idea that reason was the consequence of language rather than its origin [Cournot 1875, pp. 191-4 and 198-9].

In that context of classifications and relationships between different bodies of knowledge, the role played by psychology was under scrutiny. Cournot did not think that psychology was a science. He remarked that important bodies of knowledge, which corresponded to interesting and useful subject matters, did not have "a scientific nature." Despite their intrinsic value and usefulness, history and philosophy could not be qualified as sciences in a strict sense. Scientific practice required "the detachment of the observer from the observed object." A subject who observed her/himself put her/himself "in the worst conditions of observation and judgement." Moreover, in the field of psychological observations, the mere presence, and even worse the attentive presence, of an observer influenced the observed behaviours. That presence was a psychological phenomenon in itself: it could interfere with the phenomena under observation. Psychology could be qualified as "an observational science" in a broad sense: paradoxically enough, according to Cournot, it belonged to a set of bodies of knowledge that included "astronomy and meteorology." Nevertheless observational sciences in the domain of physical phenomena could rely on "an endowment of long-tested scientific instruments" whereas psychology had mere language at its disposal. In brief, psychology was a body of knowledge that did not deal with scientific practice but only with a partially reliable, observational practice [Cournot 1875, pp. 252-5 and 257].

The problem of the demarcation between science and other bodies of knowledge led Cournot to a more refined analysis of the concept of scientific law. A scientific law in a proper sense was something more than an *interpolation* among scattered empirical data. The label "law of Nature" itself hinted at the existence of a hypothesis that had to be simple, and at the same time in tune with "phenomena unknown to the advocate of the hypothesis itself." In other words, the scope of a law had to be wider than the set of known phenomena, in order to allow subsequently discovered phenomena to be accounted for by the scientific law. The laws of mechanics and Newton's law of gravitation were instances of scientific laws. Their status allowed scientists to look upon alleged violations of the laws as the consequence of "less general laws that masked or made difficult the acknowledgement" of the main effect. On the contrary, some empirical laws such as "Bode's law" could not be qualified as scientific laws [Cournot 1875, pp. 330-2].[4]

In the last part of the book, Cournot faced a question that lay at the foundations of scientific knowledge, namely the relationship between subject and object. He started from the widespread representation of "the mind as a mir-

4 The law made reference to the distances of the first seven planets from the Sun: it was a simple arithmetical law lacking any independent evidence apart from the existence of "a *gap* between Mars and Jupiter" that corresponded to "the asteroid belt." Bode's law did not account for the position of the planet Neptune, which had been observed almost thirty years before: it was something less than a scientific law even though it was something more than "pure chance" [*Ibidem*, p. 331].

ror that reflects the objects" of Nature, and then he followed the analogy in order to point out its shortcomings. If the mirror had the shape of a plane, the image would be "a faithful representation of the object, apart from the exchange between right and left." On the contrary, if the mirror had a curved shape, cylindrical or elliptical for instance, the image might be warped. Therefore Cournot asked rhetorically: was "the human mind like a flat or cylindrical mirror"? He answered that there was no certainty of having attained "absolute truth": truth was as unattainable as the certainty of having found the absolute rest. The adjective *absolute* had to be replaced by the adjective *probable* both at the theoretical and meta-theoretical levels. Uncertainties and perturbations in the search for a physical reference frame were structurally similar to the impossibility of attaining a perfect theoretical reference frame. Both events and theories could only be evaluated in terms of approximation and probability. Scientists had to confine themselves to "great probabilities," and probability was the keyword for the third way between dogmatism and scepticism [Cournot 1875, pp. 348-9, 354, and 359-60].[5]

Cournot found that the influence of science on philosophy had always been positive: more specifically, the great transformations in the history of philosophy had been triggered off by corresponding transformations in science. In the seventeenth century, when modern science emerged, the best philosophers had also been "the most talented scientists." During the nineteenth century, the superposition between scientific practice and a professional, philosophical commitment could not be pursued any more.[6] Nevertheless, the new philosophical attitude, which had qualified itself as *positive*, had resumed the ancient superposition: it aimed at reducing philosophy to a summary of all particular sciences. Scientific progress had led "*positivist* philosophers" to give up metaphysics, which was looked upon as an old-fashioned practice. They believed that the age of metaphysics had given way to the age of science in the same way as the age of theology had given way to the age of metaphysics. At the same time, Cournot was aware that the positivistic attitude represented a specific philosophical commitment: it was a new kind of metaphysics [Cournot 1875, pp. 371 and 373-5]. Confronted with this naïve philosophical attitude, Cournot pointed out the impossibility of getting rid of refined rational practices, and therefore metaphysics. His words were not dissimilar from the words Whewell had employed some decades before:

5 Cournot quoted a sharp passage from Pascal's *Pensées*: "La Nature confond les pyrrhoniens, et la raison confond les dogmatiques" [*Ibidem*, p. 359].

6 I remind readers that this awareness had already been expressed by Cournot in 1861, and to a certain extent in 1851 [Cournot 1861, p. VII; Cournot 1851, pp. I-II].

> We confine ourselves to remarking that every actual scientific practice consists of empirical and positive facts, together with a theory that links those facts and offers an explanation. A theory needs some ideas, where a critical attitude develops the potentiality of reason: a theory deals with what positivists label philosophy or metaphysics. In the network of scientific practice, ideas conflate with facts, and a rational principle sets in order empirical experiences. We must stress that a body of purely empirical knowledge is not a real science yet,... [Cournot 1875, pp. 375-6].

Cournot insisted on the fact that theories were necessary but also provisional: scientists needed theories even though theories changed over time, and their reliability had to be continuously under scrutiny. In the end, the contingency of history and a mild relativism entered the scene. In general, the possibility that unpredictable events appeared throughout history permeated the whole landscape of knowledge: the concept itself of science, scientific standards, and even scientific method might change over time in the same way as habits and civil law, and even ethics, had actually changed over time. He mentioned social behaviours such as "slavery of coloured people and loans with an interest rate" that, at certain times, had been forbidden or discouraged by laws after having been encouraged for a long time, or had been encouraged after having been discouraged or forbidden. Every kind of dogmatic authority, scientific authority included, naturally leant towards a conservative attitude, and science could not indulge in conservative attitudes. He found that "the authority of science" would have been "cautiously exerted," with great respect for every opposition, "even when opposition became stubbornness" [Cournot 1875, pp. 376 and 391-3].[7]

The book Cournot published in 1875 can be looked upon as a philosophically oriented history of science, where a critical historiography merged with cautious epistemological cogitations. Alongside the specific contents of his remarks on the history of science, philosophy of science, and philosophy of history, we find in Cournot a new style and a new attitude: he appears as the first scholar to have expressed a critical detachment towards scientific tradition and its mythologies, and at the same time a sympathetic attitude towards actual scientific practice that was branching out into different directions. His epistemological pluralism and the role played by contingency in his historiographical framework were not in tune with more simplified and dogmatic attitudes of other scientists and philosophers.

---

7 Cournot remarked that it would have been better not to involve God in the licence to practice slavery or in censuring financial transactions [Cournot 1875, p. 392].

2 Debates between Philosophers and Physiologists

In 1876 the physician with interests in neurology Jules Bernard Luys, who worked at the Parisian hospital *La Salpêtrière*,[8] published the book *Le cerveau et ses fonctions*. In the *Foreword*, after having mentioned the different cerebral parts and their "nervous elements," he claimed that the combinations and cooperation among those elements allowed the human brain "to feel, recollect, and react." More specifically, every manifestation of cerebral activity, psychic and intellectual functions included, could be reduced to three elementary stages. The first stage, "the stage of incidence," consisted of the arrival of a sensorial impression; the second stage, the intermediate one, corresponded to "the reaction of the interposed medium"; finally, in "the stage of reflection," the medium actively interacted with the external world. These elementary steps allowed Luys to reduce "the unfathomable domain of speculative psychology" to "regular processes of nervous activity." He stressed that, in recent times, brain physiology had become a field of research as developed as the physiology of the heart, lungs, and muscular system. This reliable and accomplished science of the brain was the only body of knowledge that could legitimately justify the existence of psychology and human sciences in general. From the medical point of view, new methods in the therapy of mental illness could only emerge from "a better comprehension of cerebral anatomy, and a more rational cerebral physiology" [Luys 1876, pp. VIII-XI].

In 1877 and 1878, the years of Cournot and Bernard's death respectively, the debate on the relationship between mind and brain flourished. The philosopher Victor Egger, *maitre de conférences* at the University of Bordeaux, published a critical review in the Journal *Revue des deux mondes*. The paper, "La physiologie cérébrale et la psychologie," was devoted to a sharp criticism of the reductionist attitudes of "the Parisian school of medicine." From the outset, Egger stressed "the insuperable hurdles that logic opposes to intrusive pretensions of physiology," and addressed his criticism to Luys, who had put forward "a physiological psychology." According to Egger, the logical mistake consisted in assuming a correspondence between "material or extended processes," on the one hand, and "psychological or non-extended processes," on the other hand: the two kinds of *facts* were intrinsically different and methodologically irreducible to each other.[9] In reality, his thesis was even more radical: psychol-

8 La Salpêtrière was a psychiatric hospital in Paris, which attracted many scholars from all Europe.

9 In order to strengthen his point of view, Egger mentioned the talk Claude Bernard had given before the French Academy in 1869 [Egger 1877, p. 195].

ogy could be set up without any reference to physiology whereas physiology had to rest upon some kind of "implicit or explicit psychology." The reason was that the brain as an anatomic entity was "visible and tangible" but its physiological functions evaded any direct observation. He remarked that the relationship between organ and function was clear for many organs, but in the case of the brain only "*imagination led by analogy*" could find a correspondence between feelings and thoughts, on the one hand, and "the motions... of the anatomic entity, namely cerebral matter," on the other [Egger 1877, pp. 193-6].

According to Egger, two main issues were at stake: in the human brain, functions were "heterogeneous to the organ," and moreover, it was difficult to associate a specific mental action with a specific region of the brain. He found that no discovery could have allowed scientists to determine the link "between a thought and a cerebral element" in the same way as such links could be established between a muscle and a contraction, or between a gland and a secretion. In brief, he claimed that an unbridgeable gulf separated cerebral anatomy and psychology. He insisted on the fact that the design of a physiological map of psychological processes was a psychological act in itself: in this sense, the physiology of the brain rested upon psychology. A scientist needed psychology "for setting down a problem of this kind, even for envisaging the idea." Therefore the physiology of the brain rested upon two pillars, anatomy and psychology, and the two components had to be carefully disentangled. In Luys' approach, he found an inaccurate superposition between anatomic and psychological elements: one of the first statements in the book, the above-mentioned "the brain feels, recollects, and reacts" was an instance of that misunderstanding. Egger also listed and sharply criticised other expressions that made reference to nervous elements or organs that were "*shaken in their sensitivity*." He could not accept the puzzling superposition between processes of different natures [Egger 1877, pp. 197-8 and 200-1].[10]

In the last part of the paper, Egger analysed the empirical and theoretical side of the heterogeneity and incommensurability between mental and cerebral phenomena. How could the anatomic-physiological correspondents of memory, imagination, deductive reasoning, fear, and hate be found? The optimistic view claimed that the brain had lost any mystery, and its functions could be explored with the same accuracy as the functions of the heart. According to Egger, the reality was different, and the correspondence between

10 It is worth noticing that the psychological character of every hypothesis concerning the human brain could not become a specific thesis against the reduction of psychology to physiology because a psychological component was at stake in every kind of knowledge. This made Egger's line of reasoning not so convincing.

functions and specific physical-chemical-biological processes in specific regions of the brain was far from being understood. He put forward a colourful analogy in order to depict the problem in a more convincing way: neither a common unit of measurement nor a shared criterion of comparison could be followed in the search for "*the sound corresponding to a given colour.*" Different levels of investigations required different methods and specific concepts and words. From the linguistic-conceptual point of view, he suggested that anatomic phenomena should be distinguished from physiological phenomena, and the latter from psychological ones. In the end, he found that no scientific physiology or psychology could be practiced without a clear separation of words, concepts, and methods [Egger 1877, pp. 209-11].

In its turn, Egger's critical appraisal was criticised by the Belgian physiologist Hubert Boëns in the journal *La philosophie positive*, in 1878.[11] In his paper, "La physiologie et la psychologie ou le corps et l'âme," the distinction between science and metaphysics was based on the opposition between "the real, the tangible, and the accidental," on the one hand, and the abstract or "the absolute, the infinite, and the incommensurable," on the other. Science was the domain of observations, experiments, trial and error, analyses, and deductions, whereas metaphysics was the domain of "pure intuition." Could there be any possible connection "between the *contingent* and the *absolute*, the *finite* and the *infinite*"? Boëns' answer was negative: the gap could only be filled by "ridiculous fairy-tales and unbelievable mysteries." He considered himself "*a sheer positivist*," and therefore did not trust in any *absolute*. As a consequence, metaphysics, or better his conception of metaphysics, was nothing else but a pointless and deceptive science. Even the cautious *rationalists* who accepted "a science of the absolute" in the context of mathematics were considered too submissive by Boëns. The absolute could not exist as an actually existing being, and could not exist even as a "universal principle." Infinite space could only be filled by material and limited entities, endowed with the features of every being, namely "extension and gravity" [Boëns 1878, pp. 345-7].[12]

11 At the time the journal was headed by the philosopher Emile Littré, and it was "the organ of the non-religious disciples of Auguste Comte" [Mucchielli 2006, p. 210]. In other words, Littré was a positivist who had not followed Comte's religious drift in the last stage of his life. In 1867 he had founded the *Revue de philosophie positive* that was published until 1883 [Benrubi 1933, pp. 23 and 25]. Boëns had been a correspondent member of the *Académie Royale de médecine* since 1862, and he fiercely opposed Pasteur's method of vaccination [*Biographie Nationale*, Académie Royale de Belgique 1956, tome XXIX, Supplément 1, p. 310].

12 It is worth stressing that Boëns' sharp separation between science and metaphysics was not structurally different from Egger's unbridgeable gap between brain and mind.

With regard to Egger's criticism of Luys' book, Boëns did not agree with Egger on the separation from the domain of physical-physiological processes, and the domain of thoughts and behaviours. He found that it was not so difficult to appreciate "the time required by a thought... to be sent to a muscle": therefore thoughts travelled through specific places over time. Boëns' radical materialism could not accept that something was assumed to be real but non-material: everything that was endowed with *duration and length* was real, finite, and material. He ventured to claim that the "*weight* and *volume*" of a human brain increased when it received an impression from outside, and its volume and weight decreased when it expressed an *idea*. He conceded that the exact correspondence between physiological processes and psychological effects had not yet been demonstrated, but also the contrary had not been demonstrated. He also conceded that psychology could be set up as an autonomous body of knowledge, but in this case it dealt only with some specific cerebral functions. He did not find it strange that thoughts could be looked upon as *things*, and that the activity of the human brain consisted in "feeling, thinking, and wishing," because the exact correspondence between organ and function and the dependence of psychological functions on brain anatomy were assumed from the outset. In reality, the opposition between Egger and Boëns was as passionate as substantially ideological [Boëns 1878, pp. 349-351].

Against "the defenders of the old spiritualist attitudes" Boëns claimed that psychology had to be looked upon as a specific section of cerebral physiology. He mentioned Bernard's references to a *vital force* or vital principle, and Rudolf Virchow's commitment to "carefully fixing the boundaries" between the domains of physiology and psychology as instances of mistaken assumptions. After having professed a radical materialism and a radical monism, he outlined a mechanical representation of the emergence of sensitivity in newborn babies. Tactile and visual impressions shook "a wide region or even the whole of cerebral matter" in the same way as "a drop of water shakes and excites a liquid mass" or "an electric spark excites a metallic wire" or "a sound wave hits the membrane of a telephone." The excitement travelled through encephalic cells, and put in vibration those cells that had already developed a small degree of sensitivity. According to Boëns, those vibrations were nothing else but "the first intellectual actions of the brain." In the end he found that, starting from those simple oscillations or mechanical motions, it was easy to understand "the operation of the human organism in the different stages of its wonderful evolution" without any recourse to non-extended entities. He did not go beyond this extremely simplified outline: probably he was not interested in putting forward a detailed explanation but only a very general hypothesis. In some way, he hinted at a mere possibility, and contented himself with defending this possibility [Boëns 1878, pp. 352-4 and 359-60].

A different line of reasoning had been followed by the Belgian philologist, philosopher, scientist and psychologist Joseph Delboeuf.[13] In 1877 he had published a paper on the propagation of variations in biological species, where we find a mild reductionism, or rather the confidence in simplified mathematical models for life sciences. In reality, in this case, we find a clearer awareness of the difference between the simplifications and approximations of a mathematical or mechanical model, and the complexity of natural phenomena. He tackled some objections to Darwin's theory: he aimed at supporting mathematically the propagation and persistence of unpredictable variations in a species. More specifically, his mathematical model showed that "the number of varied individuals could overtake the number of the unchanged ones." In his simplified mathematical model he represented a variation as the possibility of adding or subtracting a given quantity to a specific biological feature, in a symmetric way. In the model, an individual could generate n unchanged individuals, 1 individual endowed with the specific strengthened feature, and 1 with the weakened one. The number $n+2$ corresponded to the individual "*generative power.*" After some generations, the process gave birth to one set of homogeneous and two sets of heterogeneous individuals [Delboeuf 1877, pp. 672-3].[14]

After the first step there were only three kinds of individuals, which he labelled A (the unchanged individuals), $A+1$, and $A-1$. After the second step there was the opportunity of a more various descent: $A, A+1, A-1, A+2$, and $A-2$. After m steps, a symmetric set of labels was at stake: $A+1, A+2, \ldots, A+m$, and $A-1, A-2, \ldots, A-m$. Delboeuf's computation rested on the symmetry between the increasing and decreasing variations, and on the equal opportunity of life. This was a striking oversimplification that led to a striking result. Skipping Delboeuf's series of long computations, the model led to a modified population definitely greater than the unmodified one after a number of steps of the order of n. He showed that, for $n = 10$ and therefore generative power 12, only 8 steps were required in order to realise

13 In 1877 Delboeuf was elected a member of the Belgian Royal Academy of Sciences. The official biography of the *Académie Royale des Sciences, des Lettres et des Beaux-Arts de Belgique* qualifies Delboeuf as "philosopher, psychologist, philologist, naturalist and mathematician" [*Biographie Nationale*, Académie Royale de Belgique, 1969, p. 164].

14 Unfortunately, the simplified mathematical model was not in tune with Darwin's theoretical core: Delboeuf's hypothesis that death affected all populations in the same way was in contrast with Darwin's theory.

that numerical overtaking. In reality, from the biological point of view the success of a specific variation was more important than the amount of a generically varied population, and Delboeuf did not miss the point. He briefly claimed that the amount of a specifically varied population of the kind $A \pm k$ could not reach the original population "even though its relative importance does really increase." However, the results of his mathematical model also excluded the opposite outcome, namely the possibility of "recovering the primitive population." As Delboeuf pointed out, the process that led from $A \pm 1$ to A was mathematically weaker than the opposite process [Delboeuf 1877, pp. 673-7].

The paper did not contain original mathematical developments or applications, and the only generalisation consisted in writing a very general expression for the term that represented "the number of individuals of the generation $A \pm m$ after a number p of generations." It seems to me that the mathematical model is interesting in itself, and also in the context of late-1870s debates on reductionism. Nevertheless, the biological side of the model suffered from some misunderstandings because of the semantic extension of the word *evolution*. Delboeuf overlapped biological evolution and an ideal of perfection, and explicitly stated that "evolution and progress are almost synonymous." Moreover, he was interested in the cause that gave birth to a marked improvement in some species, and thought that "the cause could not be merely found in adaptation." Adaptation could only be an indefinite cause of variation: in no way could it be "a cause of progress." His mathematical model involved an indefinite variation triggered by a persistent cause of transformation, but it could not account for "a gradual improvement." He saw an intrinsic limitation of his model because he saw progress as more important than variation. In this context he stressed the role of *intelligence* as the real engine of evolution, or more specifically "the first cause of evolution." The contrast with Darwin's theoretical approach is really astonishing: sensitivity, intelligence, and freedom were not the outcome of variation and evolution but rather causes of evolution [Delboeuf 1877, pp. 674, 676, and 678-9].[15]

A network of simplified models and hypotheses, and the confidence in those models and hypotheses represented two distinctive hallmarks of that historical stage. Sometimes, radical simplifications and radical claims overshadowed more cautious and critical views.

15 He imagined that the universe, in its initial state, did contain those three features "at least in its embryonic form," in the same way as it contained matter and motion [*Ibidem*, p. 679].

3 The Radical Reductionism of an Anthropologist

We find a remarkable trust in radical reductionism in a scholar who contributed to the emergence of human sciences, a field of research where old intellectual practices like history and philosophy merged with new disciplines like psychology and sociology.[16] In 1878 Jules Soury, archivist and palaeographer with interests and serious studies in neurology and psychiatry, published a text which was something more than a booklet and something less than a book, *Jésus et les Évangiles*. From the outset, in the first lines of the *Foreword*, he claimed that he would put forward a new, radical re-interpretation of Jesus Christ's preaching: "after the god and the man" he was to take into account "the sick person [le malade]" or the patient. He claimed that if psychiatry had already emerged at the dawn of the new era, Jesus Christ would have been a psychiatric patient: he had shown "a perversion of his personal feelings," in particular towards his mother and brothers. The seat of these psychic perturbations was obviously the brain, and Jesus's brain had been afflicted by "an inveterate engorgement." The physiological imbalance had led to a series of corresponding psychical effects: the strengthening of the imaginative faculty up to hallucination, and then an extreme sensation of strength and power. At that stage, the patient Jesus had began to disconnect himself from reality: subsequently thoughts had become "absurd and frantic." The author even ventured to imagine that Jesus' irritability could have led him to explosions of violence [Soury 1878, pp. 7 and 9-11].[17]

Soury did not question his thesis in any way: once it was put forward, it was pursued in a consistent way, without any detour or critical remark. The real Jesus was a clinical case, and had to be looked upon as such. Moreover the psychiatric illness had its basis in physical transformations which the brain had experienced: the elusive behaviour of the mind was nothing else but the automatic effect of material processes taking place inside the brain. According to

---

16 The last decades of the century also saw the progressive institutionalisation of "human medical sciences" that included physical anthropology, psychiatry, and criminology [Mucchielli 2006, p. 226].

17 Soury had received a degree in humanities in 1862, and in 1865 he had began to attend the lesson of Jules-Bernard Luys and Auguste-Félix Voisin at la Salpêtrière. In the meantime he attended the private lessons of the philosopher, historian, and scholar of Hebrew and Christian religion Renan, who had published his *Vie de Jésus* in 1863. In 1867 Soury became an archivist-palaeographer, and began to publish in the journals *Le Temps*, *Revue des Deux Mondes*, and *Revue Scientifique*. For further biographic information, see Huard 1970, pp. 155-6.

this meta-theoretical assumption, the cause of the partial or total disappearance of consciousness could be found in "the consumption of elements of the cortical brain." A proliferation of fat tissues had progressively replaced ordinary cells in Jesus' brain cortex. Those parts of the cortex that had managed to preserve their normal functions had probably suffered "from blood congestion": the psychical effect was "a more or less severe delirium." The illness could have improved temporarily, and sometimes an apparent recovery could have lasted for a while, but the final consequence could definitely be forecast. The progressive weakness of muscles and intellect, and the side-effects on liver and kidney, would inevitably have led Jesus to death. This would have been the inglorious end of Jesus, if "the Jewish people had been badly advised, and Barabbas had been crucified in his stead" [Soury 1878, pp. 13-4].

Some historical remarks can be found here and there in Soury's book. He claimed that the mental illness which he qualified as meningitis-encephalitis was typical of the nineteenth century but its presence could be documented also in other centuries. The documentary evidence he put forward was based on a logical misunderstanding: cause and effect were confidently reversed. The cause was the anatomic and physiological alteration of the brain, and the effect was the emergence of peculiar ideas, which could be indifferently qualified as delirium or as political and religious commitment. Soury reversed the logical chain: the appearance of political and religious passions was assumed to be the hallmark of severe brain disease. Moreover those feelings and passions were assumed to be as strong in ancient Judea as in contemporary France: the physiology of the human brain had not changed during the last two thousand years. The universality of the science of the brain could rely on that spatial and temporal persistence in both the biological and psychological fields. Two main issues were at stake in Soury's psychiatric reconstruction: first, political and religious commitments could be identified with mental illness, and second, mental illness was the necessary result of a severe brain disorder. The historical and logical short circuit was closed by the analogy between the turmoil that had taken place in Jerusalem some years after Jesus' death, and the *Commune* which had violently shaken Paris a few years before [Soury 1878, p. 16].

It seems that Soury aimed at some kind of *global* history, however naïve it might be, where public events mixed with the physiology of individuals and the habits of races and civilisations. It is worth remarking that in 1758, in the same year that the mathematician Jean-Étienne Montucla published the first volume of his *Histoires des Mathématiques*, the French scholar Antoine-Yves Goguet published *De l'Origine des Lois, des Arts et des Sciences et de leur progrès chez les anciens peuples*. We find here the first traces of cultural anthropology, and the first traces of a *global* history of humanity that attempted to go "beyond the horizon of the history of reason devised by Fontanelle and d'Alembert."[18] Obviously, a subject matter such as the history of humanity was too wide and demanding to be mastered by a single professional profile: in the nineteenth century that ambitious commitment gave birth to a series of more specialized disciplines or human sciences, such as anthropology and sociology. The new sciences had to find their place in the wide interval "between speculative philosophy and positive sciences" [Gusdorf 1966, pp. 80 and 86].

Even genetics entered the scene in Soury's book: the apostle James, who was looked upon as Jesus' brother in the strict sense, had shared the same religious commitment and therefore had been afflicted by the same brain disease. Soury seriously claimed that this fact could hardly be questioned. Among Jesus' relatives, ancestors and descendants, a parade of "fanatics, epileptics, suicides, and drunkards" could certainly be found. Fortunately Jesus had kept himself "chaste as an ascetic," and had not engendered children who might have been idiots like him. Soury's language and concepts were quite aggressive and disparaging, but they were based on an alleged objectivity, which set apart personal feelings. The ideological disdain stemmed from a radical reductionism, which was assumed as the hallmark of every serious scientific approach. Soury stressed "the relevance of ecstasy and hallucination" in the life of men who had contributed to change profoundly our ideas or the course of historical events. With icy detachment, he claimed that Islamism, after Buddhism and Christianity, had emerged from the "visionary attitudes of an epileptic." On the other hand, foolish behaviours could even lead to positive side-effects: the hallucinations of Jeanne d'Arc "had freed France" [Soury 1878, pp. 18-21].

Soury's reductionist design was repeatedly stressed in many pages of the booklet: the origin of ideas and feelings had to be found in "the physical and material structure of man." The origin of "the most advanced expressions of

18 The historian Gusdorf saw in Goguet's 1758 book the first attempt to put together the natural history of the human species and the cultural history of mankind [Gusdorf 1966, p. 80]. This seems to me a little exaggerated even though Goguet himself, in his *Introduction*, had stressed the necessity of multiple links among different aspects of human experience. See Goguet 1758, pp. v-vi: "Je me suis proposé, en conséquence, de tracer l'origine des lois, des Arts & des Sciences d'une manière plus exacte & plus conforme à l'Histoire qu'on ne l'ait fait jusqu'à présent. J'ai cherché aussi à faire sentir l'enchainement de tous ces différents objets, & leur influence mutuelle." In 1781, at the end of the third volume of his *The History of the Decline and Fall of the Roman Empire*, the authoritative and customarily critical English historian Edward Gibbon had praised Goguet's "learned and rational work" [Gibbon 1854, p. 641, footnote 11; Wolloch 2007, p. 429].

heart and mind" was neurological. Although his assumptions were quite radical, he claimed that he had not put forward any hypothesis. He also claimed that he had confined himself to reading the Gospel without distorting the holy texts: he had only taken note of Jesus' portrait as it emerged from the Gospel itself. For a twenty-first-century reader, Soury's meta-theoretical naivety is really astonishing, but in some way his keen reductionism was consistent. He remarked that if previous researches on Jesus' life and the emergence of Christianity had overlooked that supposed link between mental disease and religion, it had to be ascribed to bias. As Soury himself stated, "we cannot find what we are not looking for." In that specific context, only the hypothesis that a religious practice had a neurological basis allowed him to appreciate the supposed neurological basis of Jesus' practice and preaching. Obviously, only the confidence in a specific hypothesis allows us to appreciate the meaningfulness of interpretations that are consistent with our hypothesis. Soury's insistence on the fundamental role played by hypotheses was not in tune with his positivistic faith, and it was not in tune with the statement that he had not put forward any hypothesis. At the end of his *Foreword*, his main assumption was briefly synthesised: "religious excitement" was the visible manifestation of "an injury of the nervous network" [Soury 1878, pp. 25-7].

In the following section, he insisted on the necessity of an objective appraisal, and stressed that he was advancing along the same pathway that had been undertaken by the historian and philologist Ernest Renan and other historians and philosophers. No disdain had urged him towards this kind of research: he was aware that Jesus was one of the characters who had deeply influenced the history of civilisation. He acknowledged that the founding father of Christianity had been one of the leaders "of our species," and the memory of the events which had studded his life had become the leading mythology of "the most remarkable part of mankind." In Soury's historical-medical reconstruction a new mythology emerged, and that new mythology rested on two meta-theoretical hypotheses. First, science was able to offer the true explanation for phenomena concerning body and mind and second, physical processes could explain mental ones. The two pillars might be labelled *scientism* and *reductionism*. Even creativity, as well as other superior activities of human beings, stemmed from some kind of burnout. Moreover, the brain could not separate the good from the bad: both creative and destructive momenta, as well as health and illness, stemmed from the same physiological root [Soury 1878, pp. 31-5].

One of the consequences of his alleged scientific *objectivity* was that some qualitative differences, which were extremely important in ordinary life, faded away when submitted to scientific procedures and classifications. The difference between health and illness was one of the most useful and convincing, but science had taught us that they were different conditions of life that were ruled by the same laws. Even the difference between faith and scepticism was a difference in attitude, or rather an anthropological difference, which stemmed from the same physiological processes taking place in the same brain. Nevertheless, this did not prevent Soury from putting forward acute ideological statements: for instance, he was convinced of "the inferiority of Catholic people of the new and old world when compared to Protestant nations." This peculiar mix of a supposed scientific objectivity and ideological commitment was one of the hallmarks of Soury's methodological approach. Paradoxically enough, the melting pot of scientific rhetoric and ideology led him to a sort of nostalgia for ancient times: the study of ancient civilisations had convinced him of the moral superiority of ancient peoples over modern ones. In the past he found moral virtues that his contemporaries had lost, and in particular "dignity, generosity, and the naïve faith in the absolute" from which saints and heroes had emerged. In the end, the search for a purely rational order had led him to long for a mythical past crowded with exceptional personalities [Soury 1878, pp. 35-7 and 42].

General history, the history of science, and anthropological theses found place in a wide-ranging framework which was at the same time a historiographical sketch and a scientist manifesto. Men had always found it hard to accept that their representations of divinity were the result of many interactions and contaminations, even though a historical genealogy could easily be outlined. More specifically, Soury found that "the Gods of Mecca, Rome and Jerusalem" had stemmed from each other, and the belief in the Virgin Mary had its roots in the cult of Isis. At the same time, those holy characters were nothing more than a re-interpretation of more ancient religions in which the Sun and other planets and stars were worshipped. However, when Soury looked at human history from the point of view of long-term physical processes taking place in the Universe, he realized that cosmic history was insensitive to "the hopes of mortal beings." Over time, nebulae had condensed, suns had flared, and then life had appeared: processes of generation and dissolution had continuously followed one another as a manifestation of "the creative chaos of the eternal Universe." In the end, the most persistent reality was nothing else but "the unity and the indifference of the whole": he had started from a supposedly *objective*, scientific approach, and ended with broad cosmological remarks [Soury 1878, pp. 46-7].

In Soury's book, the psychiatric analysis of Jesus' behaviour gave way to the interpretation of historical events which had taken place in Palestine after Jesus' death. When he took into account "the wars of Judea," he praised the Romans because of their search for harmony and peace throughout the empire, and blamed the Hebrew people for having resisted Roman attempts to restore the power of the empire in their lands. What strikes the reader is the series of adjectives which he uses for describing the resisting Hebrews: the adjectives "deaf and unmoved" preceded the description of their faces as "wild, convulsive masks" that were "monstrously misshaped by hate." That horde of "deranged people" had transformed their holy temple into "a shelter for outlaws, or rather a lair for hyenas and jackals." The Hebrews' worst crime consisted in their firm will to preserve their traditions: the psychiatric side of that political and anthropological commitment was qualified as *delirium* and *pestilence* [Soury 1878, pp. 176 and 178].

We find here an explosive mixture of a simplified scientific reductionism, a conservative political commitment, and open anti-Semitic attitudes. The last two elements were further developed in the following pages: not only was the resistance to Roman dominion qualified as *madness* but also as "a crime against civilisation." The skip from science to racism became even more audacious in Soury's historiographical sketch: the Semitic Carthage and Jerusalem had shared the same destiny because of their common anthropological impertinence. Phoenician and Hebrew had to be crushed as a dangerous *hydra* by the most developed Romans. The sack of the holy temple in Jerusalem was the triumph of the best civilisation of "the ancient world," and at the same time the triumph of the "Aryan race." When he analysed the emergence of Christianity, he remarked that the Christian religion had undermined the pillars on which that society had rested for centuries. No conciliation could be pursued between the new religion and the already existing "state religion," and moreover, the state could not accept the superimposition of "a different social structure." Christians were living inside the empire as "termites," which eroded the foundations of society from inside. Their practices had stemmed from "the Semitic theocracy": Christianity had recast that theocracy, which flourished throughout the Middle Ages until the end of the nineteenth century. Soury regretted that French society was still doomed to fighting against "the authority of Jewish tradition, and the claims of the Vicar of Christ" [Soury 1878, pp. 178-81 and 187-90].[19]

19 Soury's commitment to politics and social order was not an exception in those decades. Social order in general, and criminal behaviours in particular, underwent a process of scientific analysis on a physiological basis. Besides this "naturalisation of crime" we find a community of physiologists and physicians who claimed that medical science could rid "society of its deviants." Radical projects "for sanitary and social control" were put

According to Soury, the transformations experienced by the Hebrew religion in the transition to Christianity and the differences that still persisted between the two religions were not as meaningful as the unbridgeable gulf between foolish traditional religions and the rational religion of State and Science.

It is worth remarking that fifteen years earlier the much-praised historian Renan had put forward an anthropological approach to Jesus' life, which was influenced by his positivistic attitude, but at the same time was more refined and respectful. The book he wrote in 1863, *Vie de Jésus*, was intended as the first volume of a weighty history of Christianity, and raised a fierce debate. As a consequence of strong opposition to his historical reconstruction he was removed from the chair of ancient Hebrew at the *Collège de France*. Renan's analysis of Jesus' biography is as passionate as accurate from the philological point of view, and was based on a sophisticated historiographical framework. Unlike Soury, we do not find any aggressive and naïve scientism, but a sympathetic even though critical inquiry. He insisted on the necessity of a critical detachment: a person who had lived many centuries before could not be judged in accordance with modern standards. If behaviours that appeared "devoid of meaning or unreliable to an observer of the nineteenth century" were looked upon as the acts of "a crazy person or a cheater," this would lead to a deep historical misunderstanding. He found that Jesus' "helpless effort to set up a perfect society," and the faith in the establishment of "God's kingdom" were "the highest and most poetic expression of human progress." Our civilisation was deeply rooted in that tradition [Renan 1863, pp. 267 and 286].[20]

Renan was aware of the linguistic and conceptual impossibility of translating words and meanings from one culture to another accurately and faithfully. In some cases no definite correspondence between attitudes or ideas could be established, and in the worst case, no specific word or meaning could be found in a given cultural tradition. He stressed that the ancient use of metaphors was quite different from the modern one, wherein actual reference or literal

forward, the prevention of insane minds from reproduction included. The meaning and aim of punishments underwent a cultural change: they were rooted in "social utility" rather than "in vengeance or in expiation" [Mucchielli 2006, pp. 208-10 and 214]. It is worth remarking that the first Congress of Criminal Anthropology was held in Rome in 1885, and the second in Paris in 1889 [Hacking 1990, pp. 155 and 175].

20 Renan's book had great success but was sharply criticised by Christian scholars. Even Pope Pius IX intervened, and in the end, Renan's academic course was abolished by the Minister of Public Instruction. In 1878, after Claude Bernard's death, Renan was elected to the *Académie française*.

meaning had to be separated from metaphorical expressions. The essential condition for what he labelled "true critics," namely a reliable historical reconstruction, was the explicit acknowledgement of chronological and cultural differences: the historian had to "divest himself of automatic reactions that stemmed from an exclusively rationalist education." He reminded readers that what appeared to scholars as a legend had been written "by another race, under another sky, in a different context of social expectations." According to Renan, concepts like healthy and sick were historically determined: he regretted that "recent, widespread, and narrow-minded ideas on madness" had dramatically misled historians in their judgements. What in ancient times could be qualified as "high inspiration or prophecy," in the nineteenth century had become a mere hallucination. Every great creation entailed some kind of "acute disequilibrium" that modern physicians were proud of qualifying as "nervous accident" or at least mental instability [Renan 1863, pp. 305-6, 359, 451, and 453].[21]

Renan's refined historiography showed that a rational and critical approach did not necessarily imply the endorsement of dogmatic positivism.

4 Further Debates

Renan's cautious and attentive historiographical remarks would probably have appeared too sophisticated to the physiologist Boëns. In 1979 the latter published the essay *La science et la philosophie* with the sub-title *Nouvelle classification des sciences*. From the outset he professed a realistic and empiricist faith alongside a boundless faith in scientific progress.

> The nineteenth century will be labelled as *the century of realism*. In the end, all fields of knowledge will become subject to positive sciences. In the body of universal knowledge, we will not accept anything else but ideas stemming from observation, with direct reference to the senses. Demonstrable laws will replace hypotheses ever more, and structure and

21 In the penultimate page of the book, Renan stated that Jesus had been the person who had helped "his species undertake the greatest step toward the divine" [Renan 1863, p. 457]. It is difficult to classify Renan according to well-defined intellectual labels. Benrubi saw in Renan a complex network of influences, especially Kant, Comte, Hegel, and Darwin [Benrubi 1926, pp. 29-31]. Gusdorf focused on Renan's sincere and enthusiastic scientism or "scientific triumphalism," wherein science became something like "an eschatological mythology" [Gusdorf 1966, p. 35].

> features of different natural entities will be better understood. Nowadays some truths are firmly established, whereas the best minds of the past could hardly imagine them. Every day, the domain of real, unquestionable facts grows, new discoveries emerge, and science deploys them in order to be generalised and popularised [Boëns 1879, p. 5].

Words and concepts that he qualified as "metaphysical, spiritual, and idealistic," had to be banned in scientific practice. Among them he listed "the soul of animals, the human soul, and the soul of the world," and all the concepts and entities that could be found in the holy books of Christians, Brahmans, Hebrews, and "polytheistic philosophies from Egypt and Greece." Psychology would have deserved not to be banned only if it had been reduced to "the science of the brain's features and functions." Metaphysics was an idle body of knowledge, devoid of any "civilising momentum." And even metaphysics would have deserved preservation only if it had confined itself to "studying the fundamental laws of the universe."[22] Ethics was worthless too: private ethics had to be reduced to personal hygiene, and public ethics could be founded on the social contract of every national tradition. This was his positivistic catechism, or better the set of meta-theoretical assumptions of what he labelled "the positivistic school." In this context, philosophy was "nothing else but the *universal Science*," namely a review of all specific sciences or simply the synthesis of the whole of human experimental knowledge. According to Boëns, the ancient biases had to be replaced by truth, which imposed itself because of its unquestionable evidence. Both "willing and unwilling," people had to become acquainted with truth, and society had to adapt itself to the new trend [Boëns 1879, pp. 6-9].

Boëns insisted on the materialist monism, and on the simplified mechanistic view he had already put forward in the last part of his 1878 paper. He found that the concept of philosophy as "science of sciences" was too generic, and a more radical definition was required: he therefore asserted that "*philosophy is science in itself*." The statement sounds a bit strange because the choice of two words for the same thing appears not so wise. Moreover, it was ironic that the superposition between science and philosophy echoed an ancient attitude that the positivistic trend attempted to overcome. According to Boëns,

22 Although the passage is not clear, Boëns probably hinted at Comte's conception of philosophy as a rational summary of all sciences: this interpretation seems consistent with subsequent passages. It seems that metaphysics and philosophy were synonymous for Boëns.

from the classic age onwards, "*theology*, and then *metaphysics*" had become the leading disciplines of "*dogmatic philosophers*," and sciences had been disregarded and disdained. In his historiographical sketch, the eighteenth and nineteenth centuries had seen an astonishing turn: sciences had re-conquered "the sovereignty of the intellectual world." The second half of the nineteenth century appeared as the proscenium where *positive* philosophy replaced the two ancestors, namely theology and metaphysics or traditional philosophy. In brief, in this cyclical historiographical framework, philosophy had been identified with science at the dawn of our civilisation, then philosophy had undergone a sort of degeneration, and had given birth to theology and metaphysics; in the last decades, science had taken the lead once more, and had become the core of the *true* philosophy. The last stage was looked upon by Boëns as irreversible: science would absorb philosophy, and mankind would enjoy "the universal knowledge" forever [Boëns 1879, pp. 14-15].

The fundamental hypothesis of Boëns' positive science was "*the eternity of things, and infinity in space, time, and matter*." There was no empty space, and matter was infinitely divisible. Positive science could be identified with positive philosophy, and a well-defined borderline separated positive from speculative philosophy. Positive science could also include the old "great concepts of infinite and absolute" into a *positive* framework. Boëns outlined a naïve algebra, which would have allowed scientists to handle both finite and infinite quantities by means of "exact mathematical expressions." He made use of three symbols: $(\infty >)$, which denoted "infinitely great," $(\infty <)$, namely "the infinitely small," and x, which denoted "whatever finite object of the universe." He found reasonable to write down the following proportion

$$\infty > : x = x : \infty <,$$

namely $x^2 = (\infty >) \cdot (\infty <)$, and therefore $x = \sqrt[2]{(\infty >) \cdot (\infty <)}$. Since the number 1 was the simplest representative of "any object," the last equation became

$$1 = \sqrt[2]{(\infty >) \cdot (\infty <)}.$$

Against the possible objections of "spiritualist philosophers" he claimed that the equation was satisfied by every couple of "definite parallel numbers" or numbers "of the same numeral power, ascendant and descendent." The concept was not so clear, but he intended couples of the kind 10^n and $1/10^n$: for instance, $1 = \sqrt[2]{(10^n) \cdot (1/10^n)}$. Probably Boëns expected that this rough algebra would convince his critics of the mighty universality of positive philosophy [Boëns 1879, pp. 18-22].

His positivistic mythology was not so different from what he had labelled as bias and superstition. We find concepts and words that were in tune with religious apologetics. Positivist philosophy would not have missed "its providential mission": it would have transformed the traditional cultural landscape into "a blank slate [tabula rasa]." Civilisation and ethics would have been freed from "the mistakes gathered during the ages of illusion and ignorance." Only science would have survived, and positive philosophy would have confined itself to putting forward a "simple, methodical, and natural" classification of sciences. At the same time, the classification was intrinsically linked to a historiographical framework on the track of Auguste Comte. According to Boëns, the renowned philosopher had interpreted history as the stage where the "progressive development of civilisation" had entered the scene. He did not go beyond Comte's philosophy of history, and his classification of sciences was definitely less rational than Comte's. He put forward a three-fold partition, namely "ontology," "modalogy," and "sociology": sciences like physics, chemistry, and physiology belonged to the second section, and the third section hosted metallurgy and agriculture besides anthropology and history. It does not seem an effective and fruitful classification: it seems that, for Boëns, a strong ideological commitment in favour of a radical positivism was more important than the effectiveness and fruitfulness of specific theses and classifications. He concluded the essay with a verbal equation, which was the extreme synthesis of the essay: "*Science universelle = Philosophie*" [Boëns 1879, pp. 24-6, 28, 32, and 38-45].

Two years later, in a book he devoted to the history of ancient natural philosophy, *Théories naturalistes du monde et de la vie dans l'antiquité*, Soury put forward a milder historiographical framework, where historicism merged with a sort of scientific determinism. He definitely slackened his sharp scientism, and his mythology of objective knowledge. He claimed that "all philosophical views" had been "necessary and legitimate in their times." They had mirrored the different stages of the human mind, and had evolved together with the human mind. Even the most successful philosophy of the time being would meet the same fate: probably, in the following decades, it would be replaced by other philosophies. He did not exclude that, in the future, scholars would smile mercifully at the naivety of current theses and remarks. In the three years which had elapsed since 1878, Soury had changed his views on philosophy and history: the Comtian reference frame, which he had endorsed and interpreted with aggressive enthusiasm, appeared already outmoded [Soury 1881, p. 9].[23]

23 The book was nothing other than his doctoral dissertation in humanities, and was dedicated to Renan. Between 1879 and 1881 Soury had been deeply involved in the settlement

Looking for elements of continuity between ancient and modern science, he came across atomism. Soury looked upon atomism as a long-term conceptual stream that linked ancient natural philosophy to modern science, even though atomism was the carrier of an intrinsic dichotomy. The materialists of the eighteenth century claimed that Democritus had based his natural philosophy on "the sole authority of physical perceptions and experience," but they had failed to grasp the true nature of that philosophy. When ancient philosophers had assumed that there were "only atoms and empty space," they relied on "two absolute and infinite entities": neither perceptions nor experience could help them discover elementary components of matter or an infinite, immaterial extension. Both atoms and space were abstract concepts rather than inductions from experience. In general, he acknowledged the existence of problematic links between the domain of abstract or mathematical representations and the domain of empirical experiences. In some way modern science had deepened the dichotomy, because "the reduction of quality to quantity, and physics to mathematics" could not be demonstrated on empirical grounds. Even simple concepts like extension and solidity became problematic when scientists and philosophers explored their empirical content [Soury 1881, pp. 9-11 and 14-18].

Soury went so far as to state that science was nothing else but an "ideal science": it could not elude metaphysics. The assumption of invisible and questionable entities was practically a necessity in scientific practice. At the same time, that necessity made science closer to the body of knowledge which went by the name of philosophy. Confining himself to early Greek natural philosophy, he acknowledged that Leucippus and Democritus' atoms, as well as the atoms of the recent "materialist philosophy" were not structurally different from the Eleatic abstract *Being*. They were "objects of faith" rather than consequences of experience. Both ancient natural philosophy and modern science had to make recourse to metaphysics: from this point of view, science had something in common with what he labelled *idealism*. The commonplace that man was "the metaphysical animal par excellence" probably represented an implicit acknowledgement and a justification of the existence of some kind of metaphysical commitment in any scientific practice. In the end, neither "bodies nor minds in themselves" were directly approachable, and the mediation of human senses was as necessary as the mediation of mental abstractions. In the end, he found that any scientific practice started from the

of a chair of History of Religions at the *Collège de France*. In the end, although he had been backed by influential politicians, he failed to be appointed to the chair. For further details, see Huard 1970, p. 158.

pretension of being "an experimental science," but then it inevitably became "an ideal science" [Soury 1881, pp. 19-20].

The following year, he published a book, *Philosophie naturelle*, which followed in the footsteps of a long-lasting tradition, namely natural history: it spanned different subject matters, from the emergence of the first living cells to the history of civilisation. In the *Preface*, he came back to the foundations of science, and in particular the intrinsic tension between empirical and theoretical practices. Any scientific dogmatism appeared to Soury not so different from philosophical or religious dogmatism, and this is quite surprising when we come back to the philosophical and anthropological dogmatism he had repeatedly displayed in 1878. He accepted the existence of scientific debate on methods and aims as a matter of fact. The intrinsic tension between the natural world and consciousness could not easily be overlooked. Nature and mind appeared as the two sides of the same coin: they were different from each other, and not reducible to each other, but complementary aspects of the same reality [Soury 1882, pp. III-VIII].

Soury's intellectual pathway from 1878 to 1882 is really astonishing: in four years he had covered the remarkable philosophical distance between the most dogmatic positivism and a mild, critical attitude towards philosophical and scientific practice. It seems to me that this fragment of a specific philosophical biography could be looked upon as an instance of a more general cultural turn that took place in the 1880s. In the early 1860s, Cournot had opened the way to a historical-critical reconstruction of scientific practice, but his historical and philosophical enterprise had passed almost unnoticed. In the meantime, the material and intellectual landscape had changed. It is true that there had been a dramatic increase in the production of goods and means of transport, and in the rate of education. Some ideological certainties had been violently shaken by the social and political transformations that had followed the defeat of France in the war against Prussia and the German states, the collapse of the Second Empire, the revolutionary insurrection of the Paris Commune and the subsequent bloody repression, in the short time interval 1870-71. In the 1880s, radical positivism continued to flourish, but a more balanced and critical attitude towards science slowly developed, and more cautious epistemologies and historiographies were put forward.

The cautious epistemology that Cournot had put forward in 1875 was not explicitly debated, but certainly left some traces. Chemistry and life sciences could rely on their own foundations but also on contents and methods of mechanics and physics. No science could be completely reduced to an alleged more fundamental one. Life sciences, in particular, had challenged the concept itself of scientific explanation: Darwin's concept of natural selection appeared to Cournot not sufficient to account for the emergence of new genera and families. The analogy between individual living structures and the behaviour of communities of living beings represented another sensitive issue and a new demanding challenge for Comte's classification of sciences. From the epistemological point of view, once more Cournot's *probabilism* offered the possibility of going beyond the conception of the human mind as a mirror of nature. The active role of human reason made it impossible to free scientific practice from metaphysics and to purge scientific achievements of their intrinsically provisional and historical nature.

In the late 1870s the debate on reductionism was accompanied by attempts to apply mathematical models to life sciences and social sciences, and by more radical attempts to put forward a sharp determinism in psychology and anthropology. A wide range of philosophical attitudes emerged: Soury's extreme determinism and reductionism in social sciences stood beside a reactionary political commitment, whereas the most authoritative Renan expressed a milder positivism further mitigated by serious historical interests and sensitivity. In the end, it seems that in the early 1880s even radical reductionists such as Soury began to soften their historiographical and epistemological theses.

CHAPTER 4

Mathematics and Determinism

1 Differential Equations, Living Beings, and Free Will

In the late 1870s, the debates on reductionism were accompanied by debates on determinism, and in both cases the problematic link among the tradition of mechanics, the recently systematised life sciences, mathematical models, and philosophical commitments was at stake. In the case of reductionism, the most refined philosophical approach had been put forward by the mathematician Cournot. Even in the case of determinism, the main protagonist of the corresponding debate was a mathematician; moreover, specific mathematical questions and their questionable interpretations were involved.

In 1878, Joseph Boussinesq, a mathematician of the Lille Faculty of Science, published a remarkable essay in Paris, under the long and demanding title *Conciliation du veritable déterminisme mécanique avec l'existence de la vie et de la liberté morale*. The essay, a book indeed, was introduced by a report the philosopher Paul Janet had read before the *Academy of Moral Sciences* on 26th January of the same year, and was subsequently published in the *Comptes Rendus* of the Academy.[1] The journal also hosted a shortened version of Boussinesq's essay, which corresponded to "the philosophical section." Janet stressed that the subject matter was "very specific and technical," and he found it useful to draw the attention of philosophers to "the main idea." The core of Boussinesq's book was both mathematical and philosophical, because he described "some instances of perfect mechanical indeterminism." Some differential equations led to "*branch points* [*points de bifurcation*]," where solutions gave rise to two different pathways. From the physical point of view, a material system could evolve towards different di-

1 After having been awarded a degree in mathematics in 1861, and having undertaken a teaching career in mathematics, Boussinesq defended his PhD dissertation in Paris in 1867. He was also awarded a degree in physics in 1872, and then became professor of *Differential and integral calculus* at the faculty of Science in Lille. He spent fourteen years at the University of Lille, where he published extensively on hydrodynamics and the theory of elasticity. Afterwards he held the chair of Mécanique physique et expérimentale at the Paris faculty of Science until 1896, and then the chair of Physique mathématique et calcul des probabilités: in both cases he succeeded Henri Poincaré [http://www-history.mcs.st-and.ac.uk/Biographies/Boussinesq.html, accessed January 28, 2016].

 DOI 10.1163/9789004315235_006

rections, and the actual direction was unpredictable [Janet 1878, pp. 3 and 12-13].[2]

Different versions of the essay circulated in 1878 and 1879. In 1878 both the complete version [Boussinesq 1878a, 256 pages] and the short version [Boussinesq 1878b, 65 pages] were published: in the latter no mathematical equations appeared. In 1879 the complete version was also published in the *Mémoires de la société des sciences, de l'agriculture et des arts de Lille* [Boussinesq 1879a, 257 pages, a one-page "Erratum" included]. At that time Boussinesq had already published remarkable researches in the field of mathematical physics, and in particular fluid dynamics. In 1868 he had mathematically analysed the water flow in river bends, and in 1877 he had published a treatise on the same subject [Boussinesq 1877a]. Unfortunately the book was overlooked by the scientific community because both mathematicians and engineers were not at ease with his theoretical and mathematical approach: theoretically or mathematically oriented scholars were not interested in his practical results, and engineers did not manage to appreciate his results because of the sophisticated, mathematical approach.[3]

In the complete version of his essay on determinism and free will, we find integration among three different bodies of knowledge: mathematics (more specifically the solutions of differential equations), physics (more specifically the relationship between equations of motion and actual behaviour of natural systems), and philosophy (more specifically determinism in the context of natural philosophy). Janet pointed out that Boussinesq had explored scientific possibilities far beyond mechanical determinism: specific features of differential equations might account for the existence of living beings. This was, in reality, the most important issue at stake: could the creative power of life emerge from some kind of intrinsic uncertainty or deficiency in mathematical

2 Paul Janet had been appointed to the chair of philosophy at the Sorbonne in 1864. He was interested in psychology, and specifically the correspondence between psychological and physical phenomena. He opposed empiricism and Comte's classification of sciences because he found no continuity between natural sciences and human sciences [Benroubi 1926, p. 41; Benrubi 1933, pp. 564-5].

3 Boussinesq's 1877 treatise passed almost unnoticed whereas the less demanding papers the engineer James Thomson published in 1877 on the *Proceedings of the Royal Society* were acknowledged as an important contribution to the field [Apmann 1964, pp. 427-8 and 433-4]. Some decades ago, the American mathematician John Guckenheimer noted that Boussinesq's researches, in particular his "equations which describe fluid flow in a convecting layer" had subsequently been developed by the meteorologist Edward Lorenz in the early 1960s [Guckenheimer 1984, p. 325]. For the role played by Boussinesq in the history of hydrodynamics, see Darrigol 2009, pp. vii-ix and 233-8, and Darrigol 2002, pp. 136 and 150.

procedures and mechanical laws? This perspective appeared far more interesting than the two traditional pathways taken by scientists: the reduction of life to mechanical processes, and "the vital principle of the ancient schools of thought," which opposed the mechanical approach. The title of the book called into play the compatibility of two apparently divergent issues: mathematical determinism, and the emergence of life and thought. In reality Janet hinted at an even more sensitive issue: the possibility of a mathematical account for "human free will" [Janet 1878, pp. 18 and 21].

In the end, after a short historical review, the philosopher Janet stressed what he considered the key concept of the book: science itself could "not exclude some kind of phenomenic indeterminism," or in other words, "some degree of contingency" in the natural world. According to Janet, Boussinesq's work could be looked upon as a scientific implementation of Boutroux' philosophical thesis on the contingency of natural laws. Boussinesq had managed to fulfil the reconciliation between "two fundamental laws of our mind," namely "the law of efficient causality," and "the law of finality or progress." The former required that "everything must be explained by an antecedent," and that "the effect cannot contain more than its cause," whereas the latter required that "we add indefinitely something new, which is not intrinsically contained in the antecedent" [Janet 1878, pp. 20 and 23].

Unlike Janet, Boussinesq was a mathematician, and he was interested in explaining the mathematical aspect of that compatibility or reconciliation between "true mechanical determinism," on the one hand, and "the existence of life and moral freedom," on the other. Nevertheless, in his Avant-propos he devoted more than ten pages to a historical review and meta-theoretical remarks. He claimed that "the specific, material features of life" could be accounted for by specific solutions of differential equations, which were "seats of convergence and bifurcation of the integrals" of those equations. He reminded readers that neither recently departed nor still living "physiologists or chemists" relied on "the existence of "particular vital actions." Nor did the majority of them claim that living beings were the seat of "accelerations and chemical reaction" intrinsically different from those taking place in other "material systems" [Boussinesq 1878a, pp. 25-6].[4] However some scientists had professed extensive reductionism: ordinary physical-chemical forces must rule "all kinds of motions inside living beings," and life could not be looked upon as "a cause in itself." Boussinesq thought that those claims were in contrast with common experience, in particular the observation of "volitional motions" of muscles.

4 He listed "Alexander von Humboldt and Berzelius, among the departed," and "Claude Bernard and Berthelot, among the living" [*Ibidem*, p. 26].

Although physical and chemical processes could account for the more complex processes taking place inside living cells, they could not account for the assembling of cells and organs of specific shape. The body of knowledge of life sciences appeared as a network of assumptions and experiments which did not fit perfectly with each other, and he found that the introduction of a specific "*guiding principle*" in biological processes might be the most suitable theoretical choice [Boussinesq 1878a, pp. 30-31].[5]

He made reference to some remarks Bernard had put forward in 1867. He had assumed two kinds of *forces* inside living beings. The first had been labelled "*operating forces* [*forces exécutives*]," and they were assumed to act in the same way as in unanimated bodies, whereas the second had been labelled "*guiding or evolutionary principles*," because they had to be morphologically active. The latter did not have to be identified with the previous vital principles, because "organic morphology" was based on the usual physical-chemical forces. According to Boussinesq's reconstruction, the conflation between physical laws and the creative power of morphogenesis led to a scientific determinism which was more sophisticated than purely mechanical determinism: even some features of human freedom could stem from it. He was looking for a more comprehensive determinism, not to be confused with fatalism; determinism and freedom appeared as subsequent stages rather than alternative features of natural processes. In reality, free action could take place only during "the guiding stage," while determinism was at stake during "the operative stage" [Boussinesq 1878a, pp. 28-9].[6]

Boussinesq reminded readers that mathematicians and engineers had put forward something like Bernard's guiding principle. In 1861, the mathematician Cournot had spoken of "a principle of harmonic unity, global direction, and homogeneity," whereas in 1877 the mathematician and engineer Adhémar Barré de Saint-Venant had introduced a vanishing "trigger action [travail décrochant]," which was not so different from the small amount of force required to pull a gun trigger. However he found that new concepts and new words were unnecessary, and specified that a guiding principle was not in need of a corresponding mechanical force, however negligible it might be.

5 He mentioned Emile Du Bois-Reymond and Thomas Henry Huxley as upholders of extensive reductionism.

6 Boussinesq quoted from Bernard's *Rapport sur la marche et le progrès de la physiologie générale en France* [Bernard 1867, p. 223]. He also quoted from the second volume of the treatise that Berthelot had published in 1860, *Traité de chimie organique fondée sur la synthèse*. Berthelot acknowledged that chemistry could not account for "the level of organisation" of living beings, even though "the chemical effects of life" stemmed from "ordinary chemical forces" [Boussinesq 1878, p. 29; Berthelot 1860, p. 807].

Divergent solutions of differential equations, or in other words "bifurcations in the integrals of the equations of motion," offered a suitable mathematical model for the creative power of life, which acted in its own specific way, and "should not borrow its way of action from physical forces" [Boussinesq 1878a, pp. 31-33; Saint-Venant 1877, pp. 421-22; Cournot 1861, pp. 364, 370, and 374].[7]

Boussinesq devoted the first chapter of the book to mathematical and philosophical foundations. At first he remarked that the alliance between differential equations and physical laws for inorganic matter had been the "natural crowning achievement" of a successful scientific practice during "three centuries." On the other hand, he specified that the unquestionable scientific success in the comprehension of the inanimate world did not entail an underestimation of the specific pliability of living beings, and the creative power of the human mind. Although some scientists upheld the purely deterministic nature of life, and poked fun at the illusion of human freedom, the mathematical model of differential equations was not in contrast with the actual practice of freedom [Boussinesq 1878a, pp. 37-9]. On the contrary, to deny human freedom corresponded to denying an important feature of differential equations.

> I would like to establish that the rejection of true and active freedom, and the rejection of every influence of life on matter, are in disagreement with logic, and overlook an important mathematical fact. This fact deals with well-defined differential equations [...] that cannot be identified with finite equations, and might not yield final states in function of time and initial conditions. In fact, the process of mathematical integration [...] can lead to an uncertainty that corresponds to what mathematicians call *singular solutions* of differential equations [Boussinesq 1878a, pp. 39-40].

The existence of "singular solutions" of differential equations was the keystone of Boussinesq's scientific and philosophical design. He acknowledged that inorganic nature could unquestionably undergo the "supremacy of physical-chemical laws." From the mathematical point of view, those deterministic laws corresponded to ordinary solutions of differential equations. Singular solutions corresponded to bifurcations in natural processes, and were consistent with the emergence of life and free will. Ordinary solutions were consistent with a deterministic approach to nature, whereas singular solutions could be put in connection with unpredictable processes in the context of life and mind.

7 It is worth specifying that Saint-Venant did not make use of the expression "*travail décrochant*," even though he made use of the verb "*décrocher*."

> The existence of singular solutions, and the *pliability* that they introduce in the series of events, might offer a mathematical representation of the difference between intrinsically vital motions, mainly intentional motions, and motions that are performed under the exclusive effect of physical laws. Therefore a living being stands out from other natural bodies because of equations of motions that contain singular integrals. By acting periodically or continuously, these singularities give rise to uncertainties that call for the intervention of a specific *guiding principle* [Boussinesq 1878a, p. 40].

Boussinesq was aware that the reduction of nature to matter and motion was a meta-theoretical option, or a sort of *imagery*, rather than an inescapable necessity emerging from the natural world. The result of a rough observation by means of human vision had given birth to a representation in terms of "shapes and their changes of place over time." The subsequent mathematical approach further simplified the already simplified outcome of a specific sensory process, and replaced uneven shapes with "ideal figures and abstract quantities." What Boussinesq labelled "*mechanical determinism*" was the most sophisticated version of that procedure, and corresponded to the differential equations of motions, where "second time-derivatives of spatial coordinates" could be derived from some functions of those coordinates. In other words, the accelerations were assumed to be proportional to the applied forces, and forces consisted of "functions of coordinates, to be determined by means of observation." The future was predictable, and closely linked to the knowledge of present and past. Boussinesq labelled "general integrals" the formal solutions of the system of differential equations representing the physical system. When the actual numeric values of coordinates and velocities at a given time were inserted into the general solutions, these became "the particular integrals" or particular solution of the physical problem. Besides these ordinary solutions, which corresponded to mechanical determinism, there were those peculiar solutions that Boussinesq had already labelled "singular solutions." The name stemmed from the fact that some solutions led to infinite derivatives [Boussinesq 1878a, pp. 43, 46, and 48-9].[8]

---

8 Today the expressions *general solution, particular solution,* and *singular solution* have the same meaning as at Boussinesq's time [James & James 1992, p. 121]. Although Boussinesq looked upon *mechanical determinism* as a specific implementation of mechanics, it is worth remarking that different kinds of mechanics were at stake. The mechanical approach in terms of forces and differential equations was different from, and sometimes in opposition to, the mechanical approach in terms of matter in motion and collisions.

2 Singular Solutions of Differential Equations

When Boussinesq began to analyse some instances of differential equations, he specified that he was not in search of equations describing "*living beings*." That kind of equation would have been extremely difficult to shape, and it would have been even more difficult to integrate. More specifically, two hindrances had to be overcome "in the analytical explanation of the material phenomena of life": first, the interactions between the living being and the environment, and second, the exchange of matter between them. He therefore confined himself to "fictitious examples," which did not deal with living beings in themselves, but corresponded to some structural features of complex systems, living beings included. In reality, he started from "equations of motion of a system of points," because the mathematical model had to be as simple as possible, and "in accordance with the general principles of mechanics." In brief, he attempted to show the structural analogy between some features of biological processes and some features of the singular solutions of differential equations for simple mechanical systems. He was aware of the apparent mismatch emerging from such simplified mathematical models: "the physical-chemical instability" of living beings contrasted with "the realm of pure mechanical laws," as far as life was in contrast with death. Nevertheless, he was confident about the fruitfulness of his mathematical, scientific, and philosophical framework [Boussinesq 1878a, pp. 63-5].

The first instance he put forward was also the simplest and "the more abstract": the motion of "a tiny heavy body along a perfectly smooth curve," where any friction was excluded. Among the possible solutions, there was a singular solution that corresponded geometrically to a horizontal tangent line, and physically to a condition of equilibrium. Therefore singular solutions corresponded to points where the body was instantaneously at rest in a locally horizontal but unstable position. What the body might do afterwards was unpredictable, as it happened when a body was in equilibrium on the top of a dome. According to Boussinesq, only some kind of guiding principle was able to make the body lean towards right or left. Mathematical laws could not decide the behaviour of the body in those points: the integration of differential equations allowed the physicists only to compute the specific initial conditions v_0 that led to the conditions of unstable equilibrium. The singular points were labelled "stopping points [*points d'arrêt*]," and in those positions the body was at the mercy of a guiding principle. The state of rest could last indefinitely: the stopping point was the seat of that abstract principle of action [Boussinesq 1878a, pp. 67-70].

The number of stopping points depended on the shape of the trajectory: whenever descending paths were followed by a new smooth top, the specific initial conditions could give place to other stopping points, and once more the system was at the mercy of "the guiding principle." Many possibilities were at stake: the body could rest on the stopping point, or could descend forwards or backwards. Some curves or initial conditions could lead the body to a permanent downward pathway: in that case, "the system was like *dead*," since it was subjected only to mechanical forces. The concept of stopping point was an actual scientific concept on the borderline between mathematics and physics. On the contrary, Boussinesq's guiding principle was a questionable concept on the borderline between mathematics and philosophy. The conceptual transition between stopping points and guiding principle corresponded to a highly problematic transition between two different levels of investigation. Boussinesq's meta-theoretical design was in tune with the meta-theoretical attitudes of some contemporary theoretical physicists: he was not loath to pursue a questionable alliance between the most advanced mathematical physics and the most speculative tradition of natural philosophy [Boussinesq 1878a, p. 74].[9]

The mathematical analysis of stopping points required a finer distinction between slightly different mathematical entities, namely singular solutions in a proper sense, and "asymptotic integrals." Boussinesq undertook that analysis with great detail: singular solutions corresponded to a physical configuration wherein a body reached the dome top in a finite time. On the contrary, asymptotic integrals corresponded to a body that could reach the top only after an infinite time, at least "*from the abstract point of view*." In order to show the mathematical aspect of that difference, Boussinesq refined his model and focused on the stopping point, which was assumed as the new starting point. The mathematical derivation showed that some value of the parameters appearing in the curve equation led to singular solutions or "singular points," whereas other values led to asymptotic integrals. He then put forward the second kind of differential equation that was chosen among the best-known physical models: two bodies endowed with masses M and M_1 interacted by means of a force of mutual attraction, which depended on their mutual distance. Although the problem was "less simple from the analytical point of view," it was however "the more elementary among those dealing with the motion of a real system."

9 The daring integration among integral-differential equations, theory of probability, recently-emerged physical concepts, and cosmological cogitations that Ludwig Boltzmann had put forward the previous year might be looked upon as structurally akin to Boussinesq's daring integration among mathematics, physics, and philosophy [Boltzmann 1877 (1909, vol. II), pp. 164-6].

In that case, singular solutions "*did not correspond to stopping points but uniform circular trajectories*," because a vanishing radial velocity led to a constant angular velocity in the equations [Boussinesq 1878a, pp. 72-6 and 96-7].

When Boussinesq focused on the problematic link between mathematics and physics, he had to face two specific issues: the mathematical expression of the physical force, and the specific value of initial conditions. He wondered whether the physical force, namely "the true expression of the action between two atoms," was consistent with "the existence of one or more circular orbits." This question fell outside the field of mathematics, and no demonstration was available: he confined himself to remarking that a positive answer appeared to him as "eminently probable." The second issue had already been stressed in the first geometrical-physical instance: both in the case of a body in motion along a given trajectory, and in the case of the relative motion between two atoms, only specific initial conditions allowed physical bodies to reach singular points or orbits. Boussinesq acknowledged that such conditions appeared "very difficult to realise," but once established, the bifurcations persisted indefinitely, and the role of the guiding principle really emerged. A daring conceptual shift from mathematics to biology led Boussinesq to interpret that "perfectly unstable equilibrium," namely the indefinitely lasting motion on circular orbits, as a rough mathematical representation of the supposed instabilities that allowed life to emerge. In other words, while the material system persisted in a singular condition, it "could not die," until an external perturbation suddenly modified that condition. If life corresponded to that extraordinary condition, where a specific law of force and specific initial conditions were required, death corresponded to "the supremacy of mechanical laws." A material system was *dead* when it was "at the mercy" of mechanical laws, and it was *alive* when it could successfully elude that mechanical drift [Boussinesq 1878a, pp. 99, 105, and 115].

On the physical side of instabilities, friction and viscous dissipation represented a very sensitive issue. On the one hand, they represented a sort of watershed between mathematical idealisation and reality; on the other, dissipative effects could prevent the body from reaching the singular point, in the case of a motion along a given trajectory. Nevertheless, specific singular integrals could emerge even in the case of viscous motions, or more generally when the acceleration d^2x/dt^2 did not depend only on the coordinate x but also on the velocity $v = dx/dt$. Because of the arbitrariness of the constant of integration in the general solution, there were specific x values that led to the same value of v both in the general and in the singular solution. The existence of these values assured Boussinesq that a passage "without any discontinuity from the general integral to the singular solution" could actually take place. The physical system could experience an indefinite number of transitions from ordinary trajectories to singular ones. He remarked that, in this specific instance, the singular solutions did not depend on the choice of the initial conditions but on the analytical form of the force, in particular the analytical dependence on velocity. This fact allowed Boussinesq to widen the scope of peculiar and underestimated mathematical entities such as singular solutions of differential equations [Boussinesq 1878a, pp. 82-3 and 109-11].

He was aware that both the conceptual and mathematical links between differential equations and life sciences were quite problematic. For the time being he confined himself to pointing out that "the physical-chemical instability in a system of two atoms" could be maintained indefinitely provided that suitable initial conditions were satisfied. Nevertheless, the persistence of singular states depended on "*the external conditions of isolation*," and those conditions could be easily fulfilled in the case of a system of two atoms. On the contrary, living systems were open systems: they exchanged energy and matter with the environment. Only a wishful conjecture could lead him to guess that "a similar property," namely the existence of singular solutions, could also appear in a system which was in connection with an external environment. Furthermore, he was aware that the emergence of singular states from purely physical conditions could "open the door to the belief in spontaneous generation." In this case, Boussinesq was not able to put forward a definite answer: he confined himself to pointing out that the initial conditions which were consistent with the establishment of singular solutions were actually "as specific as to lead to a negligible probability to be produced by pure chance" [Boussinesq 1878a, pp. 113-4 and 116].

In brief, from the mathematical point of view, two contrasting features emerged from singular solutions. On the one hand, Boussinesq stressed the improbability of the initial conditions leading to specific instabilities, and on the other, the *stability* of those instabilities. The stability of singular states and their improbability were independent of each other: the astonishing combination of stability and improbability appeared as one of the essential features of living processes. He firmly rested upon this structural analogy, which was his actual methodological hallmark. Complex phenomena could not be accounted for by specific mechanical models: that complexity could only rely on the structural analogy with peculiar mathematical entities. In this context, even the above-mentioned shortcomings could be transformed into a fruitful analogy. The improbability of those mathematical states was in tune with "the experimental impossibility of spontaneous generation," and the improbability of the emergence of new biological species. The persistence of the same states was consistent with the persistence of life, and "the longevity of the species themselves." In other words, Boussinesq found that the two essential features of singular solutions could found "a physical translation" into the actual natural processes [Boussinesq 1878a, p. 117].

In the last part of the essay Boussinesq outlined a historical-critical analysis of the mathematical approach to singular integrals. Mathematicians had been astonished by this kind of solution, and they had not attempted to find a *"field of application"* in the natural world. At the same time, naturalists were not interested in mathematical models for living systems, not to mention the lack of actual mathematical competence. In brief, neither were naturalists able to handle the mathematical toolbox, nor were mathematicians interested in inquiring into the mathematical features of "those peculiar material systems which we call living bodies." During the nineteenth century, singular integrals had sometimes attracted the attention of mathematicians, but always from a purely mathematical point of view, or in connection with mechanical problems. Boussinesq mentioned Siméon Denis Poisson, Jean-Marie Constant Duhamel and Cournot's researches, and briefly commented on their texts [Boussinesq 1878a, pp. 121-30].

In the end, he synthesised the issue he had raised, and the result he had achieved: physical laws as expressed by differential equations for the motion of material systems should not be identified with "absolute determinism." Besides the deterministic drift there was "a principle which took charge of driving the systems at bifurcations," and that principle represented something more than pure chance. He hinted at the possibility that the action of the guiding principle was subject to "certain laws," and he imagined a new hypothetical *science* whose wide scope included the behaviour of both inorganic matter and living bodies, man included. Two targets appeared within reach of that peculiar science: the inclusion of "physiology into the realm of rational knowledge," and the possibility of bridging the gulf between mechanics and "social dynamics." According to Boussinesq, not only was the guiding principle of life evolution different from mechanical actions but it was also different from "free causes": it could fill the gap between the strict causality of "physical-chemical forces" and "the *principle of finality*" which was the hallmark of "wholly conscious life" [Boussinesq 1878a, pp. 133-4 and 140].[10]

3 Mathematical and Physical Aspects of Determinism

From the mathematical point of view, two issues were at stake in the context of differential equations: the existence and uniqueness of solutions, and the role

10 After his concluding remarks, Boussinesq devoted almost a hundred pages to mathematical specifications, and a further ten pages to additional remarks.

played by singular solutions. The two issues were mutually interwoven, and still under scrutiny around 1880: no systematic, conclusive, and universally accepted theory was on the stage.[11] With regard to singular solutions, some mathematicians had already come across them in the eighteenth century: among them Brook Taylor, Alexis Clairaut, and Leonhard Euler. In 1759 Euler had devoted a paper to the analysis of "some paradoxes of integral calculus." From 1776 onwards, Lagrange had inquired into the subject matter and had created the first systematic theory of singular solutions: his researches were collected in the treatise he published in 1801, *Leçons sur le calcul des fonctions*. In brief, the topic was studied in detail in the late eighteenth century, and "some results were taught at the Paris École Polytechnique." From the linguistic point of view, Lagrange labelled "complete solutions" what we call general solutions, and "particular integrals" what we call singular solutions. The label "particular solution" was referred to "a special case of the complete solution." Unfortunately a plurality of linguistic choices might puzzle the reader: Laplace used the expressions *particular solutions* and *particular integrals* in the converse senses from Lagrange. In 1806 Poisson devoted a paper to "particular solutions" and some physical applications [Gilain 1994, pp. 444; Grattan-Guinness 1990, vol. 1, pp. 155 and 227].[12] With regard to the theorem of existence and uniqueness, in 1824 and 1835 Cauchy had given the demonstration for a specific class of equations. At that time, Cauchy's researches represented a noteworthy innovation since it had previously been taken for granted that solutions always did exist. In 1868 Rudolf Lipschitz refined Cauchy's results, and showed that the conditions of existence were weaker than Cauchy's. Only in 1893 did Emile Picard offer a consistent exposition of the results of existence [Gilain 1994, p. 446; Grattan-Guinness 1990, vol. 2, p. 759].

As mentioned by Boussinesq in his essay, Poisson had hinted at the problem in his Traité de Mécanique in 1833. In the context of the differential equations of motion, he had analysed the simple case of a body in motion in a viscous medium. More specifically, he had imagined that the body did not experience gravity, and that the resistance of the medium led to a deceleration that was

---

11 Only around the turn of the twentieth century was a satisfactory systematisation achieved. Twenty years ago the historian of mathematics Christian Gilain stated that "the theory of ordinary differential equations still appears to be one of the most active branches of mathematics" [Gilain 1994, p. 451].

12 In 1772 Laplace labelled *solution* or "intégrale générale" the general solution, and "intégrale particulière" every solution "qui se trouvera de plus comprise dans l'intégrale générale." He labelled "solution particulière, toute solution qui n'y est pas comprise" [Laplace 1772, p. 326].

proportional to the square root of the instantaneous velocity. The mathematical interpretation suggested that $v = 0$ was a singular solution, and physics suggested that when velocity vanished, acceleration also vanished, and therefore "in that time the moving body must stop and then stay at rest." However he looked upon the physical example as "purely hypothetical": it had given him the opportunity "to show the necessity of taking into account particular solutions in differential equations of motion." He claimed that the corresponding processes could not happen when mathematicians confined themselves to forces that really acted in nature. In other words, actual forces were mathematical functions that could not give rise to singular solutions. The complex interplay between mathematics and physics was still waiting for a satisfactory clarification [Poisson 1833, pp. 250-1].[13]

Boussinesq reported that the problem had also been tackled in Duhamel's *Course de mécanique*, and Cournot's 1841 *Traité élémentaire de la théorie des fonctions et du calcul infinitesimal*. In reality, around the middle of the century, mathematical and physical approaches to singular solutions became different from each other, as we easily realise when we compare the mathematical treatises that Duhamel and Cournot published in 1847 and 1857 respectively, with the physical treatise Duhamel published in 1853.[14] In Duhamel's *Cours d'Analyse* the subject matter was extensively treated in the first sections of the second volume, and in Cournot's *Traité* in the fourth and seventh chapters of the second volume.[15] On the contrary, in Duhamel's physical treatise the subject matter was compressed in a short section of three pages, "Remarque relative aux solutions singulières," and the physical model he discussed was an improved version of the example Poisson had put forward twenty years before.[16] The singular solution $v = 0$ had to be added to the general solution in order to obtain "the complete solution of the problem under consideration."

13 Poisson labelled "solution particulière" the singular solution, and "son intégrale" the general solution.

14 After having attended the École Polytechnique, in 1830 Duhamel taught mathematics in the same École. Cournot and Duhamel's above-mentioned treatises were the second editions.

15 In 1841, in the second volume of his *Traité élémentaire de la théorie des fonctions et du calcul infinitésimal*, Cournot had devoted the whole chapter IV of part VI to singular solutions of differential equations [Cournot 1841, II vol., pp. 271-92].

16 However four differences between Duhamel and Poisson can be stressed: first, the former spoke of a point rather than a body, second, he made use of a more general viscous force, third, he specified that velocity had to be positive, and fourth, no mistake with regard to the inversion of functions appears.

From the physical point of view, the point was expected "to remain in the same position where it was found at that time" [Duhamel 1853, pp. 328-30].

No problem emerged on the borderline between mathematics and physics in Duhamel's analysis. It seems that the first mathematician who raised the question of *determinism* in connection with singular solutions of differential equations was really Boussinesq himself, in a brief *Note*, "Sur la conciliation de la liberté morale avec le déterminisme scientifique," he published in the *Comptes Rendus* of the *Académie des sciences* in 1877.

In reality, in 1872 Cournot had briefly discussed the case of a cone in equilibrium upon its top, and the relationship between mathematics and physics was explicitly at stake. He had found a subtle and meaningful link between unstable mechanical equilibrium and probability. He had remarked that when somebody attempted to set a cone in equilibrium upon its tip, she/he found that the equilibrium could not be attained in practice. It was "mathematically possible" but "physically impossible." The fact was that, among the infinite directions that we could choose by chance, "only one of them" corresponded to equilibrium. The probability of fulfilling the right orientation was zero, because it was given by the ratio of the number of suitable directions, namely 1, to the number of total directions, which was infinite [Cournot 1872, p. 276].

In 1877, Boussinesq started from his "mathematical definition of *determinism*," which required that second time-derivatives of the atom's coordinates were functions of coordinates themselves. It was in reality a definition at the borderline between mathematics and physics. The statement was the mathematical translation of the physical definition of conservative forces: the simplest statement of the conservation of energy required that forces, and therefore accelerations, had to be functions of coordinates. Physical laws were looked upon as "nothing else but specific applications" of mathematics, where the word *mathematics* corresponded specifically to "the differential equations of motions" [Boussinesq 1877b, pp. 362-3].

Besides "the general integrals," which offered the solutions of those differential equations and gave rise to a family of "particular integrals" corresponding to different initial states, some equations also admitted singular solutions. When such solutions appeared, the physical system underwent a transition from "a set of particular integrals to another," and the transition could take place "in infinite ways, and infinite times." Mathematical and physical determinism required that natural phenomena followed "pathways which never branch off": after having "*completely* translated the problem into equations," only one mathematical solution should emerge. According to Boussinesq, when that requirement was not fulfilled, and bifurcations emerged, the necessity of "a *guiding* principle" also emerged. The principle could repeatedly be called into play, and it could change the pathway of natural phenomena frequently over time [Boussinesq 1877b, p. 363].

The dichotomy between determinism and free will mirrored the corresponding mathematical dichotomy, which did not mean contradiction but complementarity. Determinism corresponded to ordinary solutions of differential equations, whereas free will corresponded to the domain of singular solutions. The two different domains, namely the domain of processes subjected to determinism, and the domain of processes subjected to free will, did not overlap with each other. Boussinesq stressed their independence, in the sense that freedom did not affect determinism but was complementary to it. Free will came into play when physical laws "did not manage to deduce the future from the present," and failed to prescribe "a completely definite pathway for natural phenomena." He dared to imagine a new kind of science which had guiding principles as its object, and could represent the behaviour of "a *moral* and *responsible* being." He also imagined that the singular integrals emerging from "the equations of motion for the organ of thought" could concur to set up that new body of knowledge, which was placed "at a higher level than geometry" [Boussinesq 1877b, pp. 363-4].[17]

Free will gave rise to intrinsically unpredictable behaviours: its effects could only be computed by means of statistics. When "*great numbers*" were at stake, single behaviours could undertake any direction, but only slow transformations could be expected in the social domain: more specifically, "in the mean moral state of society," the macroscopic effects could only experience gradual changes [Boussinesq 1877b, p. 364]. Boussinesq did not detail this reference to statistics that could have bridged the gap between determinism and free will, or between predictable and unpredictable events. He simply hinted at a broad analogy: the unpredictable behaviour of singular solutions and personal choices stood beside the predictable behaviour of general solutions and collective processes that involved great numbers. The fact is that the empirical side of the analogy, namely the relationship between singular and collective behaviours, was consistent with a statistical approach, whereas totally unexplored and still mysterious appeared any statistical relationship between singular and general solutions of differential equations.

---

17 In 1877, Boussinesq assumed an ambiguous position on the relationship between mathematical models on the one hand and biological and psychological processes on the other: he put forward a structural analogy but he also hinted at equations describing mental activity. In 1878, he was more cautious, and he confined himself to the structural analogy.

Boussinesq's *Note* had been presented by his mentor Barré de Saint-Venant, who did not fail to give support to his protégé. He sent a *Note* with a similar title, "Accord des lois de la Mécanique avec la liberté de l'homme dans son action sur la matière," to the *Comptes Rendus*. He started from three physical laws, which were nothing other than the three laws of conservation for linear momentum, angular momentum, and energy. He specified that in no way could free will and human actions "contradict these laws," which were "firmly established." No contradiction could emerge from the co-existence between "freedom in our visible actions" and "the invariability of physical laws that rule the subsequent motions of bodies." The existence of free will was assumed as a matter of fact. After this specification Saint-Venant took a slightly different pathway. He focused on physics, in particular explosive processes, where a little quantity of energy triggered off the transformation of huge amounts of energy. He made reference to phenomena like a little spark acting on a gunpowder box, which could blow up a fortress. In those cases, the ratio of "the work which produces the transformation of potential into actual energy" to "the amount of energy thus transformed" might be as negligible as to become zero. Living systems offered other meaningful instances of such processes; for instance, the efforts of our muscles were triggered off by "the impulse of small vibrations in the nervous system." According to Saint-Venant, nothing prevented us from imagining that physical actions in living systems could take place "without any expenditure of mechanical work" [Saint-Venant 1877, pp. 419 and 421-2].

At this point he was not so far from Boussinesq: in the realm of living beings, physical actions could be driven by some kind of guiding principle, which did not correspond to any measurable physical force. According to Saint-Venant, the mathematical representation of those processes could be found in Boussinesq's *Note*: singular integrals were "the analytical answer" to "the necessity of a *guiding principle*." In the context of physics, that principle could "extend over time the instantaneous state of rest" or make the motion restart in accordance with specific values of the general integral. The choice could be performed "arbitrarily and by free will," and "without any mechanical action corresponding to that choice." This was probably the most questionable part of Saint-Venant's line of reasoning: motions that restarted without any mechanical action, or the replacement of mechanical actions with free will, were unjustified from the physical point of view. He intertwined structural with ontological analogies. In conclusion, he found that the supposed incompatibility between free will and the laws of motion had no rational foundation: deterministic laws and bifurcations were generated by the same mathematical womb. Both general and singular solutions stemmed from differential equations, and no incompatibility was at stake. On the contrary, it was mathematics that assured the possibility of a mutual consistency between "physical laws and the free action of mind on matter" [Saint-Venant 1877, p. 422-3].

Sometimes historians and philosophers have made use of synthetic labels in order to qualify cultural trends and personal commitments. With regard to Boussinesq in particular, and the debate on determinism and free will in general, some labels such as *materialism* and *spiritualism* have willingly been used. The reduction of that debate to a competition between *materialistic* and *spiritualistic* philosophies seems to me really misleading. I do not find that a mechanical world-view should be identified with a materialistic philosophy, nor do I find that religious faith should be automatically identified with spiritualism. Therefore I disagree with theses such as "the debate started as a defense of a spiritualistic philosophy, anchored in the personal faith of Joseph Boussinesq." I also disagree with the thesis that the debate started "as a private questioning on the matter of free will, i.e. as a metaphysical problem," and ended "as a debate on the foundations of physics and mathematics." In other words, Boussinesq's research programme was looked upon as "an example of the influence of philosophy on the development of science."[18] It seems to me that the reverse is also true: that programme is an instance of the influence that scientific issues exerted on the philosophical debate in a historical context where the philosophical environment had become extremely sensitive to contemporary scientific researches [Mueller 2015, pp. 613-5].

4 A Widespread Interest in Instabilities

However, despite the heuristic power of Boussinesq's perspective, the problematic link between singular integrals and free will was still waiting to be clarified. It is worth stressing that Saint-Venant's pathway had already been

18 However, it seems to me that the author later acknowledged that the issue at stake was scientific, namely the fact that "Newtonian physics is indeterministic in some particular cases." Moreover the author softened his thesis in the following pages: he himself warned readers against any hasty labelling, and remarked that "one should be cautious in reading Boussinesq as being particularly close to any definite philosophy." I endorse this without hesitation, and I also agree with the statement that Boussinesq's main ambition was "to reconcile determinism and free will without denying determinism." In general, I find that questioning of someone's motives might perhaps be exciting from the intellectual point of view but it is definitely unfruitful for the comprehension of historical processes [Mueller 2015, p. 621-2].

undertaken by some English natural philosophers even though in a more informal way. In 1873 James Clerk Maxwell had written a brief essay that was not intended to be published: it was addressed to a club of scholars who had the habit of sharing their reflections and cogitations. He was interested in the relationship between mind and body, and instabilities and "singular points" were at stake. He found that "the soul of an animal" was not structurally different from "a steersman of a vessel" whose function was "to regulate and direct the animal powers" rather than to produce them. The text contained scattered or loosely connected issues that dealt with "the principal developments of physical ideas in modern times," but two main issues were at stake. The first dealt with "the distinction between two kinds of knowledge," which had emerged in the context of "molecular science," and which he labelled "for convenience the Dynamical and Statistical." The second issue dealt with "the consideration of stability and instability," which could shed light on the above-mentioned questions. In other words, according to Maxwell, the study of instabilities could clarify the comprehension of the problematic connection between dynamical and statistical knowledge. He defined instability as a specific condition of a system when "an infinitely small variation in the present state may bring about a finite difference in the state of the system in a finite time." The existence of unstable conditions could make difficult or impossible the prediction of future events, provided that "our knowledge of the present state is only approximate, and not accurate." Once more we find the link between instability and statistical knowledge that neither Maxwell nor Boussinesq were able to clarify [Maxwell 1873, pp. 817-9].

Maxwell saw an intrinsic connection between instability and free will: in his words, when "we more or less frequently" found ourselves "on a physical or moral watershed," we also found the same features of physical instability. In the moral state that corresponded to physical instability, "an imperceptible deviation" was "sufficient to determine into which of two valleys we shall descend." According to Maxwell, determinism was not automatically in danger in this case: a sort of compatibility or complementarity between determinism and free will could be assumed. They represented nothing other than two different interpretations of the events that took place on a watershed. On the one hand, the doctrine of free will required that "the Ego alone" was looked upon as "the determining cause." On the other hand, "(t)he doctrine of determinism" claimed that "without exception" there was an objective explanation: the resulting choice was looked upon as "determined by the previous conditions of the subject." In other words, a predictable chain of "causes and effects" was at stake in the deterministic interpretation. But *which* was the entity that was able to ascertain the chain of causes and effects? If *It* had been a human being, he/she would have been subjected to the already mentioned uncertain knowledge of the effects. If *It* had been "the Deity," Maxwell would have objected to "any argument founded on a supposed acquaintance with the conditions of Divine foreknowledge." In the end, determinism and free will were not facts but different interpretations of the same facts [Maxwell 1873 (1995), pp. 820-21].

He stressed that his inquiry into determinism was undertaken from the point of view of physics, and "theology, metaphysics, or mathematics" had not been taken into account. Obviously, this was not completely true, but the claim shows us that Maxwell looked upon these problems as actually physical problems. *Instability* was the key word and the key concept, and physics offered many instances of instability. In optics, more specifically within a biaxial crystal, when the ray direction was "nearly but not exactly coincident with that of the ray-axis of the crystal," a little change in the ray direction could produce a great change in the direction of the emergent ray. A better known instance was offered by "the conditions under which gun-cotton explodes." In all these cases, "the axiom about like causes producing like effects" was not true. What did these different processes have in common? The material system was endowed with a quantity of potential energy that could be transformed into motion. Nevertheless, the transformation could take place only when the system had "reached a certain configuration." The expenditure of work to attain that configuration might be infinitesimally small, and moreover it bore "no definite proportion to the energy developed in consequence thereof." When a little spark set fire to a great forest or a "little spore" blighted "all the potatoes" or a "little scruple" prevented a man "from doing his will" he saw the same kind of structure, namely "singular points." In those points predictions became impossible, unless we could rely on "absolutely perfect data," and "the omniscience of contingency" [Maxwell 1873 (1995), pp. 821-2].

In reality, fires, potatoes, and scruples belonged to different domains of knowledge, but all singular points shared two essential features: they corresponded to the release of huge amounts of energy, and at the same time they were isolated points, which formed "no appreciable fraction of the continuous course of our existence." However rare those points might be, singularities and instabilities could bring into question "that prejudice in favour of determinism," which was recommended by the other side of our experience, namely "the continuity and stability of things." In other words, the science of instabilities was a science of extraordinary events: the corresponding body of knowledge was negligible from the quantitative point of view but very important from the meta-theoretical point of view [Maxwell 1873 (1995), pp. 822-3].

In the same year Balfour Stewart, Professor of Natural Philosophy at Owens College, Manchester, published a book, *The Conservation of Energy being an elementary treatise on energy and its laws*, which had great success, and was repeatedly reprinted in the following years. In the last chapter, which was devoted to "the position of life," he discussed physical and chemical instabilities, and some structural analogies with life. The simplest instance of unstable equilibrium was represented by an "egg upon its longer axis." The balance could be suddenly destroyed by a perturbation "so exceedingly small as to be utterly beyond our powers of observation." Other instabilities emerged when "the force at work" was not gravity "but chemical affinity," and "the slightest impulse of any kind" might trigger off a sudden chemical reaction. The best-known instance of that process was represented by gunpowder: the slightest spark could bring about "the instantaneous and violent generation of a vast volume of heated gas." In brief, the natural world offered two kinds of "machines or structures": the former were characterised by their stability and "*calculability*," and the latter by their instability and "*incalculability*." Astronomical events represented the best instance of calculability whereas explosions, together with their sudden and violent transmutations of energy, represented the best instance of incalculability [Stewart 1873, pp. 155-9].

According to Stewart, any living being could be looked upon as "a machine of a delicacy that is practically infinite," whose motions "we are utterly unable to predict." Living beings represented the third level of instability and incalculability after the mechanical and the chemical, and their complexity exceeded at length the complexity of first-level and second-level machines. A different kind of action was involved indeed, because "the power of an animal, as far as energy is concerned," was not "creative, but only directive." In living beings, Stewart saw a shower of small perturbations that triggered off a series of energy release. It is worth remarking that he did not expect "to have discovered the true nature of life itself"; he had only confined himself to pointing out a structural sketch. Life could be associated with a peculiar kind of machinery, where "an extremely delicate directive touch" was "ultimately magnified into a very considerable transmutation of energy." He insisted on this specific issue: he was not able to offer any mechanical explanation of how a living being worked. He had simply outlined a very general analogy rather than a solution for "the problem as to the true nature of life" [Stewart 1873, pp. 161-3].

In brief, life was associated with "delicacy of construction," which entailed "an unstable arrangement of natural forces." Those forces were nothing other than "chemical forces" acting in the same way as in thermal engines, where explosions took place. Nevertheless, two important differences were at stake: unlike physical-chemical machines, living beings could rely on an internal organisation, and were "the subjects of decay." Organisation and decay of living beings were definitely outside the scope of Stewart's elementary treatise, but their existence allowed him to stress the essential feature of living entities. Stewart was aware that he was putting forward analogies rather than explanations [Stewart 1873, pp. 164-5].

In 1875 Cournot focused on the same kind of processes. He remarked that the mechanical work required for crushing and blending the components of gunpowder had "no theoretical relationship with the mechanical force produced by the explosion" of the powder itself. The mechanical energy sent out by the explosion was the consequence of the sudden release of chemical forces triggered off by the burning wick. He found that the intervention of thermal processes like explosions broke the symmetry between past and future. In mechanical processes, the knowledge of the forces acting on a physical system, and positions and velocities of the parts of the system, allowed scientists to "implicitly determine the state of the system in subsequent times." For the computation of an indefinite number of further states, only a mind "of the same kind of human mind" was required, even though endowed with greater computational power. The same procedure allowed scientists to go back through the past: the exact date of Thales' eclipse could be computed in the same way as "an eclipse that will take place within twenty centuries." On the contrary, the cooling of an unevenly heated sphere could be computed, but the knowledge of the uniform final state did not allow scientists to determine the inhomogeneous initial state. Cournot found that this kind of asymmetry was also linked to the intrinsic tension between empirical facts and scientific laws or between "the sources of information" and the "human mind." Facts or information dealt with the knowledge of the past whereas scientific laws and mental activity dealt with the prediction of future states. On the one hand, scientists knew "more things about the past than about the future"; on the other hand, "the virtual knowledge of the future" was easier than the rational deduction of the past [Cournot 1875, pp. 46, 59-62, and 64].

In this context, determinism was at stake, or rather "the absolute determinism" which was accepted "in the domain of physical-chemical phenomena." He immediately specified that even the strictest determinism did not exclude "the concept of independence of causes, the notion of accidental and unexpected, and the contribution of *chance*." Once more "the opposition between *law* and *fact*" re-emerged, and that opposition was not so different from the opposition between "essential or necessary," and contingent. In other words, determinism and indeterminism were the two sides of the same coin: both the rationality of laws and the contingency of facts were components of scientific practice. The predictions of mechanics emerged from the alliance between the rationality of mathematical laws and the contingency of initial conditions that "we do not consider necessary in virtue of a law." However the non-deterministic or contingent component of scientific practice, in other words "chance itself," could be submitted to "rational speculations." It was in this context that the theory of probability and statistics found "a legitimate application in the domain of physical sciences." Even the classification of sciences was involved in the intrinsic tension or complementarity between laws and facts: there were "*theoretical* sciences" and "*cosmological* sciences": physics and chemistry belonged to the former class, astronomy and geology to the latter. Saturn's rings and the number of Jupiter's moons dealt with "the *facts* of cosmology," but Kepler's laws were "*laws* of physics" rather than mere *accidents* [Cournot 1875, pp. 66-8 and 70-2].

Cournot envisaged a more general kind of determinism where both deterministic and non-deterministic processes were submitted to the normative role of mathematics. Moreover determinism did not have to be confined to physics, but could be extended to chemical processes. Alongside the knowledge of forces, positions, and velocities in the initial state, the knowledge of "temperatures, electric tensions, chemical affinities, and all kinds of actions" was required in order to put into action this wider determinism. Even a physiological or biological determinism could be assumed, but it was quite different from "the physical-chemical determinism." Deterministic processes involved "species and races," because their essential features could easily be forecast, but the specific features of individual plants or animals eluded any prediction. Indeterminism in individuals stemmed from the inadequacy of the initial conditions: besides a set of information about the single subject, "a detailed history of ancestors" was required. In other words, the whole history of living structures was at stake in biological processes. On the one hand, living beings were able to pursue the instinctive "accomplishment of an aim, or execution of a plan"; on the other hand, scientists were unable to determine the details of that accomplishment or execution in specific cases [Cournot 1875, pp. 113-7].

It is worth remarking that, ten years earlier, Bernard had put forward a mild and pragmatic determinism, or a determinism that called into play "the necessary condition for a phenomenon that is not forced to happen." It seems that Bernard hinted at something like the difference between necessary and sufficient conditions: a sound determinism involved the set of necessary conditions for the appearance of a phenomenon. In no way could those conditions automatically lead to the emergence of a specific phenomenon: the conditions were necessary but not sufficient. We find here different kinds of determinism. There was a strict determinism, which corresponded to *necessary* and *sufficient* conditions for a given phenomenon: it was in tune with a sort of *cos-*

mological determinism. Bernard's mild determinism was a mild or pragmatic determinism, which corresponded to the existence of *necessary* conditions. The third kind of determinism corresponded to the transformation of determinism into fatalism: it involved "the necessary emergence of a phenomenon independently of its conditions." Bernard considered his mild determinism and fatalism as two opposite world-views: in the former, a set of material conditions offered nothing more than the conditions of possibility for a given phenomenon, whereas in the latter a phenomenon could take place independently of whatever condition of possibility [Bernard 1865, pp. 359 and 383].

Cournot was aware that the concept of rational necessity [la détermination] had to be disentangled from the concept of prevision: a well-defined process could be unpredictable because of the complexity of the process itself. For instance, "perturbations of the atmosphere or Oceans" did not escape determinism but could be unpredictable in detail. On the contrary, the appearance of mammals could not be a completely deterministic process. The reason rested upon the fact that biological evolution did not follow "the philosophical or transcendental assumption" that the laws of nature were invariant through time and space. The metaphysical assumptions that scientific laws are eternal and universal "was not supported by experience": the confidence in that *postulate* suited "the determinism of chemists and physicists" but was not in tune with the determinism of physicians and naturalists. Why did "monkeys living in the New World" have thirty-six teeth while those of "the Ancient World" only thirty-two like human beings? Why did elephants live in the latter world rather than in the former? These questions dealt with contingency, and could not find any answer in the "functional harmony" of living beings or in the conditions of the environment at the corresponding latitudes [Cournot 1875, pp. 118-20 and 128].

With regard to the relationship between determinism and freedom, Cournot mentioned Descartes' remark on freedom as the awareness of "what is good and true." In other words, a greater freedom corresponded to "a more strictly determined human will." This conception was in tune with Leibniz's definition of God's freedom as "the definite determination to act in the best possible way in accordance with its nature." In other words, freedom was the highest expression of determinism: freedom could not be pursued in a world where only chaos and chance occurred. In human beings, freedom required awareness of self and a sense of responsibility. The concept of freedom was the seat of intrinsic tensions between seemingly opposite elements. It was an antinomy in the sense of Kantian philosophy. The internal tension between freedom and responsibility was accompanied by the logical and metaphysical tension between freedom and "the definite linkage between causes and effects." In Cournot's view, the philosophical antinomy could give way to a pragmatic compatibility, provided that "the cause of a phenomenon" was not confused with its "*physical conditions*." The emergence of conscious life introduced both perception and interpretation into the domain of purely physical effects, and freedom dealt with the plurality of possibilities of perceiving and interpreting those strictly determined effects. The antinomy of freedom appeared "on the threshold" that separated inorganic from organic processes [Cournot 1875, pp. 241, 243, and 249-52].

In 1844 Mayer had been puzzled by processes involving a sudden release of *force*, and in 1876 he devoted a short paper, "Ueber Auslösung," to the subject. From the outset, the two key words and concepts were "sudden release [Auslösung]" and "triggering action [Anstoß]" or impulse, and both concepts were involved in explosive processes: the latter could be looked upon as the first stage of the former. Mayer's use of the two words, and their semantic scopes indeed, were not so definite, and it is not easy to ascertain whether *Auslösung* corresponded to the early triggering action or the whole process of sudden energy release. In reality, the distinction does not seem so important in Mayer's text because the processes he described were the well-known explosive processes. Making reference to his 1842 paper, he reminded readers that in those processes no quantitative relation between cause and effect could be found, since the cause might be "an infinitely small quantity." Nevertheless, no exception to the sentence "causa aequat effectum" could be admitted: as a consequence, the expression "cause and effect" had to be used "in a completely different way" in the context of explosive processes. From the mathematical point of view, cause and effect had no common unity of measure: he ventured to assert that triggering processes lay outside the mathematical domain. They were qualitative rather than quantitative processes, and *qualities* could not be "numerically determined" [Mayer 1876 (1953), pp. 9-11].[19]

According to Mayer, triggering processes played an important role in life sciences, in particular "in physiology and psychology." Even in organic chemistry, and more specifically in the phenomena of fermentation, the *Auslösung* was at stake. Human life depended on a network of processes of the same kind, and even the motions of our limbs required such processes. He classified human motions as "instinctive, half-conscious, and conscious": in all these

19 Historians have already stressed "the influences of metamorphic conceptions (which stemmed from the context of life sciences) on physical sciences," and the role played by the concept of *Auslösung* in late nineteenth-century physics and chemistry [Guzzardi 2001, pp. 146 and 150]. I am indebted to Guzzardi for having drawn my attention to Mayer's *Auslösung*.

motions, energy thresholds and sudden discharge of electric and chemical energies were at stake, but the existence of conscious motions was "the most evident representation" of the strength and effectiveness of basic, biological processes. Conscious motions consisted of the contraction of some muscles, and the triggering action took place through the excitement of the corresponding nerves. In this case, two different processes were simultaneously involved: purely physiological reactions, and the powerful activity of the human mind [Mayer 1876 (1953), pp. 11-13].[20]

We see that in the late 1870s a scattered set of problems and remarks crossed the scientific community: it dealt with the long-standing question of free will, and the complex relationships among physics, physiology, and psychology. More specific processes such as physical instabilities, chemical explosions, and the transmission of nervous impulses were involved. The mathematical physicist Boussinesq had ventured to put forward a very general mathematical framework for that heterogeneous set of problems. No specific answer could stem from this mathematical framework because of its generality: Boussinesq could only offer a broad analogy between the essential features of those processes, and the essential features of some mathematical entities such as singular solutions of differential equations. Mathematics seemed more fruitful than previously expected: ordinary solutions of differential equations could represent the usual, deterministic, physical and chemical processes, whereas singular solutions could represent physical instabilities and the core of biological processes. It was really astonishing that the improbability of initial conditions leading to instabilities, and the undefined persistence of those instabilities, were in tune with some essential features of life and the emergence of life.

The debate on determinism and free will was as theoretically ambitious as empirically ineffective, but it cannot be denied that Boussinesq's mathematical approach represented a new and original point of view. What Maxwell, Stewart, Cournot, Saint-Venant, and Boussinesq had in common was a certain degree of confidence in mathematics, more specifically the confidence that mathematics could represent a wider domain of natural phenomena, however improbable, isolated, and *explosive* they may be.

20 Mayer's estimation of the speed of nervous impulses was "about 30 meters per second"; over the "short distances" of nerves, the transmission corresponded to a sudden process. Human volition travelled through those nerves, and triggered off "the desired action" [*Ibidem*, p. 106].

CHAPTER 5

Scientists and Philosophers on Determinism

1 A Bitter Confrontation between Mathematicians

Boussinesq's long essay I discussed in the previous chapter was immediately criticised by the renowned mathematician Joseph Bertrand, who published an aggressive and sarcastic paper in the *Journal des Savants*.[1] He immediately poked fun at "the useless array of scholarly formulae," which could "dazzle a reader who is not an expert in mathematics." According to Bertrand, that expenditure of mathematical scholarship hid a questionable superposition between the mathematical theory of mechanical systems and concepts like choice, freedom, and will. He asked himself how a mathematician could envisage "a material and inert system that was suddenly endowed with will" and could choose between two possible motions. He ironically remarked that when equations were non-determined, Boussinesq found it "necessary to compensate for their deficiency." However, when Bertrand stopped making sarcastic remarks on the weakness of Boussinesq's analogies and embarked upon a more specific criticism, he was forced to enter the meta-theoretical ground where the complex links among mathematics, science, and Nature were at stake. He started from two meta-theoretical theses: first, the results of mechanical equations could not attain absolute precision, and second, the reliability of equations could not be greater than the reliability of the principles from which they stemmed. The second statement called into play the hypothetical-deductive structure of mathematics, whereas the first echoed something like the ancient distinction between the *smoothness* of pure mathematics and the *roughness* of the natural world [Bertrand 1878, pp. 517-9].

Although Bernard had started from a sharp criticism on Boussinesq's scientific enterprise, he discussed the same physical configuration that Boussinesq had already described. He analysed the case of a body that started from a state of unstable equilibrium: he found that "an infinitely small force would bring about an endless motion," and from the mathematical point of view, the min-

1 After having lived with his uncle Duhamel, Bertrand had entered the *École Polytechnique*, and in 1862 had been appointed professor of *Mathematical Analysis* at the *Collège de France*. In 1853 he had overseen a new edition of Lagrange's *Mécanique Analytique*, and in 1874 he had been elected *Secrétaire perpétuel* of the *Académie des Sciences*. He also edited the *Journal des Savants* from 1865 to his death in 1900.

 | DOI 10.1163/9789004315235_007

imum value of that force was strictly zero. He pointed out the same problems and the same effects, but in the end, he found that nothing really important could be derived from these phenomena. Neither "mechanics appeared worried," nor "the science of soul had something to gain" from speculations on the possible link between mathematics and free will. He denied any possible connection between the two fields, and sharply concluded that "the mystery of the soul" remained unattainable. In reality, the paper did not end here, because he continued to insist on the first meta-theoretical issue he had pointed out two pages before. Among physical sciences, mechanics was "the closest to truth," but it could not reach perfect exactness. In some configurations, equations allowed the physical system to take two different pathways even though physical laws should only lead to one of them. In this sense, mathematics and physics had different natures, where the word *mathematics* made here reference to the mathematical laws of mechanics. Confronted with the uncertainty of mathematics, physics took the lead: the least amount of force could "make the ambiguity disappear" [Bertrand 1878, pp. 518-20].

The fact is that the supposed uncertainty of mathematics and the deterministic nature of physics was in opposition to the traditional view: the perfection of mathematics against the background of the coarser scientific explanations. In reality, in Bertrand's line of reasoning, the word *mathematics* had different meanings, because different kinds of mathematics were at stake: more specifically, the mathematics of the equations of dynamics could not attain the "absolute strictness of Euclid's theorems." In brief, three bodies of knowledge were involved: classical mathematics, mathematical physics, and physics. Mathematical physics, together with its toolbox of differential equations, appeared as a shaky domain when compared to the strictness of pure mathematics and the empirical certainty of physics. According to Bertrand, on the borderline between physics and pure mathematics, in particular in the mathematical representation of motion, some ambiguities emerged. Alongside the representation of physical forces as continuous entities there was the representation in terms of "subsequent discontinuous impulses acting during a finite time." He preferred a discontinuous representation of physical entities: a discrete rather than continuous representation made "multiple solutions disappear," and "the necessity of a free choice for a dithering molecule" disappeared as well [Bertrand 1878, p. 520].

Surprisingly enough, Bertrand criticised Boussinesq for his blind trust in mathematical physics: according to Bernard, Boussinesq had expected that "inert matter hesitated" when mathematical procedures led to a state of uncertainty. Boussinesq had expected that physical systems loyally followed differential equations when the latter "refused to decide." It was just the identification of physical with mathematical entities that had led Boussinesq to "class molecules among living bodies." On the one hand, Bertrand rightly pointed out Boussinesq's confidence in the possibility that mathematics could represent the essential features of complex natural processes. On the other hand, Bertrand's line of reasoning missed the point, because Boussinesq had not put forward a material analogy between singular solutions of mathematical-physical equations and processes taking place in living beings. He had confined himself to a structural analogy, where the equations that led to a plurality of pathways in the field of mathematical physics were supposed to be akin to the mathematical structures that might rule the behaviour of living systems. On the specific content of the analogy and on its global soundness, Bertrand could not agree with Boussinesq. The former confined himself to expressing his astonishment before "the eternal miracle" of "the immaterial soul" that could influence the motion of matter in living bodies [Bertrand 1878, p. 521].

Boussinesq had attempted to outline a scientific approach to that *miracle*, whereas Bertrand found that such an attempt was pointless. Boussinesq had outlined a mathematical approach to complex systems, whereas Bertrand did not dare to undertake a similar step. His last sentence, the most sarcastic indeed, consisted of a rhetorical question, which was at the same time a statement of scientific indifference and impotence. He asked: when two pathways are equally probable, "the differential equations dictate nothing," even "the guiding principle refrains from acting," and time elapses, "what can we expect to happen?" [Bertrand 1878, p. 523].

The following month Boussinesq sent a response to the *Journal des Savants*, but the journal refused to publish it: Bertrand was the editor of the journal, and perhaps this might have been one of the reasons for the rejection. As a consequence Boussinesq sent the text to the *Revue Philosophique de la France et de l'Étranger*, which published his paper under the title *Le déterminisme et la liberté* in 1879. He found that some misunderstandings in Bertrand's paper needed to be clarified. Boussinesq claimed that the main aim of his essay had been the refutation of the deterministic view put forward by "Leibnitz, Laplace, Dubois-Reymond, Huxley, etc." In positive terms, he aimed at demonstrating that "the equations of motion of a material system, *as they are assumed by classical mechanics*," could not determine "the *complete* series of motions of the system." Since Bertrand seemed in agreement with him on this specific issue, Boussinesq found it surprising that the renowned mathematician had forgotten to point out such an important issue. The readers of Bertrand's paper had rather been led to think that Boussinesq was interested in penetrating "the mystery of the immortal soul" or "the action of the soul on the body," as some quotations from Bertrand's paper testified. On the contrary, nowhere and in no way had he raised such questions: he had confined himself to criticising the "absolute mechanical determinism" which was supposed to rule "all motions that occur in the universe" [Boussinesq 1879b, pp. 58-60].

Boussinesq discussed Bertrand's meta-theoretical thesis, namely "the mysterious nuances which distinguished abstract from real," and the supposed better reliability of physical-chemical laws when compared to their mathematical language. He also commented on Bertrand's preference for a discontinuous representation of natural processes: he acknowledged that bifurcations actually disappeared from the equations of discontinuous processes, even though they could not disappear from reality [Boussinesq 1879b, pp. 60-1]. In the end, Boussinesq focused on the core of his scientific enterprise: he preferred a questionable scientific hypothesis and a rough outline of a scientific theory rather than no scientific theory. In mathematics he had found some clues as to the possibility of a third way between mechanical determinism and vital forces for the representation of elementary processes in living structures. Some mathematical solutions could suitably be associated with physical and chemical instabilities, and in their turn those instabilities showed a structural analogy with the essential features of elementary biological functions.

> Every scholar who accepts the principles of mechanics and rejects the vital forces of the old physiology cannot but acknowledge that the domain of life corresponds to the bifurcations that appear when there is an uncertainty in mathematical pathways. This is the only route left outside the domain of inanimate matter. It is a good result for mathematicians that all cases of mechanical uncertainty known until now correspond to eminently unstable states of matter. In fact, an extreme and unparalleled instability of the physico-chemical kind is exactly what chemists and physiologists consider typical of living tissues [Boussinesq 1879b, p. 62].

Since he had been sharply criticised by Bertrand because of his supposed pretension to explaining life, he specified that his simplified mathematical models could not account for life in the sense of conscious life or life endowed with intelligence. He had opened a field of possibilities: he had also specified that his models could only outline the emergence of a basic kind of life, probably "a vegetable life." In reality, in his essay Boussinesq had insisted on the structural character of his analogies: the mathematical models he had described did not correspond to specific forms of life, but aimed to offer a mathematical foundation for the creative power of life [Boussinesq 1879b, p. 63].[2] Boussinesq's spec-

2 He made reference to Boussinesq 1878a, pp. 112 and 134.

ification allows us to appreciate a meaningful conceptual shift from his 1877 paper and his 1878 book, and afterwards between the book and the 1879 paper. In the first paper, the mathematical analogy focused on free will; in the book, he stressed the essential features of physiological and psychological processes; in the last paper, he confined himself to elementary architectures of life.

In the same year Boussinesq published an unsystematic collection of essays on different issues such as "geometrical intuition," and aims and methods of "physical mechanics." However inhomogeneous, the new publication dealt with fundamental issues on the borderline between mathematics, physics and philosophy. The first two essays offered an interesting framework for the subsequent remarks on determinism. He pointed out the gap between mathematics and experience: mathematics consisted of "ideal artefacts," and it made reference to an "autonomous order of things." Mathematics required "a specific transcendental sight," or a specific frame of mind, which was quite different from that required in physics and natural sciences. On the other hand, he excluded any actual opposition between mathematics and science. In the context of mathematics, geometry represented a sort of bridge between the realm of formal structures and the realm of human experiences, where intuition and reason were mutually intertwined: geometry required "something more than pure deduction" or pure reason. Geometry contained "unexplored depths" and "infinite dark sides": the development of geometry appeared to him a demanding task which deserved to be pursued but could never be accomplished [Boussinesq 1879c, pp. 8 and 14-16].[3]

In the transition from the abstract to the empirical, Boussinesq found "some irreducibility, or so to speak, some incommensurability": the notion of space and therefore the existence of geometry filled that gap. He refused Leibniz's conception of space as "the order of co-existences," because this definition appeared too broad to him. Space was a specific order of co-existences, or better "the place where a certain order of co-existence is deployed." Once more he stressed that the nature of geometry, the science of space, was neither purely rational nor purely empirical. Even mechanics [la mécanique physique] had a mixed nature, since it could be placed on the borderline between mathematics and experience: it required a plurality of intellectual attitudes, from the most speculative to the most empirical [Boussinesq 1879c, pp. 22-3 and 48].

3 On the *mixed* character of geometry, see Boussinesq 1879c, pp. 18-19: "Il y a donc tout lieu de croire que, sans le concours apporté au raisonnement par l'intuition géométrique, les mathématiques seraient impossibles. Bien plus, nos connaissances ou notions de toute nature se trouveraient sans doute, de même coup, profondément mutilées, peut-être même anéanties dans ce qu'elles ont de précis, de scientifiques." See also *Ibidem*, pp. 20 and 21.

One of Boussinesq's essay, "Complément à une mémoire, publiée en 1878," was intended as a further elaboration of the issues he had raised in his 1878 book. He reported that some "distinguished scholars" had appreciated the content of the book, whereas others had sharply criticised him for having involved differential equations in questions which most mathematicians and scientists strongly disliked, and for having stretched mathematics beyond the boundaries of its legitimate domain. He acknowledged that the title of the book had misled some readers, because they had expected a treatise on metaphysics whereas the book was intended rather as a collection of remarks and applications dealing with a specific field of mathematics. It was an actual scientific work, "a simple mathematical-physical study" on a specific query of natural philosophy that "had involved many minds for two centuries." He had pointed out that a specific guiding principle had to be postulated in science besides matter and energy. On that ground he had built up a very abstract approach to life sciences, which followed a structural analogy between some mathematical entities and the essential features of living systems [Boussinesq 1879c, pp. 82-3]. That structural analogy had led him to conceive the possibility of simplified mathematical models for simplified living structures. Alongside matter and force (or energy), a guiding principle had to be taken into account in the most complex natural processes.

He had assumed a structural analogy between the indeterminism of integrals and "the extreme, inimitable, physical-chemical instability of a living being." The analogy was based on two essential features: the "extremely weak probability that the physical-chemical conditions for the appearance of life" emerged, and at the same time, the "indefinite persistence of life, *once established.*" Nevertheless, the confidence in that analogy required three specifications. First, the mathematical model could not account for real living beings but could only represent some essential features of very simple systems of that kind. Second, there could not be any direct connection between the domain of computation and the domain of facts: the possibility of a meaningful link was based on hypotheses and concepts that did not belong to mathematics. Third, he stressed "the practical impossibility of spontaneous generation" in the context of natural sciences [Boussinesq 1879c, pp. 84-5].

Boussinesq relied on a very strong meta-theoretical belief: natural phenomena required natural explanations, and natural explanations could be translated into mathematical laws. The existence of singular integrals, bifurcations, physical indeterminacy, and guiding principles were embedded in that general meta-theoretical framework. Moreover, he relied on another pillar of scientific tradition: the set of hypotheses and concepts that linked mathematical models to natural phenomena should not be in contradiction with the fundamental laws of physics and chemistry. Since the analogy between mathematical models and natural phenomena was structural and not ontological, Boussinesq did not dare to put forward hypotheses on the nature of the guiding principle. It could correspond to either "a higher cause, like *life* and *will*," or something akin to ordinary forces that acted on inanimate matter. In any case, no finite amount of force in the physical sense was required in order to lead matter to choose its way at the bifurcation points. He could not accept vital forces of intensity comparable with mechanical, physical or chemical forces. Once more he pointed out that the specific feature of living beings was instability, and ordinary natural forces could not account for that instability and frailty. At the same time, instability did not prevent living systems from preserving a specific kind of equilibrium, which was in reality a homeostasis or dynamical equilibrium. Only open systems could experience that equilibrium: fluxes of energy between the system and the environment had to be "exactly balanced" by simultaneous fluxes of matter [Boussinesq 1879c, pp. 90, 94, and 98].

Boussinesq stressed another specific feature of living structures: the influence of past states on present ones. History intrinsically affected natural processes: the sensitivity to history was really one of the hallmarks of life. The natural world could be represented as a hierarchy of three levels or stages: the domain of physical-chemical forces, which depended only on the current state of the system; the intermediate level of unconscious life, where the whole history of the system was at stake; and the level of "fully conscious life," where a "*principle of finality*" made the present depend on the future. In the second and third levels new causes were at stake: they could not be assimilated to physical forces, even though they "could be represented geometrically." In other words, there was a wide set of phenomena which could not be accounted for by traditional science but had to be represented by mathematical procedures. No bifurcation or indeterminism could shake Boussinesq's firm belief in the representative power of mathematics [Boussinesq 1879c, pp. 108-9].

The structural or morphological analysis of natural phenomena led Boussinesq to put forward a series of analogies between specific fluid-dynamic effects and well-known processes in life science. The germ of a living being, when placed in a suitable environment, crossed all the transient stages that led to the adult state. The process was not structurally different from a perturbation "of medium dimensions" which entered a channel with still water, when the bottom surface was horizontal and the width of the channel constant. In this case, the perturbation progressed towards "its limiting shape of *solitary wave*," which depended on its total energy and the width of the channel. Another analogy involved the phenomenon of metamorphosis, where the whole life span of insects consisted of two subsequent stages: they were similar with regard to life duration, but "very different as to morphology and ways of life." The phenomenon showed the same essential features of "a wave of great height" which travelled upstream along a rushing river. The wave could overcome the stream and propagate itself upwards until it preserved "a reasonable fraction of its original height"; below a given height, the wave was transformed into a downstream perturbation [Boussinesq 1879c, pp. 115-6].[4]

Finally, Boussinesq discussed briefly a specific differential equation, where acceleration depended on velocity. From the structural point of view, the differential equation was not different from those introduced by Poisson and Duhamel, but from the physical point of view it corresponded to an increasing acceleration rather than a viscous deceleration. Two results were emphasised by Boussinesq: first, a stopping point really existed, and its existence did not depend on the choice of the initial conditions; and second, general and singular solutions correspond to each other with continuity. Once more he stressed that the dependence of acceleration on velocity led to stopping points that could not be removed by choosing different initial conditions [Boussinesq 1879c, pp. 116-8].[5]

2 Different Attitudes of Scientists

Boussinesq and Bertrand had different attitudes towards mathematical procedures. While Bertrand can really be associated with the classical concept of *idealisation*, Boussinesq might more conveniently be associated with the concept of *structural analogy*.[6] He was not so philosophically naïve as to trust in the automatic correspondence between mathematical structures and ob-

4 I remind readers that hydrodynamics was Boussinesq's specific field of research: in 1877 he had published a treatise where he had put forward a mathematical explanation of the water flow in river bends [Boussinesq 1877a].

5 On the contrary, when accelerations depended only on co-ordinates the stopping points could be removed by changing initial conditions, and those points were nothing but "the natural positions of unstable equilibrium" [*Ibidem*, p. 118].

6 That "Bertrand argued that these differential equations are only an idealisation," and "Boussinesq also described the laws of mechanic as an idealisation of physical reality" [van Strien 2014a, p. 181], can only be accepted as a first approximation, in the sense that mathematics in general, and differential equations in particular, represented an idealisation of natural processes. A more refined analysis shows that Boussinesq did not confuse the *strength* and *truth* of mathematics: he was confident in the expressive and explicative power of mathematics, and at the same time was quite disenchanted with regard to its content of truth.

served or perceived phenomena. He only trusted in the possibility of a structural correspondence between the essential features of singular solutions of differential equations on the one hand, and some essential features of life and moral processes on the other. The correspondence was not an explanation. More specifically, he did not expect to be able to explain life, but was confident in the possibility of describing the simplified structure of some processes in suitable mathematical terms. A structural analogy might be looked upon as something weaker than an idealisation and at the same time something stronger, depending on the point of view. A structural analogy is weaker than an idealisation in the sense that it does not require any belief in the superiority of ideal models. On the other hand, a structural analogy requires confidence in the reliability of formal or mathematical representations, in the sense that mathematical language manages to catch some essential features of natural processes. While an idealisation involves firm confidence in philosophy, a structural analogy involves firm confidence in mathematics.

In December 1878 Maxwell published a three-page paper in the journal *Nature*, which was formally a review of the book *Paradoxical Philosophy* which Balfour Stewart and Peter Guthrie Tait had recently published.[7] On the second page Maxwell went back to his previous cogitations on singular points or "singular phases" where "a strictly infinitesimal force" might determine the course of a system towards "any one of the finite number of equally possible states." He mentioned Stewart's book *The Conservation of Energy* with reference to the physical side of the problem, and reminded readers that Saint-Venant and Boussinesq had analysed "the corresponding phase of some purely mathematical problems." The difference between living and dead matter involved neither matter nor "that more refined entity" called *energy*. A third level was involved, where "the application of energy may be directed without interfering with its amount." It dealt with a directive power, and the power of mind was a suitable instance of that peculiar power. He mentioned "the engine driver, who does not draw the train himself" but rather "directs the course of the steam" in order to "drive the engine forward or backward, or to stop it." It was only "*in general*" that the present configuration and motion could determine the whole course of the system. There could also be "certain isolated and singular phases" where a tiny impulse could change the pathway of a huge amount of energy [Maxwell 1878 (1890), pp. 760].

The following year, in a letter to the mathematician Francis Galton, Maxwell mentioned Boussinesq and Saint-Venant once more: the former was infor-

7 The book was intended as a sequel to their successful *The Unseen Universe*, which had been published some years before.

mally labelled a scholar "of hydrodynamic reputation" and the latter a scholar "of elastic reputation." Maxwell credited Boussinesq with having accomplished "the whole business by the theory of the singular solutions of the differential equations of motion" in his 1878 essay. In just a few words Maxwell managed to grasp the essence of Boussinesq's research programme: "when the bifurcation of path occurs" — he wrote — the material system "ipso facto invokes some determining principle." That principle was definitely "extra physical" although in no way "extra natural," and allowed the system to determine which of the two paths it had to follow. Maxwell also managed to grasp the difference between Stewart and Saint Venant's *physical* approaches, and Boussinesq's *mathematical* one. The first two scholars had assumed the existence of "a certain small but finite amount" of force or energy, whereas the third had "managed to reduce this to mathematical zero." Stewart and Saint-Venant had made reference to actual physical processes whereas Boussinesq had made reference to mathematical models. Those models stemmed from actual physical processes that could be looked upon as mere starting points for further, more general, analogies. In fact Stewart's "trigger-work" or Saint Venant's "travail décrochant" were conceptually different from Boussinesq's guiding principle. Finally, Maxwell appreciated the philosophical scope of "Boussinesq's method" inasmuch as it was "a very powerful one against metaphysical arguments about cause and effect," and offered a better alternative to "the insinuation that there is something loose about the laws of Nature." In other words, Maxwell acknowledged that however questionable Boussinesq's research programme might be, it had the advantage of relying on mathematics and natural laws, and therefore could retrieve extraordinary or *singular* events within the domain of a scientific theory [Maxwell 1879 (2002), pp. 756-8].

In 1880, Du Bois-Reymond commented on the debate that had been raised by his 1872 lecture. He stressed the "impossibility of understanding the nature of matter and force," and the impossibility "of explaining consciousness on mechanical foundations." He deployed seven queries or open problems [Schwierigkeiten]: the nature of matter and energy, the cause of motion, the origin of life, "the apparently intentional and purposive disposition of Nature," the origin of sensorial perception, the existence of rational thought and language, and the existence of free will. The last query was extensively discussed by Du Bois-Reymond: he started from the question "whether human beings were really free to act or were bound and determined by inescapable constraints." After having discussed some philosophical approaches in ancient times and in the Middle Ages, he pointed out what he considered to be the answer offered by current science. The conservation of matter and energy prescribed that the present state of the world, our brains included, was "the definite mechanical effect" of its state in a previous instant, and was "the definite mechanical cause" of the following state. According to a "monistic point of view," volitions were "necessary and undisputed epiphenomena of motions and rearrangements of brain molecules [Hirnmolekeln]." The universe was seen as a mechanism, and in a mechanism there was no room for free will [Du Bois Reymond 1880, pp. 65, 74-6, 79-80, and 82].

In the last part of his lecture, the renowned physiologist mentioned "the late mathematician Cournot," Boussinesq, and "the Parisian Academic Saint-Venant." They were credited with having claimed that motion could be produced, or a change in the direction of motion could be realised, "without any expenditure of force." In particular, Cournot and Saint-Venant had introduced "the concept of the triggering action [Auflösung] or *décrochement*." According to Du Bois-Reymond, Boussinesq had followed a slightly different pathway: he had pointed out that "some differential equations of motion" led to singular solutions, and "ambiguous or completely undetermined" states of motion followed. It was "a kind of mechanical paradox" that had already been noticed by Poisson. The main aim of Boussinesq was rightly grasped by Du Bois-Reymond: the possibility of a better comprehension of "the multiplicity and uncertainty of organic processes." At the same time, "the German school of physiologists," he himself included, relied on "a particular kind of mechanism," and could not be satisfied with such an approach. Underneath Boussinesq's guiding principle German physiologists saw the old and unreliable vital forces, in spite of Boussinesq's reference to the French physiologist Claude Bernard [Du Bois Reymond 1880, pp. 88-9].

According to Du Bois-Reymond, a theoretical approach to simple forms of life, or "unconscious living beings," could be pursued without any reference to bifurcations in integrals or guiding principles. He found that the difference between crystals and living beings lay in the conditions of equilibrium: inorganic matter was "matter in a state of steady equilibrium" whereas organic matter was in a state of "completely unstable equilibrium." With regard to great amounts of energy which could be delivered by a little triggering action, he conceded that there was no quantitative relationship between the former and the latter. At the same time he thought that the latter could "not decrease under a given threshold," and the threshold could not be zero. If a zero value for energy was ineffective for triggering off a transformation of energy from potential to actual, "no guiding principle of immaterial nature" could steer "a material point on the top of a knoll." However he specified that his seventh query did "not necessary entail the rejection of free will or its representation as a deception": every query involving free will could rather be looked upon as a *transcendental* riddle in the Kantian sense. In the end, no specific solution was offered, and he let us believe that even his mechanistic meta-theoretical option was *transcendental.* The last word, "*Dubitemus*," seems more the acknowledgement of an intellectual stalemate than an actual answer to very demanding questions [Du Bois Reymond 1880, pp. 89-92].

Boussinesq's mathematical and philosophical approach puzzled the physiologist Du Bois-Reymond no less than the mathematician Bertrand. Only the physicist Maxwell appeared more sympathetic towards it. However, it seems that Du Bois-Reymond managed to grasp the specific weakness of Boussinesq's research programme better than Bertrand. In unstable equilibrium, when the transition from ordinary to singular solutions took place, mathematics was silent and philosophy even too talkative, but physics imposed a definite answer: a negligible but non-zero amount of energy was in order. However, the link between mental activity and its physical effects appeared more problematic. A nervous impulse carried energy and could trigger off a much greater amount of energy, but it was doubtful whether the purely mental act that prompted the nervous impulse did really require energy. As Maxwell had pointed out, it was the connection between mind and body that could probably find a suitable representation in Boussinesq's mathematical and philosophical interpretation of singular solutions.

Cournot, Maxwell, Stewart, and Boussinesq's remarks have been interpreted as a common "concern about the irreducibility of life and the mind to physics" but also as "reactions to the law of conservation of energy" [van Strien 2014c, pp. 1 and 3-4]. The fact is that the four scholars belonged to different generations and had different agenda. Moreover, apart from Boussinesq, they did not put forward a systematic research programme on the problematic link among the determinism of physical law, the emergence of life, and the practice of free will. In some cases, for instance Maxwell and Stewart, we find scattered and sometimes informal remarks. Quite different appears the case of Boussinesq: he consciously undertook a third way between vitalism and mechanical reductionism. Rather than some kind of dualism I see a sophisticated and unified approach, where mechanical determinism on the one hand and the emergence of life and free will on the other represented the opposite poles of a wide set of mathematical structures. If something may be found in common among the four scholars from Cournot to Boussinesq, it is a general commitment to unification. They explored the possibility of a wide-scope picture, where different processes, from the physical to the historical, could find place. Sharp statements such as affirming that Maxwell and Boussinesq "regarded an antimaterialist and dualistic metaphysics as essential for free will," or "Cournot and Boussinesq can both be counted as vitalists," or "they had a dualistic conception of life," do not help us to understand the critical commitment of those scholars [van Strien 2014c, pp. 8, 11, 16, and 18-19].[8] Simplified philosophical frameworks, such as the already mentioned opposition between materialism and spiritualism, are too naïve and misleading. In different ways, those scholars ventured to approach the complexity of the natural world, the complexity of scientific practice and its history, and the complex interactions between science and philosophy.

In the context of Boussinesq's researches, both determinism and indeterminism played an important role in the natural world, even though he preferred not to have recourse to the word *indeterminism*. He put mechanics, or deterministic mechanics, on the one hand, and life and free will, rather than indeterminism, on the other. Determinism corresponded to predictable and stable trajectories: physical stability had its mathematical counterpart in ordinary solutions of differential equations. Life and free will corresponded to mechanically unpredictable and unstable trajectories: physical instability corresponded to singular solutions.[9] In the case of life and free will, the correspondence was formal or structural: from 1878 onwards, he never spoke of or hinted at anything like the equations of living processes or free actions. The correspondence consisted of a structural analogy which was not based on specific material similarities but on wide-scope mathematical structures. He was not interested in defining what determinism was: determinism corresponded to ordinary mechanics, and ordinary mechanics corresponded to ordinary solutions of differential equations. Boussinesq's research programme realised an integration among different traditions of research. A set of different problems that emerged from mathematics, physics,

8 With regard to Poisson, Duhamel, Boussinesq and Bertrand's approaches to "indeterministic systems" [van Strien 2014a, pp. 167 and 170], I must stress that Poisson, Duhamel, Boussinesq and Bertrand did not share the same view on determinism. Moreover, with regard to Poisson and Duhamel, who explored the mathematical side of what we nowadays call determinism, it is worth stressing that in no way were they interested in determinism. On Poisson and Duhamel's supposed commitment to indeterminism see also van Strien 2014c, p. 13. For a synthetic reconstruction of those debates, see Bordoni 2015b, pp. 29-32 and 34.

9 According to Boussinesq, the non-uniqueness of solutions allowed two complementary domains to emerge: the domain of determinism, which dealt with ordinary solutions, and the domain of life and free will, which dealt with singular solutions. A complementary representation of the natural world also emerged, where both dead and living beings found room. In this context, I found questionable that "for these authors, … whether or not there was determinism in physical reality did not necessarily depend on whether the equations of physics had unique solutions." At least for Boussinesq, indeterminism did indeed depend on such non-uniqueness, but depended in a way that is different from that expected by twenty-first-century philosophers of science [van Strien 2014a, pp. 168 and 179]..

physiology, and philosophy found an original synthesis and a provisional equilibrium.

3 A Tentative Dialogue between Philosophers and Scientists

Structural analogies between physical and biological processes, the questionable link between mathematical physics and free will, and determinism in general attracted the attention of philosophers after having raised some debates in the scientific context. Philosophers made use of words and concepts very different from the words and concepts belonging to the tradition of mathematics and mathematical physics. Boussinesq, Barré de Saint-Venant, and Bertrand had spoken of singular solutions of differential equations, bifurcations and determinism starting from a specific field of mathematics, and making reference to specific physical phenomena. The reference to explosive chemical processes, to complex biological processes, and even more complex psychological processes rested upon a mathematical analogy. Philosophers occasionally made reference to specific natural processes, and widened the scope of the debate by focusing on very general themes. In general, it does not seem that those scientific debates managed to open new perspectives in philosophy, even though sometimes mathematical and scientific issues became instrumental in criticising or upholding traditional philosophical theses. Different attitudes towards science emerged, but two main attitudes can be singled out. On the one hand we find philosophers who were acquainted with recent developments in science and acknowledged the philosophical meaningfulness of specific contents and theories. On the other hand we find philosophers who underrated the scope of scientific enterprise and expected that no scientific achievement or research could be philosophically meaningful.

In reality, in the second half of the nineteenth century philosophical debates on determinism had already flourished. In 1872 a French philosopher with interests in politics and sociology, Alfred Fouillée, had published the book *La liberté et le déterminisme* with the explicit aim of a conciliation between them. He found that "the method of conciliation" was better than the method of refutation, in the same way as "liberalism in the social context" was better than having recourse to repressive means. No philosophical system could encompass the whole truth, which did not stem from the exclusion but rather the inclusion of different, even opposite, trends. According to Fouillée, "the present state of science" was not consistent with a radical determinism: the net of causes and effects, which was one of the hallmarks of scientific practice, called into play the problematic link between reason and experience, or more specifically the expectations of reason, and the autonomous development of natural events [Fouillée 1872, pp. v-vi, 5-6, and 8-10].[10]

He focused on the different meanings of the word *determinism*, and on the difference between a coarse and a sophisticated determinism. A sound determinism was consistent with the achievements of contemporary science, but a blind determinism was not so different from "the lazy sophism of oriental fatalism." The latter led human beings to believe that "phenomena occur despite causes," whereas the former implied that phenomena occur "according to causes." In the sophisticated version of determinism, the subject and his/her mind entered the scene, and they played an important role in the chain of rational connections between causes and effects. In other words, the intervention of the subject, or even her/his thoughts, could change the development of events. For instance, the awareness of the existence of the blood flow in the human body did not directly affect the flow itself, but beliefs "in the necessity of defending us from danger" were among "the causes of the motion of our limbs." There was an interaction between consciousness and free will on the one hand and material processes on the other: that interaction assured that no incompatibility could exist between the freedom of human beings and the necessity of natural laws. In a certain sense, the compatibility had to be assumed from the outset because determinism was just the result of that interaction [Fouillée 1872, pp. 11-12].[11]

In 1878, Charles Renouvier, a renowned philosopher who had never held an academic position, sharply criticised Balfour Stewart. In his paper, "Des notions de matière et de force dans les sciences de la nature," he made reference to a French translation of Stewart's 1873 booklet, and his criticism left no room for appeal. He focused on the transformations of energy, in particular the application of "the principle of living forces" — namely the principle of conservation of mechanical energy — to biological and physiological processes, where he found that "some vague, indefinite forces, which cannot be measured" were at stake. Renouvier had nothing to say about the quantitative conservation and transformations of mechanical energy "that was able to pro-

10 In 1872 Fouillée became maitre de conférences at the École Normale Supérieure in Paris. In the same year, the publication of the book on freedom and determinism and another book on Plato's philosophy allowed him to obtain a PhD in philosophy.

11 Benrubi qualified Fouillée as an eclectic who, despite "his very definite idealistic attitude," aimed at "carrying on Leibniz's great work of reconciliation." More specifically he attempted to bring into harmony idealism and positivism, science and philosophy, and free will and determinism. His opposition to empiricist positivism stood beside his eclectic and conciliatory attitude [Benrubi 1926, p. 148; Benrubi 1933, pp. 610-11].

duce work," but he was firmly convinced that every application of the principle to other domains was useless. He firmly stated that "an obscure metaphysics superposed to physics" could not be accepted. He did not enter into details but confined himself to blaming Stewart for having confused "the real object of scientific researches" with indefinite concepts. In reality, Renouvier denied the possibility of a fruitful cross-fertilisation between scientific concepts and procedures on the one hand, and the investigation of mind on the other [Renouvier 1878, pp. 168-70].[12]

In 1879 the Swiss philosopher Ernest Naville published the paper "La physique et la morale" in the journal *Revue Philosophique de la France et de l'Étranger*. In the first three passages of the paper we find many occurrences of words like "la pensée," "la morale," "l'ordre spirituel," "faits spirituels," phénomènes spirituels," and "phénomènes psychiques." In spite of this spiritualistic-oriented language, he was seriously interested in recent scientific developments: he focused on science, and pointed out an essential tension in recent scientific practice.[13] On the one hand, he saw the emergence of deep connections among different sciences and the explicit acknowledgment of those connections, in spite of the process of professionalization and specialisation that had taken place in the second half of the century. He mentioned the close links between physics and physiology, and between physiology and psychology. On the other hand, natural philosophers and scientists had continuously to confront the gap between "material facts as experienced by the senses, and mental facts as experienced by the mind." He undertook a conciliatory pathway, where a mutual influence between matter and mind could not be excluded, and any kind of sharp reductionism was rejected. Mind and thoughts could not be imagined as the results of mechanical processes, even though "molecular motions, or waves" could offer the condition of thinking. In particular, he found unacceptable the radical reductionism of the British

12 After having entered the *École Polytechnique* Renouvier was influenced by Comte's philosophy, and subsequently went back to Kant. His criticism was more radical than Kant's: philosophy had to be replaced by mere critique, and no *thing in itself* could be found beyond phenomena [Benrubi 1933, pp. 298-303]. In 1872 he had founded the influential journal *La critique philosophique*. He never held an academic position but was looked upon as an authoritative philosopher by his contemporaries.

13 Naville was Professor of Philosophy at the University of Geneva, and theologian and minister of the Evangelical Protestant Church. He aimed at the reconciliation between science and religion, and in 1883 he published *La physique moderne: études historiques et philosophiques* (in 1884 the book was translated into English). This book will be analysed in the following chapter.

philosopher Herbert Spencer. Although human perceptions could be looked upon as the effect of a chain of mechanical processes, in no way could the act of thinking and the content of thought automatically be reduced to matter and motion [Naville 1879, pp. 265-7].[14]

He found that a radical answer to the most radical determinism had already been given by Renouvier: physical principles of conservation might not be universally applied to all kinds of processes. Nevertheless, Naville was unsatisfied with this perspective, and was to take another pathway. He did not find good reasons to give up the principle of conservation of energy: he was rather more interested in showing that the existence of specific principles of conservation did not collide with "the commitment to defend moral freedom." In the context of "a positive and cautious" scientific practice, it could be assumed that both "physical laws, and specific laws for living beings" were at stake in life processes. He reported Bernard's conception of a specific "living force" which did not oppose ordinary physical laws: he stressed that the renowned physician had specifically spoken of "a *legislative* living force, which was not *executive*." In other words, living beings could be looked upon as the seat of specific actions, but those actions stemmed from a *driving* rather than *creative* power. Those "plastic forces" were not able to oppose the physical laws of conservation, because they acted at a different level. His philosophical perspective stemmed from a different ground and a different agenda but it was actually in tune with what Bernard had repeatedly stated [Naville 1879, pp. 273-7 and 281-2].[15]

In the end, Naville found "no conflict between physics and ethics." Human beings could not create energy [force], but could make use of the amount of energy at their disposal in accordance with the laws of physics. Human beings could freely decide how their energies were to be spent: they could choose "for good or for evil." Ethical principles, or the guiding principles which acted in accordance with free will, were placed "outside the domain of mechanics." He specified that the laws of mechanics could be applied to everything, but they

14 He insisted on the problematic link between the physical bases of life and the activity of the mind [*Ibidem*, pp. 272-3]. He also mentioned and widely quoted from the French edition of Spencer's *The First Principles*, which had first been published in 1862.

15 He mentioned Claude Bernard's *Leçons sur les phénomènes de la vie communs aux animaux et aux végétaux*, and *Rapport sur le progrès et la marche de la physiologie générale*. The former had been published in the same year (1879), after Bernard's death, whereas the latter had been published in 1867. In the scientific context, the word *force* and the principle of conservation of *force* had already been replaced by the word *energy*, and by the principle of conservation of *energy*.

did not explain everything. Nevertheless the existence of some kind of logical or philosophical link between mechanical effects and free actions deserved to be explored. According to Naville, his conciliatory option could save both the determinism of physical laws and the freedom of moral judgment. In the last passage of his paper, he claimed that his perspective was not so different from Boussinesq's, who had arrived at similar conclusions "by means of mathematical considerations" [Naville 1879, pp. 284-6].

Two years later, the polymath Delboeuf published a paper on the same subject in the same journal *Revue Philosophique de la France et de l'Étranger*. The paper, "Determinisme et liberté — la liberté démontrée par la mécanique," appeared in three different parts, in two subsequent issues of the journal. Probably because of his multiple training as a philologist, mathematician, and psychologist, he was able to grasp both the core of Boussinesq's research programme and its scientific and philosophical frailty. On the borderline between science and philosophy he attempted to cope with the problem of freedom, which he found at the same time "fascinating and discouraging." From the outset he pointed out certain paradoxes which stemmed from the alleged opposition between freedom and determinism. He found that, on the track of a sharp determinism, a determinist who upheld his thesis was "nothing else but a puppet in the hands of fate." In other words, a coherent determinist should accept that her/his determinism was not a free belief, but a consequence of determinism itself. If determinists applied determinism to themselves, they would fall either into a contradiction or a vicious circle. On the other hand, according to Delboeuf, a naïve conception of freedom would be pointless, since it would allow ourselves to subvert every scientific procedure, and would transform axioms "into a mere illusion" [Delboeuf 1882a, pp. 453-5].

Delboeuf was aware of the debate on determinism which had emerged from the scientific and mathematical context: he mentioned Cournot, Boussinesq, Janet, Saint-Venant, Bertrand, and the German physiologist Emile Du Bois-Reymond. In particular, he found that Boussinesq's essay on the mathematical roots of indeterminism was "rich in scholarly formulae and clever metaphysical remarks," but was "quite confused." He qualified Boussinesq's guiding principle as a "*deus ex machina*," and criticised the mathematician for having looked upon specific mathematical abstractions as facts.[16] Moreover he found unconvincing that a material sphere in unstable equilibrium could be put in motion without any force: however small or negligible it might be, only a force

16 This is not completely correct, but probably Delboeuf made reference to Boussinesq's firm confidence in his structural analogy.

rather than a guiding principle could put it in motion. The concept of freedom could not be associated with mechanical processes: a sphere could fall in equally probable directions, because those directions were "*totally independent* of each other." From the mathematical point of view, there was a perfect symmetry; from the ethical point of view, if ethics made sense in this context, there was a sort of *indifference*. On the contrary, in human experience freedom called into play something more complex: when freedom could really be practised, it transformed the landscape of facts and feelings. In the end, Delboeuf did not trust in the *ingenious* attempts Cournot, Saint-Venant and Boussinesq had put forward, because he considered those mathematical and philosophical solutions as "artificial and unworkable" [Delboeuf 1882a, pp. 467, 469-70, 475, and 477-8].[17]

In the second part of his paper Delboeuf tackled the conceptual link between freedom and time: freedom depended on the possibility of making use of time. If a person could delay some actions, he/she would be free, and any forecast would become impossible. The explosion of a gunpowder box produced a given amount of energy, but a person could choose whether to burn it today or tomorrow. The effects of the choice between two different times might be quite different: for instance, "a useful task today, and the death of centuries of people tomorrow." The same thing happened in the case of a gun trigger. According to Delboeuf, the freedom to decide when a given action should be performed could really overcome Laplace and Du Bois-Reymond's determinism [Delboeuf 1882a, pp. 613-16 and 623].

He addressed specific physical contents with a certain degree of competence, but the flow of thoughts and words followed a tangled pathway. He wandered through a crowd of instances, which encompassed physics, astronomy, and ordinary life: a stone which broke away from a mountain and then rolled, leapt, and struck various hurdles, or a little ball in motion in the reference frame of a moving ship, or an asteroid in motion around the sun, ... In the end, he insisted on the specific feature of intentional motions in living beings, which he identified with the key-word *discontinuity*. As an instance of that kind of discontinuity he mentioned the sudden and intentional attack of a wild animal on its prey. In conclusion, a specific "kind of power" was at stake in non-deterministic processes. That power was involved in the volition of living

17 He criticised both Boussinesq and Cournot. See, in particular, Delboeuf 1882a, p. 478: "L'être libre n'est pas dans un monde à lui, il est dans le monde, et, au moment où sa liberté se déploie, il donne une physionomie nouvelle à la scène qui s'y jouait." In short, he thought that the domain of deterministic processes and the domain of free actions could not be separated.

beings, and it was different from the "initial forces that had triggered off and maintained the motion of the universe." In short, he found that Boussinesq' guiding principle could probably be applied to intentional actions rather than physical processes or biological processes in general [Delboeuf 1882a, pp. 626-9 and 632].[18]

In the same year, in the same journal and in the same volume, Fouillée published a paper on the same subject, "Les nouveaux expédients en faveur du libre arbitre. Expédients logiques et mécaniques." He actually criticised Naville and "the learned mathematician and psychologist Delboeuf," and reported Bertrand's criticism of Boussinesq, but his remarks about physics were quite vague. He also criticised Saint-Venant's interpretation of explosive processes: he stressed that an increasingly vanishing [aussi petite qu'on veut] force was different from *no* force at all. He analysed the concept of guiding principle and concluded that it led to improbable consequences. He found that the principle was in contrast with the principle of action and reaction: as a consequence, "the conservation of centre-of-gravity motion," the conservation of linear momentum, and "the principle of areas" (the principle of conservation of angular momentum) were in danger. At the same time, in Fouillée's reconstruction, the gun trigger was represented as a suitable implementation of Boussinesq's bifurcations, but the network of remarks does not seem so consistent. In this context he mentioned "the possibility of a *clinamen*," which he attributed to Epicure and Descartes: the *clinamen* violated the principle of conservation of energy, and allowed men to create "motive force." This reference seems even more misleading: it is at least doubtful whether Boussinesq's guiding principle had anything to do with Epicure's *clinamen* [Fouillée 1882, pp. 585, 600-2, 604, and 608-9].

In the next issue (January-June 1883) of the same philosophical journal Fouillée published a three-page *Note* which followed another three-page *Note* by the French engineer and historian of mathematics Paul Tannery. The two *Notes* were published under the common title of "Le libre arbitre et le temps," and were specifically devoted to criticising Boussinesq, Naville and Delboeuf. Although the content of Tannery's text was in general agreement with Fouillée's theses, his language and his general style appeared as definitely less radical. Tannery acknowledged that, in accordance with the principle of causality as it was "intended nowadays," determinism could be considered as "its preferred logical consequence." He stated that he was "in perfect agreement

18 He ventured to label "theorem" or "axiom" his definition of determinism, and "corollary" the intrinsic link between determinism and continuity [*Ibidem*, pp. 627 and 630].

with Fouillée," and was satisfied with "Fouillée's conception of free will," but he found that future states of natural processes "could not be exactly foreseen" because they were not the consequence "of a mere mechanism" [Tannery 1883, p. 85]. On the contrary, Fouillée found that "logic, physics, and psychology" imposed strict constraints on mathematical models, and the evolution of the natural world was more deterministic than Tannery was willing to acknowledge. To explore the boundaries of mechanics in order to account for the existence of free will appeared to him as pointless, because mechanics was "the domain of the utmost determinism and *passivity*." Neither bifurcations nor time could be called into play, and Delboeuf's remarks on the possibility of time delays were nothing other than an act of faith in "a sort of mechanical miracle" [Fouillée 1883, pp. 86-8].

4 Philosophy Took the Lead

In 1883 Renouvier went back to the question of free will, and criticised the theses Fouillée had put forward the year before. He put forward a more radical approach which excluded a fruitful dialogue between science and philosophy. He mentioned a list of scientists and philosophers who had attempted to "explain the possible existence of a certain degree of indetermination in the natural order." He put scientists and philosophers together but this must not deceive us: with regard to some specific issues such as free will, he found that only philosophers were allowed to speak. With regard to "some important questions," which had been debated for "two thousand and five hundred years," and still represented meaningful philosophical issues, he claimed that science really did "*know nothing*." According to Renouvier, determinism was a typical philosophical issue that must not be tainted by scientific procedures or mathematical models. Determinism did not deal with the domain of actual experiences: it could not be tested by experience, nor could the conception of determinism be changed by a series of better observations. Some facts or events could be submitted to experiments and computations, but other entities could not, because they were not "observable facts" but general laws "encompassing the totality of facts of a given kind." In brief, determinism was a very general, regulative principle rather than a specific, scientific principle [Renouivier 1883a, pp. 371-2 and 375].[19]

19 It is not easy to follow Renouvier's line of reasoning because of the fragmented and branched structure of his texts, and the exceedingly argumentative style: sometimes his criticism blurred into mockery.

In a subsequent paper, "Les objections de M. Fouillée contre la conciliation du libre arbitre avec les lois du mouvement," Renouvier focused on a more specific issue, namely the breakdown of equilibrium in explosive processes. He conceded that those processes offered a mechanical representation of "the sudden transformations of energy that actually take place in living beings." The question was whether those discontinuous transformations could be triggered by a merely mental action, without any expenditure of mechanical energy. He leant towards a positive answer. Despite the incommensurability between "the mental force" and the transformation of mechanical forces that followed the breakdown of the equilibrium, a "*law of correspondence*" between them could be assumed. Nevertheless, that correspondence remained quite mysterious because the connections among natural processes really were also quite mysterious. Not only were "the fundamental natural laws unexplainable" but also the relation of causality was unexplainable. In the end, a sceptical point of view emerged. He did not claim that mental acts did not require any expenditure of physical energy; he confined himself to "showing that it might be so." He found that the assumption of a mere *possibility* did not distance him from Cournot, Saint-Venant and Boussinesq, but the relationship between his mental force and Boussinesq's guiding principle was not further clarified [Renouvier 1883b, pp. 389, 391, 396, and 398-400].

Although the debate among philosophers was substantially inconclusive, the theses of French philosophers offered a reference frame for a debate that overstepped the boundaries of French-speaking countries. In 1884 the American philosopher and psychologist William James addressed the students of the Harvard Divinity School delivering a lecture on "The Dilemma of Determinism." From the outset he credited Renouvier, Fouillée, and Delboeuf with having "completely changed and refreshed … the form of all the old disputes." He stressed the pliability of the world towards our attempts at interpretation. The world had "shown itself, to a great extent, plastic to this demand of ours for rationality," and that pliability could bridge the gap between the ontological and epistemic level. At the same time he underlined the mythological nature of every belief, rational beliefs included. Both scientific and philosophical ideals were nothing else but "altars to unknown gods." He took care to separate the "old-fashioned determinism," which he labelled "*hard* determinism," from the recent "*soft* determinism," which repudiated "fatality, necessity, and even predetermination." The latter could easily be identified with "true freedom." James' conclusion was essentially conciliatory but also surprisingly detached, as if the intellectual trend had already faded away. In the end, a pragmatic scepticism emerged [James 1884, pp. 145, 147, and 149].

The debate continued in the philosophical journals but the specific issues raised by the debate between Boussinesq and Bertrand slowly faded into the background. In 1885 Renouvier published a critical analysis of the history of philosophy, and devoted some pages to the relationship between science and philosophy. That relationship deserved "specific attention" because sciences "had always had the honour of suggesting new metaphysical pathways." However, he cast doubt on the possibility that something like "*the* science" and a single scientific method could really exist: he saw a plurality of specific sciences endowed with their specific methods. When he focused on physics, he noted that from Descartes onwards, "the universal mechanism," namely the mechanical world-view, had been identified with "absolute determinism," and that determinism had been extended to the whole of the natural world. He credited Laplace with having imagined the physical world as something that "could be expressed *a priori* by an equation of rational mechanics." Renouvier did not rely on that possibility, and even more questionable appeared the possibility that "science could encompass the whole" of human experience. He stressed the hypothetical nature of scientific foundations, and the intrinsic uncertainty of scientific simplifications and generalisations: scientific theories could not be freed of "their uncertainties." For this reason, he found that scientists should abstain from "absolute and unverifiable statements going beyond any possible experience" [Renouvier 1885, pp. 286-8].[20]

In general he saw an unbridgeable gap between theory and experience, and in particular he questioned the empirical value of general principles such as the principle of conservation of energy. It belonged to "the domain of mathematical principles" rather than empirical laws. In brief, he saw science as a three-level enterprise, which consisted of three different stages: empirical, theoretical, and metaphysical. The last stage brought science closer to philosophy, and even closer to ancient natural philosophies and *mythological* traditions. The relationship between the theoretical and metaphysical levels was as problematic as the relationship between the empirical and theoretical levels. There was no automatic connection between specific physical theories and the world-views that could be drawn from them: any world-view represented a questionable extrapolation rather than a logical necessity. The most meaningful instance was the problematic link between "mechanical physics" and the mechanical world-view based on the reduction of "every kind of phenomenon in the universe" to matter and force. Renouvier found that mechanical physics

---

20 On Laplace's determinism, I make reference to my words in the second chapter: the alleged radicalism of Laplace's determinism was overestimated by Renouvier.

became *mechanism* in the sense of a metaphysical option when it claimed that not only did "perceptions correspond to motions," but also that perceptions were "motions in themselves." The metaphysical option that he had labelled "absolute determinism" was looked upon as not so different from "a mythological body of knowledge" [Renouvier 1885, pp. 289-91].

In 1889 the philosopher Henri Bergson published the book *Essai sur les données immédiates de la conscience*, where the gap between rational and empirical practices was at stake from the outset.[21] The dialogue between the two domains required "an unnatural translation from spatially extended objects to an entity without extension." In other words, the discontinuous structure of things had to match the continuous structure of thought. That problematic link also corresponded to the awkward and misleading connection between spatial extension and temporal duration. In Bergson's book, the dichotomy or essential tension between the two domains was also expressed by means of other linguistic and conceptual couples, as for instance "the extensive and the intensive," or "the external world and the internal states" of our consciousness, or "matter and mind," or "sequence and simultaneity." The fact that some kind of correspondence was "useful in ordinary life, and even necessary in scientific practice" did not soften that tension. In the end, Bergson found that the problematic link between rational and empirical practices, and therefore the problematic link between space and time, had given rise to a multiplicity of misunderstandings, in particular in debates on determinism. In reality, a correspondence between the intrinsic structure of space and the intrinsic structure of time could be put forward, but it could only involve past time. Currently flowing time could not be associated with a spatial structure. Freedom, or the free choice [l'acte libre], could only take place in the context of flowing time [Bergson 1889, pp. VII-VII, 168-70, and 172].

He found in Kant a serious mistake with regard to time: the supposed homogeneity of time led to the possibility of reproducing the same events. Here Bergson found the root of a strict determinism, and therefore the impossibility and incomprehensibility of freedom. Bergson's 1889 book put forward an

21 Bergson was really interested in the foundations of science. It is worth stressing that his interests in science and mathematics could be traced back to his first publication in the *Nouvelles Annales Mathématiques* in 1878, when he had not yet decided to undertake a philosophical career. After having studied at the *École Normale Supérieure* in Paris between 1878 and 1881, he taught in high schools in Angers and Clermont-Ferrand. The *Essai* corresponded to the doctoral dissertation he submitted in 1888. In 1897 he returned to the *École Normale* as a Professor of Philosophy, and in 1900 he was called to the *Collège de France*.

extremely rigid separation between the spatially structured world of external experience and the temporally structured world of internal experience: such a sharp separation involved the impossibility of that fruitful interaction between science and philosophy which had been wished for by Cournot and Boussinesq. The sharp separation between physics, where the same causes always produced "the same effects," and psychology, where a given cause never led to the same effect, seems more a parody of physics and psychology than a serious attempt to grasp the actual physical and psychological processes. The representation of science as the domain of absolute certainty, and the domain of consciousness as the domain of unpredictable events, allowed Bergson to find his solution for "the problem of freedom" but opened an unrealistic gulf between physical and psychological processes, which are both natural processes [Bergson 1889, pp. 153-4].[22]

In 1890 Naville published a book on free will, *Le libre arbitre. Études philosophiques*, but only the third chapter was devoted to determinism: we find explicit references to Bayle, Spinoza, Condillac, Spencer, Schopenauer, Fouillée, and Hegel's philosophical theses, but no reference to the heated debate triggered by mathematicians appears. The processes of professionalisation and specialisation made the pursuit of a cross-fertilisation between different bodies of knowledge ever more difficult, and discouraged further attempts in this direction.

The debate on determinism between mathematicians was also a debate on the rate of confidence in mathematics and its explanatory power. Boussinesq believed that in the wide range of mathematical procedures, different kinds of natural processes could find a suitable representation or, at least, a broad structural analogy. Singular integrals and bifurcations of mathematical solutions could be put in correspondence with physical instabilities and Bernard's guiding principles in living beings. Unlike the mathematician Bertrand and the physiologist Du Bois-Reymond, who preferred a quiet stalemate to a problematic theoretical sketch, the mathematical physicist Maxwell appreciated Boussinesq's attempt to address complex and elusive problems. In brief, some mathematicians and scientists were worried by the complexity of certain natural processes, whereas others praised a mathematical approach to complexity.

---

22 As Benrubi wrote in the 1920s, "Bergson's philosophy is not to be pigeon-holed under the label of any existing *ism*"; he opposed dogmatic positivism but leant towards an equally dogmatic cult of freedom. His critical attitude did not manage to clarify the complexity of scientific practice, and the complexity of its relationship with philosophy [Benrubi 1926, pp. 170-2 and 175].

Philosophers were attracted by those attempts and those debates, but in general they did not manage to grasp the meta-theoretical aim, namely the new confidence in the possibility of a mathematical inquiry into complex natural processes. Paradoxically enough, it was a philosopher and theologian, Naville, who understood the importance of Bernard's guiding principles and Boussinesq's mathematical bifurcations for the enduring debate on free will. Naville found that scientific researches and hypotheses could help renew the search for the compatibility between the relative determinism of natural laws and the relative freedom of moral judgements. In the late 1880s Bergson emphasised the distance between the *spatial* domain of scientific enterprise and the *temporal* domain of philosophical practice: he saw determinism on the one side, and indeterminism on the other. This simplified account of science and philosophy led to the thesis of an alleged incompatibility. Fortunately, more refined analyses of scientific problems and foundations led to more interesting outcomes, as I will show in the following chapter.

CHAPTER 6

Naïve versus Sophisticated Meta-theoretical Frameworks

1 A Sophisticated Historical and Epistemological Framework

Apart from specific issues such as reductionism and determinism, the intellectual landscape of French-speaking countries in the 1880s deserves to be further analysed. The landscape included philosophers who were deeply interested in the latest developments of science, and were also interested in putting forward a sophisticated philosophical approach to science. In 1883 a Parisian publisher sent a book by the Swiss Naville to the printing press. The book dealt explicitly with physics and its history: the title, *La physique moderne: études historiques et philosophiques*, pointed out a historical and critical reconstruction. In the first lines of the short *Avant-propos*, the author expressed his main concern: although physics had fast developed in recent times, and had increased "the power of man on nature," both the foundation of recent theories and their philosophical consequences had been neglected. We must stress that, at that stage of the history of science, when the professionalization of scientific practice was accomplished and the boundaries between different sciences began to be firmly established, a reliable definition of *physics* could really be put forward. Naville separated physics "in its strict sense" from a broader field of physical sciences, which included mineralogy and some sections of geology. Making reference to recent debates on the classification of sciences, he found questionable whether chemistry should be included into physics or should enjoy an autonomous status. Obviously, the study of "living beings was excluded": the field of physics started from its boundaries with "logic and mathematics" and ended at the dividing line with biology. He stressed the suitability of the expression "modern physics," because every science had its own history, where different stages could be identified. The adjective *modern* had obviously a relative meaning, because what appeared modern in a given time would have become old in a subsequent stage of history [Naville 1883, pp. 1 and 3-4].[1]

1 Naville informed the readers that the first part of the book had already been published in 1872. Although Naville's philosophy was qualified as "spiritualist positivism" or simply *spiritualism*, he attempted to combine physical determinism with the existence and practice of freedom [Benrubi 1933, pp. 566 and 568-9].

 | DOI 10.1163/9789004315235_008

Naville found that the foundations of physics could be approached from three different points of view, namely "scientific, logical, and aesthetic." From the scientific point of view, with regard to "the intrinsic nature of phenomena," he found five specific differences between modern and ancient physics. The first essential feature of modern physics was "the mechanical nature of phenomena," namely their reduction to matter in motion. That reduction required a series of additional hypotheses: among them, the existence of "*molecular* or *atomic* motions" and "*aethereal* motions" besides the ordinary macroscopic *mechanical* motions. In no way could the existence of aether stem from "direct perceptions": it was a hypothesis, and it was justified just by its fruitful explicative power in the context of optics and the theory of heat. The second feature was the unitary nature of all kinds of matter: the great variety of natural substances could be looked upon as a combination of a definite number of simple elements. This second feature was consistent with the first one: in some way it specified what kind of matter was in motion, and allowed physicists to overcome the traditional distinction among solid, liquid, and gaseous states [Naville 1883, pp. 5-6 and 8-11].[2]

The third feature was labelled "the *transformation of motions*": Naville preferred his label to the most widespread expressions "*correlation of forces*" or "*transformation of forces*." He insisted on the concept of motion rather than force because the actual experimental determination of force corresponded to nothing other than "a real or virtual motion." It is worth stressing that he identified the foundation of physics with its mechanical foundations, which began to be questioned exactly at this time. The fourth feature specified that transformation required conservation, and in particular the conservation of both matter and motion: the two principles of conservation mirrored a sort of symmetry between the two entities. When visible motion disappeared, it was transformed from the macroscopic into the molecular or aethereal kind. In accordance with the above-mentioned identification, Naville interpreted energy as "the cause of real or virtual motions": he interpreted the two expressions *actual energy* or "living force," and *potential energy* in this perspective. The last essential feature was in reality a meta-theoretical option, namely the possibility and necessity of a mathematical explanation of natural phenomena. Although mentioned as the last feature, it affected the other four: mathematics

2 On the states of aggregation Naville wrote: "Pour la physique moderne, les états solides, liquides, gazeux sont considérés comme pouvant appartenir à tous les corps sans exception. La liquéfaction de l'oxygène récemment obtenue par MM. Raoul Pictet et Cailletet a confirmé cette manière de voir." [*Ibidem*, p. 11].

offered the suitable language for the whole body of knowledge [Naville 1883, pp. 11-12, 14-16, 18-19, and 21].

When Naville proceeded to explain the differences between modern and ancient physics, he stressed that the latter was "closer to direct experience" than the former. At the same time, in ancient physics he saw a lack of unity: for "every new fact … a new explicative principle" was required. On the contrary, the mechanical world-view corresponded to a really unifying trend in modern physics. In reality, Naville's assessment of ancient science appears as commonplace rather than the result of a careful analysis, because the commitment to unifying all fields of physical sciences was a permanent process in the history of science, which survived the great transformations experienced by that body of knowledge. However, he found that the fulfilment of the mechanical programme could not be completely accomplished because "the science of matter," unlike the science of motion, could not be reduced to mechanics. As a consequence of the broad definition of physics he had put forward in the first pages of his book, physics encompassed both mechanics and part of chemistry ("the science of matter"), and the latter could not be reduced to the former. The same difficulty affected cosmology, which was looked upon by Naville as "the most daring part of the complete programme of physics." The research programme of reduction of chemistry and cosmology to mechanics was far from being accomplished. In some sense, the wide extent of physical sciences made the programme, and therefore the mechanical world-view, quite difficult to put forward [Naville 1883, pp. 22-4 and 28].

In the section on the inertia of matter, Naville discussed and criticised one of the most meaningful concepts of modern science, which also represented a dramatic watershed between ancient and new science. He found that the concept of inertia could offer a unifying conceptual framework for the whole body of knowledge of modern physics. Inertia encompassed both the concepts of matter and force: in the presence of two bodies, each of them represented "a force with regard to the other," in the sense of a cause of the change of motion. Matter could not be defined "independently of its connections" or forces which linked it to other parts of matter. Inertia was nothing else but the force that opposed the displacement of matter itself, and as such it was "a real force." He quoted and endorsed Newton's passage "*Materiae vis insita est potentia resistendi*," where inertia was looked upon as the inner power of matter to resist other forces. However Naville found that the concept of inertia could not be derived by experience; it was a concept, and the principle of inertia was a theoretical law. It was a peculiar theoretical law indeed, because it eluded any actual experimental check. The impossibility of an empirical detection depended on three specific practical impossibilities. First, both in astronomic and microscopic fields, a real state of rest was impossible to ascertain; second, the existence of gravitation implied the unquestionable existence of accelerations; third, an isolated body did not exist. The principle of inertia could not be justified by "direct observation" but by the fruitfulness of some "results corroborated by experience" [Naville 1883, pp. 28, 32, and 35].[3]

When he took into account "the logical features" of modern physics, he started from the dynamical status of hypotheses: when a hypothesis received a certain degree of experimental confirmation, it could become a physical law endowed with a relative degree of certainty. The "degree of probability" of physical laws was a key concept in Naville's historical-critical reconstruction. Concepts like *degrees* and *probability* were introduced in order to exclude the interpretation of laws as "absolute laws revealing the eternal and necessary nature of things." The potential existence of specific facts in contradiction with a physical law required such a cautious attitude. In many pages he stressed "the experimental character of theories," and guarded against the faith in an absolute and necessary order. A consequence of that belief was a reductionist attitude, which had led some scholars to extend mechanism to social sciences. A dogmatic faith in physical theories could also lead to disregarding facts in contradiction with such theories, and could "destroy the foundations of moral order." With regard to the necessity of avoiding that "dangerous delusion" he mentioned and quoted some passages from Robert Mayer, Joseph Bertrand, and Claude Bernard [Naville 1883, pp. 41, 44-7, 50, and 52].[4]

It is worth remarking that at this stage Naville did not point out specific differences between laws and theories, and moreover he labelled as theory even the reductionist meta-theoretical attitude. It was only after having pointed out the necessity of an anti-dogmatic approach to science that he explored the

3 The status of the principle of inertia as a hypothesis, even though "a strongly confirmed hypothesis" rather than "a simple induction" from experience, was also stressed in a following section of the book [*Ibidem*, p. 43]. However, we find here an ambiguity that Naville and other critics overlooked: the principle of inertia could not be derived from experience and was rather in contradiction with common experience, but at the same time, it could lead to consequences that were in accordance with other experiences, at the end of a chain of deductions. Naville's analysis of the concept of inertia echoed some remarks Cournot had put forward in 1861 and 1872 [Cournot 1861, pp. 162-4, 179, and 181-3; Cournot 1872, pp. 269-70, 278-9, 282-3, and 285].

4 We find here another similarity with Cournot's epistemological attitude, namely a mild fallibilism or *probabilism* [Cournot 1851, vol 1, pp. 171-2; Cournot 1872, pp. 1-5; Cournot 1875, pp. 348-9, 354, and 359-60].

differences among laws, theories, and principles. Laws were identified with "experimental laws," which stemmed from facts and described them, whereas theories aimed at a more general explanation of facts and laws. In their turn, theories dealt with "the nature of phenomena," whereas principles were mental devices that led to the creation of theories. Apparently we find a definite hierarchy that went from the empirical to the theoretical level, and then to the meta-theoretical level. In reality, in Naville's sketch, the interdependence among experiments, theories, and principles was more complex. A physical law did not necessarily call for the formulation of a theory and was in some way independent of it: the existence and the validity of optical laws neither required a complete theory on the nature of light nor had led to the endorsement of the emissive theory rather than the wave-theory. During the profound transformations that had taken place in the history of science, the relative independence between laws and theories had allowed the scientific body of knowledge to survive: optical laws had survived "the subsequent hegemony of the two theories" [Naville 1883, pp. 52-3].

According to Naville, theories represented the pivotal stage in scientific practice: he claimed that "the elimination of theories would represent the elimination of science." He saw a complex relationship between laws and theories: not only could laws lead to the emergence of theories but laws could also cast doubt on theories or even overturn them. From the opposite side, theories could suggest new experiments, and also "lead to the discovery of new facts and new laws." In its turn, the building up of theories was influenced by guiding principles, even though sometimes the role played by principles was not "perceived consciously." The tacit presence of principles in scientific practice, and their persistence over time, were repeatedly stressed by Naville. Principles corresponded to meta-theoretical commitments or general ideals: a typical instance was the principle "of order or harmony." Laws and principles could enjoy a relative stability. With regard to laws, he remarked that Newton's mathematical law of gravitation had survived the debates on the nature of gravitation. With regard to principles, he pointed out that different kinds of scientific practice had in common the search for order, unity and harmony. In contrast with the stability of mathematical laws and the persistence of regulative principles, theories represented the most problematic component of scientific enterprise: the history of science showed a continuous "fluctuation of theories," namely their emergence, success, fall, and potential re-emergence [Naville 1883, pp. 53-4].

Theories occupied the "intermediate region" between the base of experimental laws and the top of the ideals or principles: that intermediate position was consistent with their nature of "changeable and provisional" entities. According to Naville, the history of science had shown the existence of "a continuous progress towards a deeper knowledge of the universal order," and theories were the most dynamical component in the pursuit of that progress. Physical laws corresponded to an empirical necessity, and guiding principles corresponded to a rational necessity, whereas theories could only rely on a problematic correspondence between the empirical and the rational domains [Naville E. 1883, pp. 55-4]. The dynamic process of emergence, development, and replacement of physical theories was the essential feature of scientific progress: it was just the caducity of theories that protected science from involution and decadence.

> Scientific theories flow but science remains. Scientific practice is the search for order and harmony, and therefore its main commitment is consistent with the emergence and fall of theories. The emergence of a theory shows the necessity of the human mind to find an order that accounts for facts. The fall of a theory stems from its intrinsic inadequacy, and invites scholars to search for a more satisfactory order. The fast replacement of theories by other theories might raise some doubts [about the reliability of scientific practice]: the confidence that present theories might escape the fate that has overturned many others would be a strange blunder. Nevertheless, the history of science shows us that the fall of a theory is followed by the emergence of a firmer and wider theoretical reference frame [Naville 1883, p. 55].

We find here a sophisticated conception of science as a dynamic body of knowledge rather than a collection of empirical procedures and rational truths. On the track of Cournot, whom Naville mentioned only occasionally, we find the conflation of a pliable and dynamic epistemology with an equally pliable and dynamic historiographical framework.

With regard to "the aesthetical features" of modern physics, Naville stressed that ancient science had been "expressed poetically": in their search for a universal harmony, ancient scholars naturally merged physics (in the ancient sense) with poetry. From the outset modern physics had been expressed in prose, and apparently it had destroyed the ancient poetry of Nature. In reality, when the old poetry had been destroyed another had been created: the rational order of laws disclosed a new kind of harmony and a new kind of poetry. He claimed that the great progress in science had been accompanied by an evolution in aesthetic sensitivity. At the same time he specified that in no way should scientific aesthetics be identified with the glorification of Nature or the glorification of science in itself. In other words, scientific practice did not have to be transformed into mythology or into some kind of natural theology, which he qualified as "an aesthetic mistake" [Naville 1883, pp. 57-63].[5]

In the fourth part of Naville's book, even ethics was at stake. He started from the fact that physiology had recently been influenced by physics, and psychology by physiology. A radically reductionist attitude had led to the belief that psychology could be reduced to physiology, and physiology to physics. Two potential risks had emerged, and had worried some scholars: the possibility that physics could endanger the foundations of "spiritual order," and the existence of an intrinsic incompatibility between physics and ethics. He found that a sharp reductionism was scientifically unnecessary, and no serious opposition between physics and ethics was actually at stake. His first step consisted in showing the inconsistency of the identification of human thought with some kind of matter in motion. The British philosopher Herbert Spencer had put forward a very general theory of mutual conversion among the different forces in Nature: in accordance with this intellectual framework, human thought stemmed from a specific transformation of mechanical energy, in the same way as sound, light and heat could be transformed into each other. Naville stressed that there was no evidence of such a direct transformation. The relationship between the domain of material transformations and the domain of thoughts appeared to him definitely more complex [Naville 1883, pp. 211-16].

2 On Determinism and Reductionism Once More

Naville's reconstruction of recent debates on the borderline between science and philosophy cast light on the close relationship between reductionism and determinism. Since "the existence and the manifestation" of mental processes required "the motion of matter" as necessary support, reductionists inferred that ethical and spiritual attitudes stemmed necessarily from "their material conditions." This reductionism automatically involved a radical determinism. The determinism that pertained to the material domain was automatically handed over to the moral domain: the deterministic laws of physics which ruled the material domain had to rule the moral domain. Within that framework, freedom was looked upon as an illusion because free will would have endangered the universal determinism of phenomena. Naville pointed out that,

5 According to Naville, in recent times the philosopher Auguste Comte, the physician Rudolf Virchow, the natural philosopher John Tyndall, and the authoritative physiologist and physicist Hermann von Helmholtz had leant towards various kinds of natural mythology. He quoted from their texts, and discussed some passages [Naville E. 1883, pp. 63-5].

in reality, freedom did not damage determinism. The hidden flaw in determinism was not represented by the actual practice of freedom, because it could not be demonstrated that an action was freely performed rather than performed under constraint. The most sensitive issue was "the idea of freedom": how could the existence of the concept of freedom be explained in a deterministic way? Which was the deterministic cause of the idea of freedom? The idea really did exist, and this was a matter of fact. Determinism could show that freedom was a delusive concept, but where did that concept stem from? Was it "an idea without cause," or a concept without a deterministic explanation? If ideas without cause had to be accepted, then why not "motions without cause"? According to Naville, determinism was tottering under its own weight [Naville 1883, pp. 223 and 227].

Naville's criticism of determinism then focused on a specific feature of the principle of conservation of energy: he found that the latter suffered from space- and time-indeterminism. Space-indeterminism depended on the fact that energy was a scalar quantity, which did not depend on "the direction of motion": if only the direction changed, kinetic energy did not change. Time-indeterminism was based on the fact that the conservation of energy did not depend on the duration of processes: transformations of energy in accordance with the principle could take place at "different times," and could require different time spans. Ethics and free will did not require the creation or destruction of energy but only the possibility of making use of energy where and when the subject found it useful or right. Unfortunately Naville did not offer further details: he confined himself to pointing out that the processes that preserve total energy were insensitive to space- and time-translation, as well as to time duration.[6] The practice of freedom did not change the amount of energy but only allowed the subject to decide the occurrence of physical processes through space and time [Naville 1883, pp. 228-9].[7]

Not only did human consciousness enjoy that intrinsic indeterminism, but so did biological processes in general. He mentioned Claude Bernard and his conception of a "living force" which corresponded to a specific "*legislative* power" in living beings. In other words, biological systems could *direct* but not *create* energy: they could "make different uses of physical motion although

6 It is doubtful whether Naville made reference to the insensitivity of kinetic energy to direction or he underestimated the fact that the change of direction in a given velocity required a force, and that such a force had to be provided by the subject or another physical system.

7 Naville acknowledged that Cournot had put forward similar remarks. We have already met this line of reasoning in Delboeuf, in the previous chapter [Delboeuf 1882a, pp. 613-6 and 623].

its amount does not change." It seems that Naville did not manage to disentangle carefully the concept of direction in a physical sense, namely direction in space, with the concept of direction in the sense of biological purposiveness. The fact that he let this ambiguity propagate throughout his text does not facilitate the comprehension of his line of reasoning. The consequence is that his concept of time-indeterminism appears consistent but the concept of space-indeterminism remains more problematic. However, at the end of the section he devoted to this subject only the role of time was in prominence. He concluded that "no opposition between physics and ethics" really took place: human beings could not create force, but they could "make use of the amount at their disposal" at a specific time: they could choose "for good or for evil."[8] He also mentioned the book Boussinesq had published in 1878, and stressed that the French mathematician had arrived at the same conclusions: according to Naville, this fact reinforced the reliability of those conclusions [Naville 1883, pp. 235-6, 239, and 241-2].[9]

In the last part of his book, which he devoted to "philosophical consequences of modern physics," a dynamic epistemology and a dynamic historiography were at stake once more. A correct relationship between science and philosophy required a careful analysis of scientific method and its history. Galileo had based his new science on a powerful alliance between reason and experience. If Bacon had stressed the role of experiments, Descartes and then Leibniz had emphasised the role of deductive procedures, and the history of physics appeared as a series of actions and reactions of meta-theoretical nature. The eighteenth century had seen "a reaction in favour of empiricism," then rationalism had gained its hegemony at the beginning of the nineteenth century, and afterwards positivism could be looked upon as a reaction of empiricism to the idealistic philosophy of Nature. What remained after "the violent oscillations" between "absolute rationalism" and "strict empiricism"? The key to the scientific method was nothing other than the continuous search for a dynamic balance between the two poles. He remarked that the influence of "specific sciences" on philosophy had become a matter of fact in the late nineteenth century: both theoretical developments and technical applications of science had become "prominent intellectual hallmarks of the time." At the same time he warned against the recent trend towards transforming the influence of science on philosophy into new philosophies based on physics or some

8 The same concept had already been expressed by Naville in 1879 [Naville 1879, pp. 284-6].

9 Naville made explicit reference to Boussinesq's structural analogy between singular solutions of differential equations, and bifurcations in biological processes.

other specific science that continuously oscillated between the two poles. He found that philosophy dealt with something more general and more universal, and no direct transition between the two domains was philosophically legitimate [Naville 1883, pp. 243-5, 247-50].

According to Naville, no field of intellectual activity really aimed at the unity of knowledge more than philosophy, but the search for unity clashed with the intrinsic duality between mind and matter. In the knowledge of matter, the mind expressed itself as a subject that could not be reduced to its object. The logical and physical dualism was really profound and could not be easily solved. Nevertheless, the gap could be spanned by a bridge, and the bridge was the existence of physical laws. Laws were rooted in the human subject rather than in physical phenomena, and mathematics played an important role in the process because it highlighted "the *a priori* nature of reason." At the same time mathematics gave prominence to the essential tension between reason and experience: the ideas that lay at the basis of mathematics could not be developed without the influence of experience, but their content did not deal with experience. According to Naville, physics was deeply affected by the tension "between facts and thoughts," but at the same time it was a meaningful instance of the possible harmony between them: only that harmony could make the world comprehensible. The existence of physics showed that some kind of correspondence between phenomena and laws of thought could really be pursued. In the end, Naville relied on a pragmatic solution. Since that correspondence was "the condition for the potential existence of science," and science really did exist, that existence showed that such a correspondence could fruitfully be pursued [Naville 1883, pp. 258, 263-5, and 267].

In the last pages of the book he pointed out what he considered the "legitimate consequences" of modern physics: among them, "the demolition of universal scepticism" and the demolition of materialism [Naville 1883, pp. 268-9]. The presumption that modern science had managed to dismantle permanently both scepticism and materialism was an evident overstatement, and this shows that his firm belief in a non-dogmatic approach to scientific practice contained slightly dogmatic nuances, even though his critical and historical analysis managed to grasp the complexity of scientific tradition.

It is worth remarking that another critical and historical analysis of physics was published in Paris the following year. In 1882 a German-born scholar, Johann Bernhard Stallo, had published a book in the United States on the foundations of physics. Although he was an outsider, the book enjoyed some success and was re-published in 1884. In the same year the Parisian publisher Alcan sent the French translation of a book to the printing press. The book was prefaced by the French chemist and mineralogist Charles Friedel, who drew the attention of readers only to one specific issue discussed in the book: the atomic constitution of matter, and the author's corresponding criticism. Side by side with the debate on atomism and the kinetic theory, it seems to me that Stallo's criticism about the concepts of inertia, gravitation, and elasticity also deserves to be briefly reported because there is an interesting similarity with what had emerged in France from Cournot to Naville. The fact that an American book had undergone an almost immediate translation shows that its content was in tune with some contemporary intellectual trends. I will confine myself to briefly outlining Stallo's position, and then Friedel's comments.[10]

Stallo focused on "the molecular or atomic constitution of bodies," and on what he labelled "the atomo-mechanical theory." In his reconstruction, that theory was based on four assumptions that had rarely been expressed in an explicit way: first, "*elementary units of mass*" did exist, second and third, those elementary units were "*hard and inelastic*" but also "*absolutely inert and therefore purely passive*," and fourth, potential energy could be reduced to kinetic energy, and any kind of force could be replaced by a system of masses and motions. Obviously, the theory corresponded to the most radical mechanism and reductionism, and it was not shared by the totality of physicist. Stallo's criticism was therefore addressed to the combination of radical mechanism with atomism.[11] He found that the concept of inertia as passive resistance to forces could not bear close scrutiny: a force could only be resisted by another force. Newton's definition of inertia as *vis insita* was in tune with the deep correlation between "*inertia* and *force*." The passive concept of inertia was linked to the concept of "isolated existence of a body," but bodies existed "solely in virtue of their relations," and in no way could gravitation be removed. He also found that the deep connection between inertia and force had not been satisfactorily explained by *dynamical* theories that had reduced atoms or molecules to "cen-

10 The book had something in common with the more famous book by the German-speaking physicist Ernst Mach, *Die Mechanik in ihrer Entwickelung historisch-kritisch dargestellt*, published in 1883, which was to be translated into French only in 1904. The title of Stallo's book, *The Concepts and Theories of Modern Physics*, was translated into French as *La matière et la physique moderne*. When compared with the original American edition, both the French title and Friedel's *Introduction* to the French translation stressed the sensitive issue of the atomic structure of matter.

11 Apart from Helmholtz and William Thomson's hypothesis of vortex-atom, Stallo mentioned lesser-known hypotheses on the nature of matter. The Italian astronomer and physicist Angelo Secchi had revived Louis Poinsot's (French mathematician) model of elastic collision undergone by inelastic atoms endowed with a rotational motion: rotation could transform an inelastic body into an elastic one [Stallo 1882, pp. 44-6].

tres of force," and looked upon force as the most fundamental entity [Stallo 1882, pp. 28-9, 39, 43-5, 68, 83, 152, 161-3, and 205].[12]

Even the elasticity of elementary atoms appeared questionable to Stallo because elasticity involved motion of parts, and the possibility of parts was inconsistent with the assumption of elementary units of mass. On the other hand, if molecules had been "wholly inelastic or imperfectly inelastic," inevitably their motions would have come to an end. Elasticity was also required by the principle of conservation of energy: inelasticity would have led to the dissipation of mechanical energy at the microscopic level. Moreover, the recently-developed kinetic theory of gases required some kind of elastic scattering between gaseous molecules, and/or elastic scattering between the molecules and the walls of the container in order to account for gas pressure. Another paradoxical conclusion emerged when scientists assumed that the essential features of atoms had to correspond to the features of ordinary matter. In this case, the impenetrability of atoms would have implied the impenetrability of ordinary matter: how could the diffusion of gases be explained? Furthermore, if the behaviour of ordinary matter had to be explained by the features of atoms, there was a vicious circle: atoms were endowed with the feature that they had to explain. When the elasticity of gases was explained by elastic atoms, the effect was transformed into a cause, and even worse, the cause of itself [Stallo 1882, pp. 31-2, 34-5, 40-2, and119-22].

In the *Introduction* to the French translation, Friedel focused on the status of the atomic theory of matter, but the specific subject matter gave him the opportunity to put forward wide-scope methodological remarks. He found that a fruitful and widely accepted scientific theory should be submitted to a careful scrutiny after a distinguished service "for the advancement of science." He considered it worthwhile to criticize that theory "in its foundations," but he was aware that neither a serious objection, nor a whole set of objections, might "completely destroy a reliable scientific conception." It was far more probable that a theory had to be more or less deeply modified. In other words, a questionable theory was better than no theory at all. Chemists were aware of the flaws of the atomic theory, and some of them were waiting for "a mechanical explanation" of the processes that had been labelled "*atomicity* or valence of atoms." Nevertheless they should have continued to make use of a theory that

12 Remarks on the interdependence between matter and force had already been made by Helmholtz and Maxwell. Yehuda Elkana pointed out Kant's influence on Helmholtz, and Peter H. Harman pointed out the influence of Whewell on Maxwell's *dynamical* conception of matter [Helmholtz 1847 (1889), p. 5; Maxwell 1878, pp. 166-8; Elkana. 1974, pp. 167; Harman 1998, pp. 28-36 and 190-4].

had helped them to collect "a great number of facts," and had led them "to discover new facts every day," even had the objections been more convincing than they actually were [Friedel 1884, pp. VII-IX and XI-XII].

With regard to the structure of scientific theories, Friedel endorsed a simplified version of the meta-theoretical analysis Naville had put forward the previous year. The starting point of every scientific enterprise could be found in "observation and experience," and a simple hierarchy of entities stemmed from that basis of perceptions. Immediately above he placed scientific laws, which allowed scientists "to collect an increasing number" of those observations and experiences by means of "a mathematical connection." At the top of the hierarchy he saw "simple and general principles." The intellectual path that led from the empirical basis to the first principles involved the transition from a higher to a lower degree of certainty. In other words, scientists could rely on their observations much more than on general principles. In the transition from experiences and experiments to principles, the process of increasing abstraction was intrinsically linked to the increasing "risk of being misled." In his remarks, the chemist Friedel was definitely less subtle and detailed than the philosopher Naville, but he showed a remarkable sensitivity to the complexity of scientific practice [Friedel 1884, p. VIII].[13]

3 An Optimistic Scientism

A completely different meta-theoretical attitude can be found in the book the authoritative chemist and politician Marcelin Berthelot published in 1886. He was an important character of the French Third Republic: as he himself explained to his readers, not only was he committed to science, but also to "specific applications to industry and national defence, public lecturing, and general politics." The book was a collection of papers he had already published in various journals in the 1860s and 1870s. The scattered collection of texts on different subjects was however unified by the presence of specific "philosophical views," which could be looked upon "as a sort of intellectual and moral biography of the author." Among the four main subject matters, "scientific philosophy, history of science, public teaching, and politics and national defence," the

13 Friedel had studied with Louis Pasteur at the Sorbonne, and after having worked in Charles-Adolphe Wurtz's laboratory, he became professor of mineralogy at the Sorbonne. In 1884, he exchanged this position for the chair of organic chemistry that had been held by Wurtz.

first two deal with the context I am exploring, and the first essay of the book focuses exactly on what we might label history and philosophy of science. In reality the qualification might be too ambitious because the renowned chemist confined himself to an apologetic and simplified history of science, and was committed to an equally simplified philosophical analysis of the science of his time. From the formal point of view, the first essay was an open letter he had addressed to another intellectual father of the Third republic, the historian Ernest Renan [Berthelot 1886, pp. II-IV].[14]

In the "Préface" to the book, which was arguably written in the same year as its publication, Berthelot outlined some basic theses of his "scientific philosophy" which he had already put forward in 1879 in his "grand work" *Essai de mécanique chimique fondée sur la thermochimie*. He mentioned the identity "in principle and in fact" between chemical processes in living beings and in inanimate bodies, the reduction of chemistry "to the most general laws of mechanics," and the perfect homogeneity between the macroscopic domain of cosmological events, and the microscopic domain of atomic processes. He carefully stressed that his celebration of contemporary chemistry and physics did not lead him to underestimate the ancient body of knowledge, which had spawned some kind of "intermediate half-mystic and half-rational sciences." Some subject matters, for instance "alchemy and astrology," had greatly contributed to "the evolution of the human mind."[15] He focused on "positive science," which started from facts, and connected them by means of "immediate relations." Since science dealt with observable facts, it was not able to attain "the first causes or the fate of the material world." Nevertheless science had managed to lead mankind to "the explanation of a huge number of phenomena" merely on the basis of "the coarsest facts." We find here both faith in a continuous and unbound scientific progress and confidence in a simplified scientific practice, which could smoothly and successfully lead from ordinary observation to very general scientific laws [Berthelot 1886, pp. V, VII, and 4-5].

He was willing to offer readers a simple and interesting instance of that *positive* practice: the explanation of the functioning of a torch or lamp, or in other words the answer to the question "why a torch lights up." He insisted on the evidence of chemical processes: the fact that the torch contained carbon and hydrogen was a direct result of "observable facts." It seems that Berthelot

14 Berthelot was a member of the political and academic establishment in France: an influential chemist with serious interests in the history of science, a professor at the *Collège de France*, and a moderate republican, he was also a member of Parliament and a Minister.

15 In this context he reminded readers that he had recently published the book *Les origines de l'Alchimie*, and some excerpts were there reproduced.

made use of the adjective observable in the sense of "scientifically explained" rather than in the sense of actually observable: in his view, current scientific explanations had to be necessarily endowed with the hallmark of evidence. Scientific practice had to be associated with the disclosure of evidence rather than a more complex practice of interpretation and translation of facts into a series of rational entities, concepts, and statements. In spite of these conceptual and linguistic overlaps, Berthelot continued to rely on the chain of evident *facts*, such as the *fact* that the combination of oxygen with the elements of the torch, namely carbon and hydrogen, generated heat. However, in the end he acknowledged that he had passed from the observation of facts to "more general notions," and specifically that those notions were "more general than the specific facts" from which he had started. [Berthelot 1886, pp. 5-7].

Not only could "physics and chemistry be reduced to mechanics," but also the process of reduction was an empirical necessity rather than a rational option. According to Berthelot, the reduction did not stem from "confused and uncertain insights, or *a priori* reasoning" but from "unquestionable notions," which were "always based on observation and experience." Great advantages and great expectations emerged from that reduction: great experimental results, very general laws in conformity with "the nature of things," and "a simple and invariable method." He relied on a circular practice, which started from the perception of facts by means of "observation and experience," then went on with the establishment of relations, (which stemmed from the comparison among simple facts, and led to "more general facts"), and ended with the empirical check of those general facts by means of observation and experience. In Berthelot's essay, the polysemy of the word *fact* is really striking because he also considered as facts what he qualified as "progressive generalisations deduced from previous facts." In other words, scientific practice consisted in discovering an increasing domain of *facts*. The empirical nature of science was "one of the principles of positive science": no actual knowledge could be "established by means of reasoning." Centuries of philosophy were swept away by that flood of facts, and by a new, naïve philosophy. His philosophy was in tune with a radical empiricism: the conclusions drawn from theoretical cogitations could only be "probable and never certain," whereas certainty was really attained by "direct observation" [Berthelot 1886, pp. 9-11].[16]

16 I find that a short passage deserves to be quoted: "Une généralisation progressive, déduite des faits antérieures et vérifiée sans cesse par de nouvelles observations, conduit ainsi notre connaissance depuis les phénomènes vulgaires et particuliers jusqu'aux lois naturelles les plus abstraites et les plus étendues." [*Ibidem*, p. 10].

The history of science was reconstructed and presented to readers by means of some brush strokes which roughly depicted the pathway that had led from the ancient natural philosophy to "the solid principle" on which modern sciences rested. At the dawn of civilisation, "Indian wise men" had relied on meditation, and Greek philosophers — in particular neoplatonic philosophy — had relied on the power of speculation. He found that even "the advancement of mathematical sciences" had supported such a delusive trend. Only modern scholars, and in particular "Galileo and Florence academicians," had managed to understand that the axioms of mathematics had to be deduced from observation, and the conclusions had to be checked "by means of the same observation." That a circular process of this kind was not consistent with the actual development of mathematics seemed outside Berthelot's intellectual horizon. According to his historical reconstruction, the sixteenth century had seen the first achievements of "the forefathers of positive science," and the eighteenth century had seen "the triumph of the new method." The "ultimate aim" of the nineteenth century was the establishment of a social order "in accordance with the principle of science and reason" [Berthelot 1886, pp. 11-13].

The appearance of *reason* in this context seems quite strange, because Berthelot had continuously discouraged the use of reason in favour of ubiquitous observations and experiences. In reality this seems a rhetorical slip, because he immediately specified that "the moral domain" and "the material domain" required the same methodology based on facts and observations of facts. The method that solved "every day the problems of the material and industrial world" could solve the "fundamental problems which emerged from the social organisation." Surprisingly enough, the starting point of social sciences was "the primordial fact of human nature," namely "the feeling of good and evil." This fact was accompanied by other essential primitive facts: the existence of a sense of duty and the desire for freedom. The semantic scope of the word *fact* was thus further widened in order to encompass human feelings in general [Berthelot 1886, pp. 13-5].

According to Berthelot, positive science had gained an unquestionable authority, which was based on "the necessary conformity between his results and the intrinsic nature of things." Neither the conformity nor the corresponding necessity required a justification or at least an explanation: they were *facts*. Every man endowed with a basic education would have been able to appreciate "the results of positive science as the only gauge of certainty." An optimistic trend marked the last stage in the history of science: the ancient attitudes, which had frequently stemmed from ignorance and imagination, were fading away in favour of new conceptions based on the observation of Nature. Even the semantic scope of the word *nature* was widened in order to host "moral nature" besides physical Nature. The essential features of positive science were reliability and steadiness, which differed from the ancient philosophies that had ceaselessly changed over time. The reliability of science was closely linked to technological power: material success ensured that the new body of knowledge would never be overturned [Berthelot 1886, pp. 14-6].[17]

The *positive* method had to be applied always and everywhere, and from the outset that option automatically excluded any supernatural entity. We are facing here a great reductionist design, where all human practices, feelings and desires pivoted around a powerful *positive* science. The domain of historical researches offered some resistance to the reductionist design, for two reasons. First, the knowledge of the past was intrinsically incomplete, and second, every kind of experimentation on the past was intrinsically precluded. At the same time, history had shown "the continuous advancement of science, material conditions of life, and morality." The existence of scientific and social progress was confirmed *a posteriori* by historical studies. Therefore history could not be submitted to the positive method, but it could show the positive effects of that *positive* method. Berthelot endorsed the Galilean rhetoric of science based on sound experiments and certain demonstrations. Unlike ancient methods, which were dogmatic, the new method rested on the acknowledgment of "individual opinions and freedom." How to combine freedom with "the intrinsic certainty" of positive science was a mystery as unfathomable as the unquestionable certainty that spontaneously emerged from the collection of individual observations [Berthelot 1886, pp. 32, 34-8, and 40].[18]

A similar intellectual optimism towards science can be found in the texts of the authoritative historian Renan. I have already mentioned his sophisticated historiography, where different historical ages and different cultural contexts required different standards of evaluation. In 1890 he published a book that contained remarks and reflexions he had written down in 1848 but had left unpublished for a long time. He thought that those pages, which had stemmed from the mind of "a frank young man," deserved to be published even after more than forty years, because he still shared the same ideas, and because he found that those ideas were still up-to-date. In the *Foreword* he confessed he

17 The possibility that ancient people and ancient traditions, and in particular ancient scholars, had been involved in careful observations of Nature, and the possibility that the role of observations was not so pivotal in the emergence of modern science, were not taken into account by Berthelot.

18 In reality Galileo's epistemology was more sophisticated than Berthelot's, because the former had explicitly acknowledged the necessity of two elements in the scientific enterprise: mathematical procedures and a suitable experimental practice. On the contrary, in Berthelot's epistemology we only find the exclusive trust in observation.

had only changed his mind slightly since he had begun "to think freely": he professed a well-defined religion, which consisted of "the advance of reason, that is science." He continued to believe that "only science could improve the bad condition of man," even though he did not see "the solution as close as he had seen at the time" [Renan 1890, pp. v, vii, and ix].[19]

He found that, unlike *positive* religious commitment, traditional religious zealotry was the worst danger for human society: the "blind faith of ancient times" had transformed human beings into "a fanatical crowd." Definitely, "immoral people" had to be preferred to "fanatical people." Luckily, science had advanced, and "apart from some disillusions," it had proceeded along the pathway he had imagined. He saw the universe as the seat of a continuous, magnificent progress: the fact that "neither preternatural facts nor mysterious revelations had ever taken place," and the fact that "inequalities among races" had been firmly established were listed among the most recent scientific achievements. On the other hand, he did not expect that science could offer answers to the most enduring and demanding questions: "neither the will of Nature nor the aim of the universe" could be grasped. Although the relentless work performed during the nineteenth century had allowed scientists to increase astonishingly the knowledge of man and Nature, he acknowledged that the ultimate fate of mankind had become "more mysterious than ever." Science could satisfy "the noblest ambition of human nature," namely curiosity, and at the same time it could "supply man with the only means for the improvement of his destiny." Nevertheless science could not lead to truth: it could only "protect against mistakes." That negative feature was of remarkable help indeed, because it prevented human beings from being led astray [Renan 1890, pp. x-xiv and xvi-xix].

A critical and sceptical nuance distanced Renan from Berthelot: Renan's optimistic scientism was tempered by a finer historical sensitivity, and a finer philosophical attitude.

4 A More Critical Trend

Definitely more critical was Boutroux, who published a book on the concept of natural law in 1895. The book, *De l'idée de loi naturelle dans la science et*

19 Benrubi saw in Renan a complex network of influences, especially Kant, Comte, Hegel, and Darwin. A close friend of Marcellin Berthelot, Renan was confident in the healthy effects of scientific progress in the cultural and material domain, but his sceptical attitude distanced him from Comte [Benrubi 1926, pp. 29-31; Benrubi 1933, pp. 34-6].

la philosophie contemporaines, contained both historical and critical remarks, and reproduced the text of fourteen lectures he had held in the academic year 1892-3.[20] The first lesson, "The problem of the meaning of natural laws" started from the role played by Bacon and Descartes in the emergence of modern science. Boutroux credited both of them with having given scientific laws "the distinctive feature of universality and reality." They had in common the specific ambition to know reality in a definite way, even though Bacon had followed the empiricist and Descartes the rationalist pathway. Rationalism could not overcome the dichotomy between physics, namely the "field of efficient causes," and ethics, which was "the field of final causes." Nor had empiricism solved the query because it had simply reduced *external* laws to the internal laws of the subject. Descartes had found it hard "to link the actual to the universal," whereas Bacon had found it hard "to link the universal to the actual." From the philosophical point of view, it was not very easy to imagine natural laws that were "universal and actual" at the same time: modern science seemed to have succeeded where philosophy seemed to have failed. It had managed to combine mathematics with experience in order to yield "tangible and intelligible laws" [Boutroux 1895, pp. 5-7 and 9].

Science had become independent of philosophy, and every specific science, namely mechanics, physics, chemistry, and life sciences had striven to attain the same certainty as mathematics. The existence of a series of specific sciences led to a series of specific questions: did the different sciences require different foundations? Did the transition from one science to another require "the introduction of a philosophically non-reducible principle"? Did sciences represent elements of reality or were they simply collections of symbols? Were they absolutely true or only true in a relative sense? Eventually Boutrox asked whether determinism was an essential feature of Nature, or simply the way "by which we must link things, in order to make them objects of mind." He was aware that he had listed ancient questions that had been debated at length, but he would have attempted to frame those questions into "a current perspective" [Boutroux 1895, pp. 10-11].

Following a positivistic trend, he discussed scientific languages and contents in accordance with a predictable classification: at first logic, and then mathematics, mechanics, physics, chemistry, biology, psychology, and sociology. Strict determinism and necessity ruled the laws of logic. At the same time, logic offered a minimum of objectivity because of the existence of "an unbridgeable gap" between logic and reality. Mathematics also enjoyed a status

20 After having taught at the *École Normale Supérieure*, in 1888 Boutroux was appointed to the Chair of *History of Modern Philosophy* at the Sorbonne.

of relative certainty, even though logic and mathematics were quite different in their nature, and corresponded to very different ways of reasoning. If logic assumed the existence of entities to be linked together, mathematics built up its own entities. In mathematics, the principle of recursion aimed at merging induction with deduction, and it was a procedure that went far beyond pure logic: it was a sort of "apodictic induction." Mathematical laws entailed a very complex structure, where both analysis and synthesis were at stake, as well as *a priori* and *a posteriori* lines of reasoning. The existence of logic and mathematics showed that the human mind was in need of some kind of rational steadiness, and that a certain degree of similarity between mind and external world could be assumed [Boutroux 1895, pp. 19-24 and 30].[21]

According to Boutroux, among "the laws of the real world," the laws of mechanics were the most akin to the laws of mathematics. The most important among mechanical laws was the principle of inertia, which marked the difference between ancient and modern natural philosophy because it stated the equivalence between motion and rest. Although that equivalence had led Descartes to the abolition of the concept of force, Newton had shown the necessity of a new concept of force, which was definitely "an extra-mathematical element." Descartes had attempted to reduce physics to geometry, whereas Newton had pointed out the impossibility of that reduction. However, the mathematical character of natural laws corresponded to "the effort to fit things to our mind," and that effort was the keystone of modern science. The essential features of mathematical laws were *continuity* and *immutability*, whereas actual observations took place at different, separated times, and the natural world (living species, for instance) changed over time. Boutroux therefore stressed that continuity and immutability were not essential features of natural events in themselves [Boutroux 1895, pp. 30-32, 34, and 37-8].

The intrinsic tension between the experience of actual events and their mathematical representations was pointed out by Boutroux in the same way as in 1874. It was the specific philosophical weakness of modern science, but the philosophically problematic alliance between experiments and rational

21 See in particular, p. 30: "Les lois logiques et mathématiques témoignent du besoin qu'a l'esprit de concevoir les choses comme déterminées nécessairement; mais l'on ne peut savoir *a priori* dans quelle mesure la réalité se conforme à ces symboles imaginés par l'esprit: c'est à l'observation et à l'analyse du réel qu'il appartient de nous apprendre si la mathématique règne effectivement dans le monde. Tout ce que l'on peut admettre, avant cette étude expérimentale, c'est qu'il y a vraisemblablement une certaine analogie entre notre nature intellectuelle et la nature des choses. Autrement l'homme serait isolé dans l'univers."

explanations was also the key to its success. Boutroux was aware that both the weakness and success had to be historically analysed.

According to his historical reconstruction, from Descartes onwards, mechanical laws had suffered from *dogmatism*; he found that even Leibniz and Newton had been *dogmatic*, even though Newtonian science should not be confused with "Newtonian metaphysics." Concepts like "homogeneous space devoid of any quality" and "extended and indivisible atom" stemmed from Newtonian metaphysics, and they were contradictory concepts. The representation of the natural world as a mechanical engine, whose behaviour was deterministic and completely defined by equations and initial conditions, was more a mythology than a matter of fact. The mathematical mythology had survived until the last decades of the nineteenth century, when a more critical attitude had emerged. Boutroux reminded readers that the mathematical physicist Boussinesq had pointed out that sometimes "initial conditions cannot define completely the way taken by the phenomenon." As Boussinesq himself had explained in 1878, bifurcations and "the action of a guiding power" were at stake [Boutroux 1895, pp. 39-42 and 46].[22]

In the end, mechanical determinism appeared to Boutroux as the source of an unnatural separation between laws and phenomena. Once more he asked himself whether physical laws were "a specific case of mechanical determination" or whether they possessed "their own originality and meaning." This question concerned the whole hierarchy of positivistic sciences: logic, mathematics, mechanics, physics, chemistry, life sciences, psychology, and sociology. Could scientists and philosophers rely on a comfortable reductionism, where mathematics stemmed from logic, mechanics from mathematics, ..., and psychology from the sciences of life, or did every science require specific contents and methods? As he had already done in 1874, he criticised every reductionist attitude, and focused specifically on physics. Even more specifically, he mentioned the main issue at stake in the context of thermodynamics: the apparent incompatibility between the reversibility of purely mechanical processes and the irreversibility of actual physical processes, where mechanical work was dissipated into heat. Moreover thermodynamics required a sort of qualitative hierarchy of energies, namely "an element of differentiation and heterogeneity": the quality of ordered mechanical work was higher than the quality of disordered energy or heat. The second principle of thermodynamics dealt with

22 It is worth remarking that in 1895, when Boutroux published his book, Boussinesq held the Chair of *Mécanique Physique et Expérimentale* at the Faculty of Science in Paris, which had been held by Henri Poincaré until 1886.

a patent impossibility in nature, the impossibility of a spontaneous passage of heat from a body to a warmer one. The negative content of the principle pointed out the impossibility of "determining phenomena in a definite way" [Boutroux 1895, pp. 51-2, 54, and 56-7].

Boutroux then drew readers' attention to Stallo's criticism about atomism: homogeneity, hardness, and inertia of atoms were accepted by chemists despite the logical and physical difficulties that emerged from those assumptions. However Boutroux acknowledged that the atomic theory had been useful, and had given chemistry "a precious notation." He agreed with Friedel on the provisional role of every scientific theory, and at the same time, on the necessity of a fruitful theory, its logical and physical faults notwithstanding. In reality, scientists relied on the wave theory of light, even though they were aware of the contradictions in the conception of the luminiferous aether. In the end, after a short historical report of atomism in the context of the eighteenth- and nineteenth-century history of science, Boutroux concluded that atomism was an extremely pliable hypothesis. It could "account for everything," provided that the atom was endowed with "the features which it should explain." In brief, Boutroux pointed out the same vicious circle Stallo had criticised more than a decade before [Boutroux 1895, pp. 64-5].[23]

In the conclusive chapter of his collection of lessons, Boutroux set the emergence of modern science, and then nineteenth-century scientism and reductionism, against the background of the history of philosophy. In particular, he compared the alliance between mathematical language and experiments, which had been the hallmark of the scientific tradition between the seventeenth and the nineteenth centuries, with "the ancient philosophy," which had rested on the corresponding dualism. In other words, ancient philosophers had hesitated in front of the problematic link between logic and mathematics, on the one hand, and experiences, arts and crafts, on the other. In more philosophical terms, they had stepped back before the gap between "the realm of eternal and necessary," where *truth* had its seat, and the variable and shaky realm of phenomena. Modern natural philosophers had dared to overcome

23 Stallo had made the same remarks, and had emphasised the "delusions that the elasticity of a solid atom is in less need of explanation than that of a bulky gaseous body." His criticism was definitely sharper than Boutroux.' See Stallo 1882, p. 128: "It may seem strange that so many of the leaders of scientific research, who have been trained in the severe schools of exact thought and rigorous analysis, should have wasted their efforts upon a theory so manifestly repugnant to all scientific sobriety — an hypothesis in which the very thing to be explained is but a small part of its explanatory assumptions."

that dualism: in philosophical terms they had trusted in the possibility of conflating "the science of being" with "the science of becoming." Logic and mathematics offered the component of necessity, whereas observations and experiments offered empirical reliability. According to Boutroux, the faith in that alliance was the root of "modern determinism" [Boutroux 1895, pp. 135-6].

He underlined that the close alliance between the formal structures of reason and the body of empirical practices rested upon a simple *belief*, or something that was not *scientific* in itself: it was something like a faith or a mythology. The chain of subsequent reductions, which started from logic and led to sociology, was extremely weak even in the first rings: mechanics had "elements that were not reducible to purely mathematical determinations." In the end, Boutroux made two conclusive remarks. First, the reductionist approach to Nature had stemmed from the ignorance of the "incommensurability between reality and mathematics," but it had had "pleasant effects," namely the emergence of modern science itself. Second, the concept of one science and one scientific method that encompassed all specific sciences was misleading: it was nothing more than a mere abstraction. There were many specific sciences, endowed with "their specific features [physionomie]" and methods. Following the series of different sciences, which went from astronomy to "the study of life and mind," a long series of assumptions could be identified. The complexity of the assumptions dramatically increased from physical sciences to human sciences: the assumptions were few and simple at the beginning of the hierarchy, and "ever more plentiful and unfathomable" at the end [Boutroux 1895, pp. 136, 138-9, and 141].

According to Boutroux, naïve reductionism and dogmatic determinism were the consequences of underestimating the problematical link between reason and experience. At the same time, philosophical naivety and dogmatism had given birth to modern science, a new body of knowledge that had modified the intellectual and material landscape of our civilisation. Ultimately, he acknowledged that the philosophical weakness of science was closely linked to its effectiveness. In its turn, the effectiveness was intrinsically linked to the plurality and pliability of scientific methods, and that plurality had to be explicitly acknowledged.

At the end of this partial exploration of the different attitudes towards scientific practice that emerged in French-speaking countries between the 1860s and 1890s, it is worth coming back to the historical reconstruction put forward by the historian of philosophy Isaac Benrubi. In 1926, he classed French philosophers in three main streams: empiricist positivism, critical-epistemological idealism, and metaphysical-spiritual positivism. In 1933 he slightly modified the first label, which became "empiricist and scientist positivism." Important characters of the emergence of history and philosophy of science in France such as Bernard, Cournot, Poincaré, Duhem, and Milhaud were included by Benrubi in the second stream. In the first stream, he placed Comte and his followers, and in the third Janet, Fouillée, Boutroux, Naville, and Bergson [Benroubi 1926, pp. VI-VIII and 13-4; Benrubi 1933, p. 4].[24]

The substantive that Benroubi associated with the second intellectual stream — which would deal with some of our characters — namely *Idealism*, seems unsuitable and potentially misleading whereas the adjectives *critical* and *epistemological* are in tune with the actual commitments and texts of the scholars I have discussed in the previous chapters. Benrubi's interpretation of their theses as an attempt to explore the limits of science and a commitment to go beyond "exaggerated beliefs in the omnipotence and self-sufficiency of exact science" seems in tune as well. He also stressed that those scholars and philosophers had emphasised "the part played by the intellect in the formation of exact science," and this might explain his choice of the word *Idealism*. Moreover, when Benrubi claimed that Kant could be looked upon as "the chief pioneer of this movement," and that Kant had dethroned "dogmatism and scepticism at the same time," it is clear that the label *Idealism* should be intended in the sense of Kantian idealism. The reference to Kant is certainly appropriate but his last linguistic choice, namely "Critical School in France" appears definitely more appropriate [Benrubi 1926, pp. 84-7]. Benrubi stressed both Kant's anti-empiricist attitude, and the anti-empiricist attitude of Bernard, Cournot, and philosophers like Naville and Boutroux. At the same time, he stressed Kant's insistence on the limits of "the demonstrative procedures of reason," and the importance of conjectural and intuitive practices. In this context, he repeatedly pointed out the influence of Pascal; more specifically, he traced back the complementarity between pure reason and practical reason to the complementarity between "the esprit of finesse and the esprit of geometry" [Benrubi 1933, pp. 294-5]. It seems to me that the influence of Pascal was probably more marked in the above-mentioned French scholars than in Kant, but the philosophical line of descent is definitely plausible.

In reality, scattered but sharp and critical remarks on the history of natural philosophy, and likewise critical accounts of the foundations of the emerging

24 In 1926, Benrubi published a summary of French philosophy in the last decades: the book, *Contemporary Thought of France*, was published in London. In 1928, an enlarged German version, *Philosophische Strömungen der Gegenwart in Frankreich*, appeared in two volumes, and finally a French version, *Les sources et les courants de la philosophie contemporaine en France*, was published in two volumes in 1933. He taught history of philosophy at Genève University and Bonn University.

new science, can be found everywhere in Pascal's writings. In the context of the new science, he had devised a treatise on fluids and vacuum that should have included both the *Traité de l'équilibre des liqueurs* and the *Traité de la pesanteur de la masse de l'air*, the two treatises that were published after his death in 1663. He only managed to write some introductory notes to the planned all-inclusive treatise (*Fragment d'un traité du vide*), and in those notes we can read insightful remarks dealing with what nowadays we label philosophy of history and philosophy of science. Other remarks on the same subjects can be found in his scattered *Pensées*, and in a collection of letters he addressed to fictive interlocutors in the context of the theological debate between Jansenists and Jesuits (*Les Provincials*). Pascal was a *modern* who did not reject the most important achievements of *ancients*. He faced two apparently contradictory processes: the unpredictability of historical events, and the existence of an actual progress in the history of mankind. In some sense, the interaction and superposition of individual events could give rise to a cumulative and progressive effect. He also focused on the complex relationship between rational and experimental procedures, and on the necessity of a clear separation among the outcomes of senses, reason, and faith. He was aware that logic, mathematics, and rational procedures in general had to be accompanied by a network of rational choices that involved conventions and intuitive skills [Shea 2003, pp. 187-93, 206-7, 209-15, and 222-3].

I will confine myself to reporting some introductory notes to Pascal's planned treatise on fluids and vacuum, which was subsequently published as *Fragment d'un traité du vide*. Against the principle of authority in natural philosophy, he defended the innovative momentum of the new science. He stressed the dynamic nature of scientific knowledge: reason and experience could not but lead to an increase and improvement of knowledge. Scientific research required a continuous exploration of new lands. In some way, the *esprit* of research was an essential feature of human anthropology, and distanced human beings from animals, who could only rely on the repetitive procedures triggered off by their instincts. *Progress* became Pascal's key word, and progress involved both preservation and increase of knowledge. Although progress was realised by specific individuals, it was a collective accomplishment, and mankind could be looked upon as a single living entity that continuously learned and advanced on the pathway of knowledge. The complex relationship between reason and experience, or rather between deductive and experimental procedures, was highlighted by the intrinsic asymmetry between them. The universal value of rational deductions did not correspond to the universal value of experimental proofs, because a finite number of actual experiments could never lead to a positive certainty. On the contrary, a single experimental event could "refute a general statement," and therefore experiments could potentially lead to some kind of negative certainty. The impossibility of demonstrating the validity of empirical knowledge whenever and wherever entailed the intrinsically provisional and incomplete nature of scientific knowledge [Pascal 1872, pp. 159-62].

From the point of view of my historical reconstruction, the separation between critical idealism (Benrubi's second intellectual stream), and what he labelled metaphysical-spiritual positivism (the third intellectual stream) is certainly less meaningful. From the point of view of philosophical and historical researches on scientific tradition, scientific practices, and scientific method(s) put forward in the last decades of the nineteenth century, a reasonable separation between naïve and sophisticated approaches seems to me more suitable. Naïve approaches can be associated with a simplified version of scientific practice, the celebration of a simplified scientific method, and the uncritical mythology of scientific progress. Sophisticated approaches managed to grasp the complexity of scientific practice, the complex interaction among rational, empirical, and intuitive components in scientific research, the plurality and pliability of scientific methods, and the existence of different scientific traditions which had developed through history.

My separation criterion between naïve and sophisticated historiographical and epistemological frameworks overcomes the separation between critical idealists and spiritualists, which I find partially misleading. The fact is that Benrubi's separation makes sense in the context of very general philosophical commitments but it becomes unsuitable, and definitely less meaningful, in the more specific context of history and philosophy of science. There are meaningful similarities between Cournot's criticism and Boutroux' alleged spiritualism.[25] In the end, I must warn against any rigid implementation of my criterion: we find in Renan a naïve scientism but also a sophisticated historiographical framework. He was more sophisticated than Berthelot, but we will see slight changes in Berthelot's attitude in the next chapter. Naïve and sophisticated are pragmatic and relative parameters that help us understand the historical roots of the professional history and philosophy of science.

25 It seems to me that Benrubi himself softened the separation between critical idealism and metaphysical-spiritual positivism when he remarked that going back to Kant had been the essential step for "the impressive development of critical and epistemological idealism, and in part, spiritualism" [Benrubi 1933, p. 298].

CHAPTER 7

Histories of Ancient Science and Mathematics

1 Curiosity and Erudition on Ancient Alchemy

In 1885 Marcelin Berthelot undertook a demanding task: a history of the ancient alchemy as an introduction to understanding the roots of modern chemistry. The latter was then a young science: it had emerged in the last decades of the eighteenth century, and therefore it had a short history. On the contrary, alchemy could rely on a long history that covered many centuries until the end of the eighteenth century.

It is worth remarking that there was a tradition of research on the history of science in France: the mathematician Jean-Étienne Montucla had published *Histoires des Mathématiques*, whose first volume appeared in 1758, and the astronomer Jean Sylvain Bailly had published *Histoire de l'Astronomie ancienne depuis son origine jusqu'à l'établissement de l'école d'Alexandrie* in 1775, followed by *Histoire de l'Astronomie moderne depuis la fondation de l'école d'Alexandrie jusqu'à l'époque de 1780* in two volumes in 1779, and other two volumes in 1782 and 1787.[1] After Bailly, other histories of astronomy had appeared: the French astronomer Jean Baptiste Delambre had published a two-volume *Histoire de l'astronomie ancienne* in 1817, then *Histoire de l'astronomie du moyen Age* in 1819, and the two-volume *Histoire de l'astronomie moderne* in 1821. The first half of the nineteenth century also saw the publication of histories of natural sciences or life science in general by the authoritative naturalist and zoologist Georges Cuvier, and by the physician and naturalist Henri-Marie Ducrotay de Blainville. Finally, a *Histoire des Sciences médicales* was published by the physician and historian of medicine Charles Daremberg in 1870: he was the first scholar to hold the Chair of History of Medicine at the *Collège de France*, a Chair that had been created in 1864 [Gusdorf 1966, pp. 89 91, and 103].

---

1 Gusdorf traced back the emergence of the history of science in France to Bernard le Bouyer de Fontanelle, around the turn of the eighteenth century, and the emergence of philosophy of history to Voltaire, after the turn of mid-century. He looked upon Fontanelle as the first scholar to have put forward a history of science as "a history of human esprit." More in general, Gusdorf found that "the awareness of the historicity of human life and culture" had emerged in the eighteen century. It is certainly true, even though it was not a hegemonic cultural trend. In 1879, after the Revolution, Bailly was also mayor of Paris, but in 1793 he was guillotined because of his alleged conservatism [Gusdorf 1966, pp. 56-7, 65, 67, and 70].

 | DOI 10.1163/9789004315235_009

Berthelot was an authoritative scientist and politician, and therefore could rely on the collaboration of other scholars in the fulfilment of his historical reconstruction. Historical, philosophical, and philological skills were required, and in some way that reconstruction was a collective enterprise. The result was a book wherein Berthelot's naïve and radical positivism was softened by the acknowledgement of the existence of a neglected cultural tradition. He started from the statement that "a purely rational conception" could be traced back to the time of the ancient Greeks, and since then science had laid claim to "the material universe." Nevertheless, the time interval between Greek rationalism and modern chemistry had been filled with a mixed practice where both magic and "a well-defined positive attitude" overlapped. In Comtian terms, that time could be qualified as "half-rationalist and half-mystic": although chemistry was "probably the most positive among sciences," it had emerged from a melting pot of sound observations and "extravagant imagination." Alchemy was an intermediate body of knowledge endowed with multifarious connections with the philosophical traditions that had gained hegemony in the first centuries of the Christian age. It rested upon a meta-theoretical commitment that was not uncommon in subsequent scientific practice: the hypothesis of "the unity of matter." In other words, Berthelot acknowledged the existence of meaningful analogies between "the deep views of the first alchemists," and modern conceptions on "the structure of matter" [Berthelot 1885, pp. VI-VII, IX, and XIV-XV].[2]

Berthelot divided the book into four parts: first the sources, and then the characters, which were followed by "facts and theories." In the *Introductions* to the different parts and chapters we find Berthelot's personal remarks and reflections, whereas many other information and passages reveal the specific skills, mainly philological skills, of the other scholars. He stressed the empirical component of chemistry, and the fact that its fast progress had contributed to the transformation of "material civilisation" in the nineteenth century. On the other hand, unlike other sciences such as "geometry and astronomy," chemistry had not existed in ancient times: it was the result of a profound intellectual transformation of "the debris of a previous body of knowledge," which he surprisingly qualified as *scientific*. In reality, the adjective appears less surprising when Berthelot remarked that the body of knowledge known as alchemy could rely on a wide domain of empirical practices and "practi-

2 Gusdorf remarked that Berthelot had started from the conception of alchemy as a false science that had faded away, and had left many pieces of theoretical debris of different value. Alchemy was looked upon as "a sort of mental archaeology" [Gusdorf 1966, p. 109].

cal discoveries" that dealt with arts in general, and more specifically "metallurgy and medicine." To some extent, those practices had been inherited by the more recent scientific practices that had converged on the emergence of chemistry. Berthelot's empiricism, and the alleged empirical nature of chemistry, projected the qualification of scientific knowledge on to ancient empirical practices. He ventured to extend the qualification to the theoretical side: "since its first days" chemistry had aspired to become "a philosophy of nature," and that ambition could be found from the outset in the tradition of alchemy. According to Berthelot, alchemy was deeply rooted in three different traditions: arts and empirical practices of ancient Egyptians, speculative theories of Greek philosophers, and "mystic daydreams of Alexandria's scholars and Gnostic philosophers" [Berthelot 1885, pp. 1-2 and 4-5].

Alongside the close entanglement among practical arts, philosophy and magic in alchemic tradition, Berthelot also stressed the difference between ancient and modern attitudes: in ancient cultures, practices like metallurgy and magic appeared as different implementations of the same art. Although he hinted at a more ancient Babylonian tradition to which Egyptian tradition could be traced back, he did not offer a detailed reference frame from this historical and historiographical point of view. He mentioned the known secondary literature, and offered many pages of quotations. He also insisted on an alleged connection between alchemy and the philosophical tradition from the Milesians to Plato and Democritus, but he did not succeed in clarifying the frequent mention of Democritus in the literature.[3] Other hints at the contemporary emergence of alchemy in China, and the Juidaic Gnostic influences on Greek-Egyptian alchemy, also remained undeveloped and unexplained. In Berthelot's sketch, different elements and different traditions were brought back to a common, general attitude, which he qualified as "global syncretism." In the end, he found that chemistry had existed in ancient times, and corresponded to an empirical practice, but that practice was overwhelmed by more mysterious and less scientific practices. At the same time he ventured to point out a structural analogy between the wide and broad network of symbols of alchemy and the more definite network of symbols of modern chemistry. The insistence on this analogy does not seem in tune with Berthelot's radical positivism [Berthelot 1885, pp. 17, 19, 29, 36, 46, 52, 54-7, and 65-6].

3 Probably the Democritus here mentioned was Bolus of Mendes "from the Nile delta in Egypt" who "wrote under the name of Democritus": sometimes he is qualified as Democritus Bolus of Mendes or the pseudo-Democritus: it seems that, "in the late second century BC," he was "a crucial figure in shaping the subsequent development of Graeco-Egyptian alchemy" [Kingsley 1994, pp. 5-9; Russo 2013, p. 195].

Hidden among many digressions, we find the main historiographical thesis of the book: a continuous line of descent led from the third century to the late Western Renaissance through Byzantine and Arabic culture. Alongside the usual three-folded convergence of arts, philosophical traditions, and "mystic imagination," and the frequent references to Plato and Pseudo-Democritus, we find a synthetic reconstruction of ancient Greek philosophy, where broad and simplified analogies emerged. Heraclitus' world-view was associated with modern "transformation of forces and the mechanical theory of heat." Alchemists had been confident that their theories could be based on a more ancient scholarly tradition in the same way as Berthelot was confident in the semantic analogies between that tradition and the foundations of modern sciences. He singled out two different lines of descent: the first went from Pythagoras to Georg Ernst Stahl's "definition of phlogiston" through Plato's Timaeus and Stephanus of Alexandria's "mystic kinship between alchemy and astronomy." The second one started from Leucippus and Democritus, and led to both alchemy and chemistry. He confined himself to specifying that mentioning Democritus in Alexandria's alchemy or in mediaeval alchemy did not mean an endorsement of "the atomic theory" [Berthelot 1885, pp. 78-9, 248, 250, 252, 262-5, 267, 271, and 275].

The most interesting part of the book is the second chapter of the fourth section, where Berthelot discussed some analogies between alchemic theories and modern chemical ones. Underneath "mystic explanations and symbols" he found "a little number of clear and credible ideas" that showed a striking analogy with the science of his time. The endless and cyclic transformation of the alchemical tradition could be put in connection with the chemistry of metals, where iron minerals in their natural state could be heated up until they became metallic iron; in its turn, iron could be transformed into an oxide by atmospheric agents, reaching a state that "was akin to the primitive one." The material foundations of alchemy consisted of "positive facts and industrial practices" that had survived the transformation of alchemy into chemistry. According to Berthelot, some empirical practices of alchemists were continuously reproduced in modern laboratories. Nevertheless there was an important difference between the two practices: in modern chemistry no scientist attempted to go behind the level of "chemical elements." No simple or basic element could be transformed into another. However he found that the modern classification of chemical elements, and the hypothesis that all atomic weights were multiples of a basic weight unit were in tune with the Pythagorean idea of a close connection between "the real features of beings, and the mysterious properties of numbers" [Berthelot 1885, pp. 278-80, 283-5, 287, and 291-2].

The search for such analogies rested upon a certain degree of relativism and historicism that stemmed from Berthelot's empiricism. Science was mainly based on a sound empirical practice, and every empirical practice inevitably led to updating or replacing every scientific theory.

> Our current theories on atoms and aethereal matter will probably appear fanciful to future scholars in the same way as the ancient natural philosophers' theory of mercury appears odd to present-day scholars [Berthelot 1885, p. 321].

This historical perspective mitigated Berthelot's previous positivism: historical studies and the professional relationship with historians, philosophers, and philologists had probably softened his naïve conception of scientific practice.[4]

2 A Sophisticated History of Greek Science

After two year the engineer and mathematician Paul Tannery published a very different history of ancient science. Although he had spent his career in the corps of engineers of the French state factories, he had pursued fundamental researches in the history of ancient science: he could rely on mathematical competence, philosophical sensitivity, and the painstaking study of ancient languages [Duhem 1905, p. 216]. He had made a great effort to become acquainted with historical and philological issues. The accuracy of his historical reconstructions, the careful and detached analysis of original texts, and the presence of a cautious but definite historiographical perspective, make his history of science a milestone in the intellectual landscape of the late nineteenth century. As I have already remarked in the *Introduction*, Tannery did not manage to gain an academic position in France even though he was acknowledged as one of the most competent European scholars in the history of

4 Berthelot had started from an extremely naïve historiographical framework, where prescientific bodies of knowledge were looked upon as purely mythological or partially scientific but essentially unreliable. The alchemic body of knowledge had probably challenged that simplified approach because of the different traditions and attitudes converging on it: philosophy, empirical practices, and mystic attitudes. See Ryding 1994, p. 121: "Wrapped in the shrouds of allegory and speculative metaphysics, alchemy belonged to a conceptual universe that deliberately defied the conscious, linear and logical mind."

ancient science.[5] He contributed to the establishment of a modern history of science, where the adjective *modern* means a history of science that does not retain traces of hagiography or justification of present-day scientific theories and beliefs, and does not confine itself to a list of successes. His determining contribution rested upon the critical analysis of primary and secondary sources, and the necessity of a rational disentanglement of historical events [Brenner 2003, pp. 184-5].[6]

Tannery's historical researches fell within a meaningful tradition of histories of mathematics that had began with Montucla, had gone on with the Göttingen poet and mathematician Abraham Gotthelf Kästner, who published a four-volume history of mathematics in the last years of the eighteenth century, and then had been developed by the German Moritz Cantor and the Danish Johan Ludvig Heiberg and Hieronymous Georg Zeuthen.[7] Tannery considered himself a follower of Comte, even though he should not be listed in the number of scholars who followed Comte "in the messianic adventures" of the last part of his life. Tannery acknowledged the influence of Comte on the French cultural environment and even "on the esprit of the civilised world." At the same time, he acknowledged some faults of Comte's conceptions: among

5 For a reconstruction of the events that prevented him from being appointed to a Chair of "Histoire Générale des Sciences" at the *Collège de France*, see Milhaud 1906, p. 14, Sarton 1948, p. 30, Gusdorf 1966, pp. 43-4, 62, 98-101, and 104-6, Canguilhem 1979, p. 63, Stoffel 1996, p. 416, Brenner 2003, pp. 5 and 101, Locher 2007, p. 217, and Chimisso 2008, p. 85, fn. 1. It is worth remarking that Berthelot had supported Tannery's nomination. I briefly remind readers what I have written in the *Introduction*: the chair of *History of science* was looked upon as "a fiefdom of the positivist school (or church)" [Sarton 1938, p. 690].

6 Gaston Milhaud stressed the international standing of Tannery's historical researches: in 1906 he claimed that "c'est encore son nom qui évoque le mieux et le plus justement, devant le monde savant tout entier, la part de notre pays dans les travaux d'histoire des sciences" [Milhaud 1906, pp. 11 and 14]. In the late 1930s, Pierre Boutroux stressed two essential features of Tannery's intellectual enterprise that can be considered as hallmarks of the professional history of science: the acquaintance with the languages of primary literature, and the critical analysis of secondary literature [Boutroux P. 1938, p. 693].

7 Cantor had published *Mathematische Beiträge zum Kulturleben der Völker* in 1863, and the first volume of his monumental four-volume *Vorlesungen über Geschichte der Mathematik* in 1880. After having published many papers on the history of mathematics, Zeuthen published the volume *Geschichte der Mathematik im Altertum und Mittelalter* in German in 1896 (the Danish version had appeared in 1893). After having discovered previously unknown texts of Archimedes, since 1880 Heiberg published editions of Archimedes, Euclid, Ptolemy, Apollonius, and Hero's works. After Tannery's death, Zeuthen and Heiberg took care of the edition of Tannery's collected works [Boutroux P. 1938, p. 695].

them, the underestimation of some advances in the domain of life sciences, and the fact that Comte's classification was definitely unsuitable for ancient and mediaeval natural philosophies. Moreover, with regard to the alleged *law of three stages*, Tannery stressed the misuse of the word *law* in the context of historical researches, a misuse that had stemmed from naive "attempts to imitate scientific methods." He also remarked that the alleged law could not be extended to other times and other civilisations [Gusdorf 1966, pp. 89-91 and 113; Tannery 1905, pp. 410-14].[8]

In 1887 Tannery published a vast study on early Greek natural philosophy under the title *Pour l'histoire de la science Hellène*. Although the book was an almost four-hundred-page volume, the time interval he explored was really narrow: the sub-title specified that he would discuss the ancient scientific tradition "from Thales to Empedocles." From the outset he explained how the word *science* was questionable in the context of early Greek philosophy; he also pointed out that the practice of writing histories of science could be traced back to classical Greek tradition. Two important contributions had stemmed from the Aristotelian school: Theophrastus of Eresus' history could be qualified as a history of natural sciences, whereas Eudemus of Rhodes wrote a history of mathematics. Tannery discussed the meaning of the term *science* during the centuries from Thales to Ptolemy and Galen: apart from mathematics and medicine, no *science* in the modern sense had managed to become an independent and easily identifiable body of knowledge. He found that ancient science could naturally be divided into four stages: "*Hellenic* science" in the strict sense, then "*Alexandrine* science," "*Graeco-Roman* science," and finally "*the age of commentators*" that was an age of *decadence*. The book focused on the first stage, which had been "the object of most studies" but was also "the most obscure" stage. He regretted that the history of the early representations of Nature had been embedded in the general histories of philosophy: those representations deserved and required a specific inquiry [Tannery 1887a, pp. 1-5 and 7-9].

He stressed the necessity of separating "philosophical history" from "scientific history" since they involved "totally opposite methods." He found that the first philosophers were more "naturalists [*physiologues*]" than philosophers in the modern sense. Unfortunately, historians of philosophy had naturally leant towards an abstract classification of theories in accordance with modern lin-

8 Milhaud emphasized the distance between Comte and Tannery's historical and philosophical views: "la lecture de Paul Tannery est certainement une de celles qui ont plus contribué à me mettre en défiance contre la philosophie scientifique de Comte" [Milhaud 1906, p. 13].

guistic and conceptual standards. They had tacitly assumed a sort of ideal continuity between different conceptual contexts in order to safeguard the unitary structure of philosophy. On the contrary, a history of science required the analysis of both historical filiations and discontinuities. Tannery preferred a chronological order rather than the more abstract "order of schools" typical of the tradition of philosophical history. Another question involved the sources: the original texts had been lost, and the history of subsequent loans and influences had to be analysed. In reality his history was both a history of original ideas and a history of historical reconstructions. The first histories of sciences, namely Theophrastus and Eudemus' histories, had been written under the influence of the incoming Aristotelian tradition, and subsequent histories composed under their influence showed relevant misunderstandings. Tannery's history of science was also a history of the transmission of those misunderstandings. A complex network of direct and indirect genealogies, and broader influences, had to be explored [Tannery 1887a, pp. 10-11, 14, and 18-9].[9]

Some historiographical theses emerge from Tannery's historical inquiry. Both the influence of *barbarous* (according to Greek linguistic tradition) contributions, and "the unquestionable originality of Greek genius" were at the basis of the development of Greek mathematical and astronomical sciences. More specifically, he claimed that Greeks had learnt arithmetic from Egyptians: this was true at least for Thales, and even Thales' cosmology was "absolutely identical to the cosmology that can be found in the most ancient Egyptian papyruses." He looked upon Anaximander as the first scholar who had put forward new views, even though he had been credited with later and extraneous scientific knowledge. He found misleading the meaning Aristotle had given to Anaximander's απειρον: it corresponded to an *indefinite* substratum rather than a more abstract *infinite* extension. According to Tannery, Anaximander had assumed a rotatory motion of the world, and "a rotation extended to infinite" was a patent contradiction. It had been Xenophanes who had later assumed an infinite world, but at the price of "excluding the dogma of the revolution." However, unlike the infinite extension of space, which involved kinematic problems, the infinite extension of time, and "the eternity of the world" had never been questioned "on Greek land" [Tannery 1887a, pp. 53, 61, 65, 71, 87, 94, and 96].

Even Anaximenes of Miletus did not manage to go beyond the Chaldean cosmological reference frame, but Tannery found two interesting novelties in

9 On the influence of German history of philosophy on Tannery, and more specifically on the philosophical background of the conception "of history of science as *complementary* to history of philosophy," see Catana 2011, pp. 517-23.

what remained of his original texts: a new "clarity and simplicity of style," which was remarkable when compared to Anaximander, Pythagoras, and Heraclitus' "poetical and obscure" fragments, and the association of the five known planets with the Sun and the Moon rather than the fixed stars. He found in Anaximenes the first clear assumption of "the unity of matter, or rather a substance endowed with perception, intelligence and will." Tannery found that the confidence in the existence of a universal substratum was not so structurally different from the modern confidence in the existence of aether. In both cases, neither a formal demonstration nor direct or indirect experiences could justify that confidence: the unity of matter was nothing more than an hypothesis [postulatum] both for Anaximenen and modern scientists. Tannery took inspiration from his historical reconstruction for stressing the persistence of very general hypothesis in the history of science despite the fact that "the logic of facts" had transformed our empirical body of knowledge. After Anaximenes, scholars had relied on the meta-theoretical belief in the unity of matter "with unshakable stubbornness" in spite of the set of facts that appeared "consistent with pluralism." He found another firm belief in the context of Pythagorean tradition, namely the unshakable faith in the intrinsic dichotomy between the two opposite "material principles": the definite, limited matter [περας], and the indefinite or infinite substratum [απειρον] [Tannery 1887a, pp. 150, 158, 162-3, 178-80, and 202].

According to Tannery, the great problem at the borderline between science and philosophy that concerned the origin of the world, or the possibility of an "indefinite series of worlds" over time, could not be solved in the context of "positive knowledge." The gap between rational speculations and actual experiences had emerged when Parmenides had put forward a rational representation of the world. The alleged absolute rest of a finite world had emphasised the "patent opposition between the consequences of reason and sensorial experiences." For the first time in the history of human civilisation, at least in the written form, the existence of an intrinsic tension between the formal structures of reason, and the appearances of senses had explicitly been acknowledged. That opposition had been expressed in different ways over time but it had proved to be one of the most long-lasting and sensitive issue in human knowledge. The opposition between *being* and *becoming* in Xenophanes, or *truth* and *opinion* in Parmenides were different implementations of what Tannery qualified as a "chasm [abîme] that cannot be filled." According to Tannery, the ancient commentators, starting from Aristotle and Theophrastus, had failed to grasp the close link between cosmological and philosophical issues in Xenophanes and Parmenides texts [Tannery 1887a, pp. 100-1, 125-7, 130, and 134].

He claimed that Parmenides' texts contained "the foundation of what we call theory of knowledge." He found a structural analogy between Pythagorean physical dualism, and Parmenides' epistemological dualism between truth and opinion, even though Tannery acknowledged that his view could not rest upon a reliable Pythagorean tradition before Philolaus. That analogy could explain the fact that some astronomic and geographical theses such as the roundness of the Earth had been attributed by commentators to both Parmenides and Pythagoras. On the other hand, the immobility of the Earth at the centre of the World suggested some kind of proximity to "Anaximander's pure doctrine." In brief, he found in Parmenides "one of Anaximander's reasonably loyal disciples … as to physics" alongside "decidedly Pythagorean elements" that could be traced back to a common intellectual and geographical environment. This fact suggested to Tannery that the Pythagorean school did not rely on its own "physical system," and "the Ionian doctrine" still represented the common physical background of some Pythagorean circles. From the point of view of the history of natural philosophy, Tannery traced back the emergence of *dynamism* to Pythagoras and Heraclitus' fragments, namely philosophical traditions imbued with a theological background. The roots of modern mechanism had emerged afterwards, when the atomistic school had rejected that "well-established dynamism" [Tannery 1887a, pp. 223, 225-6, 229-30, and 234-6].[10]

Sometimes Tannery's theses were not in tune with the received view on ancient Greek philosophy. With regard to the Eleatic school, he repeatedly claimed that Zeno of Elea had been misinterpreted, and "the true aim of his lines of reasoning" had not been suitably grasped. Although Zeno had probably not been a mathematician in a proper sense or a naturalist [physicien], he had been one of the scholars who "had done more for the principles of mathematics." Far from having denied motion in itself, he had simply shown the contradiction between motion and "the pluralistic view." The historical tradition that went from Aristotle to Simplicius through Eudemus of Rhodes and Alexander of Aphrodisias had missed the point, and had propagated a mistaken reconstruction of Zeno's paradoxes. In the Pythagorean tradition, a geometrical entity consisted of "a sum or plurality of points" in the same way as a number consisted of "a plurality or sum of unities." In spite of the discovery of irrational numbers, Pythagorean scholars had gone on with the representation

10 Some years later, Pierre Duhem paid homage to Tannery's historiographical perspective. According to Duhem, Tannery combined critical with creative attitudes, and this effective alliance had allowed him to put forward both "detailed analyses and ingenious inductions" [Duhem 1905, pp. 219 and 221].

of geometrical shapes or physical bodies as "arrangements of points." Zeno had shown the intrinsic contradictions of that representation: a body could not be a sum of points, time could not be a sum of instants, and motion could not be "a sum of simple passages from point to point." According to Tannery, Zeno's paradoxes had played an important role in the theory of knowledge. After Parmenides' separation between logical intelligibility and experience, Zeno had identified "a borderline between the mathematical and the empirical points of view" [Tannery 1887a, pp. 248-51, 253-6, and 258].[11]

In the context of the Eleatic school, the separation between the domains of reason and experience became even sharper in Melisso of Samos' fragments, wherein Tannery saw a *monistic* form of idealism. In reality, I find questionable whether we can actually find in Melisso "the seeds of the doctrine of Ideas": it seems to me that the epistemological dichotomy between the formal structures of reason and the body of empirical knowledge does not deal with idealism. From Parmenides onwards, the awareness of that dichotomy was the tacit hallmark of the whole of Greek civilisation, and it found the most explicit clarification in Aristotle's *Analytica Posteriora*. However Tannery remarked that a plurality of interpretations could stem from the heterogeneous collection of Melisso's original fragments, and in the end, the Eleatic school had undertaken an intellectual pathway that had led natural philosophers astray [Tannery 1887a, pp. 261-3, 267].

3 Historical and Critical Reconstructions

In the same year Tannery also published a book on Greek geometry that collected the researches he had undertaken since Spring 1885, which had been already published as single papers in the journal *Bulletin des Sciences mathématiques*. The book was intended as "a critical essay," and the subtitle specified that the author had explored sources, transmission, and later receptions of ancient mathematics. Once more the French scholar conceived a history of

11 Tannery's researches can be looked upon as the starting point of modern debates on Zeno's paradoxes. Bertrand Russell, who was not interested in historical and historiographical issues, did not mention Tannery but made reference to subsequent French literature. He stressed the philosophical soundness and fruitfulness of Zeno's "immeasurably subtle and profound" arguments. According to Russell, Zeno did not "prove that the continuum, as we have become acquainted with it, contains any contradiction whatsoever" [Russell 1903, pp. 347-8 and 355]. On Zeno's paradoxes, and on the role played by Tannery, see Barnes 1982, pp. 182-216, 495, 527, and 540, and Fano 2012, pp. 15-7.

science that was a history of the specific ideas and theories contained in the original sources or commentaries, and at the same time a history of the network of historical reconstructions. In other words, he was interested in how the traditions had developed, and how they had been transmitted. He wrote for educated people in general but, at the same time, he addressed a definite community of historians, and faced a definite received view. He was aware that, in every epoch, there was a body of shared knowledge and "a set of shared ideas that are consequently hegemonic"; he was also aware that "the chance of success for a new conjecture" depended on the degree of agreement it could gain in the professional community. On the track of Darwin's "theory of *evolution*," and on the track of a broader concept of evolution, "the general conception of the history of mathematics" rested upon the meta-theoretical pillars of "an obscure origin," and a subsequent continuous or indefinite progress. He found that this historiographical framework was too naïve, and did not manage to grasp the actual historical development, wherein "sudden transformations" had taken place. More specifically, a reliable history of mathematics had to account for "*the events and the causes*" that had led to stages of "*past decadence*" [Tannery 1887b, pp. V-VI, 4, and 8-9].

With regard to the sources of Greek geometry, for centuries Proclus had represented the most important source. Nevertheless, he had lived in the fifth century, and historians agreed on the fact that he had never had direct access to "any geometric work before Euclid" nor to Eudemus of Rhodes' history which he had frequently mentioned. The question was therefore "with what loyalty" the history had been transmitted from Eudemus to Proclus. A complex network of filiations had to be rebuilt, and intermediate characters had to be identified. Among them, the Stoic philosopher Geminus had played an important role in the first years of our age. The historical relationship between Geminus and Proclus was so important that Tannery devoted to it the first chapter of the book. His historical and philological analysis led to the conclusion that Proclus was "a more or less faithful echo of authors like Geminus" for the first part of his commentary, and "Pappus for the last part" of it. At the same time he found that other authors had to be excluded from the list of Proclus' sources. Among them, Eudemus, Speusippus, Apollonius, and Heron of Alexandria. Although Tannery's analysis was quite detailed, he specified that his research on sources allowed him to put forward conclusions that had to be looked upon as merely provisional. Another issue involved the classifications of sciences and mathematics in ancient culture. Different systems of classification were at stake, and the semantic content of words like *mathematics* and *geometry* had changed over time. The borderline between abstract methods of geometry and computational practices or *logistic* were crossed by Diophantus of Alexandria, who followed an "analytical pathway" which filled the gap between two bodies of knowledge that had traditionally been separated [Tannery 1887b, pp. 15-6, 21, 24, 28, 38-47, and 51-2].

From the point of view of Tannery's historiographical perspective, it was not so important to establish what ancient mathematicians probably knew but rather to ascertain how the tradition concerning their achievements had developed over time. The analysis of sources had convinced him that "a treatise of geometry under the name of Pythagoras" had probably been in circulation since the mid-fifth century B.C., and the text had "the same structure as Euclid's *Elements*." According to Tannery, the mathematicians who played the most important role in the development of geometry between Pythagoras and Euclid were Eudoxus of Cnidus and Theaetetus of Athens. More specifically, he found that Eudoxus' theory of proportions represented "a crucial step" in the process that profoundly reshaped Greek geometry. In general, the postulate of parallels and "the principle of similarity" were the keystone in the emergence of "geometrical abstraction." Tannery insisted on the importance of Eudoxus' achievements in the context of geometry and astronomy, which was another section of mathematics in the ancient classification. He also claimed that Eudoxus had been the first to put forward "a mathematical representation of the world," and "almost the totality" of the classical texts before Ptolemy dealing with the shared body of astronomical knowledge could be traced back to him [Tannery 1887b, pp. 94-5, 98-9, and 127].

With regard to the reception of Euclid's *Elements*, Tannery was interested in ascertaining when they had become "a classical text for teaching geometry in ancient times," and when the practice of writing commentaries to that text had really begun. He was convinced that Heron had "actually written a commentary on Euclid" that was probably "complete and detailed." It had been used by Porphyry in the second half of the third century, by Pappus in the first half of the following century, and subsequently by many Arabic-speaking commentators. Although Heron's work on Euclid still existed at Proclus' time, the latter had probably made reference to Porphyry and Pappus' subsequent commentaries. Tradition had also attributed to Heron the essay *Definitions of terms in geometry*, but Tannery's philological and mathematical analysis led to the conclusion that the text was to be attributed to Geminus rather than Heron. In the end, it seemed to Tannery that, after Heron's time, Euclid was to be looked upon as "a classic without any hesitation." From then onwards the Euclidean text together with its first commentaries were handed down to subsequent generations as a stable body of knowledge [Tannery 1887b, pp. 165-6,

174, 176, and 178-81].[12]

Six years later Tannery published another comprehensive book on ancient science, *Recherches sur l'histoire de l'astronomie ancienne*. From the outset he explicitly avowed his aims: an analysis of Ptolemy's *Almagest* more detailed and correct than those previously put forward, and an inquiry into "the antecedents" as far as ancient sources made it possible. More specifically, he was interested in clarifying Ptolemy's scientific debt towards Hipparchus of Cyzicus. His conclusions were not in tune with the received view because he found that Hipparchus' role had been "oddly overestimated." He stressed that the development of astronomy in ancient times was consistent with a substantial continuity between Eudoxus and Hipparchus. Ptolemy's theory represented the third stage in the development of Greek astronomy, and the different stages could be associated with three different words and concepts. The first stage of "Hellenic *astronomy*" corresponded to "questions of calendar" whereas the second stage corresponded to *astrology*, which was to be understood as a twofold practice, namely cosmography and the divination of the future. The third stage was to be associated with the expression "*mathematical composition* [συνταξις]," and corresponded to "subsequent improvements in computation and observation" [Tannery 1893, pp. V-VIII].

Tannery remarked that afterwards, in the Middle Ages, the words *astronomy* and *astrology* were to be looked upon as "synonymous terms." The term astronomy could not be seen as the proper linguistic translation of what ancient Greek meant for the science of the cosmos. The Greek word νομος, namely *law*, implied "a human institution," and it was opposed to φυσις, namely *nature*. In other words, celestial bodies moved "according to their *nature*": in this sense, they could "not follow any *law*" in a proper sense.[13] The attention to

12 Pierre Boutroux summarised Tannery's historiographical achievements into three main theses. First, Greek mathematics flourished during almost the five centuries from the emergence of Pythagorean school to Hipparchus and then Hero of Alexandria. Second, the main discoveries were put forward, and the most innovative perspectives in the field were opened, between Plato and Archimedes' lifespans. Third, from European Renaissance onwards, the sources of Greek mathematics consisted of "second-hand literature," namely Ptolemy, Diophantus, and Pappus' compilations: they had been written when the classic body of knowledge still survived but the intellectual momentum had faded away [Boutroux P. 1938, pp. 694-5].

13 Apart from these two words and concepts, he was also interested in a philological analysis of terms like *star*. According to Tannery's interpretation, among "the most ancient Greeks," the astronomer was a man who forecast the weather: he was more concerned "with inspecting the horizon rather than the stars" [Tannery 1893, p. 19].

philology and cultural context led Tannery to other remarks on the meaning of constellations and the measure of time. Constellations represented a suitable solution for the practical problem of determining the hours during the night. Nevertheless, the Greek concept of an hour was different from the modern one: *hour* did not mean a constant time interval but a variable fraction of day or night. The hour could change in the same way as the duration of day and night changed according to the seasons [Tannery 1893, pp. 1-4 and 9-10].

Tannery remarked that Aristotle's mention of "*astrology* and astrological theorems" made reference to Eudoxus and his system of homocentric spheres, and to Callippus of Cyzicus, who had reformed that system. Eudoxus had put forward a mathematical representation of the Sun, Moon, and planets' motions by means of an ingenious combination of circular uniform motions. The weakness of that geometric model was due to the prescribed constant distance between Earth and planets. Aristotle had managed to grasp the two essential features of that *astronomy-astrology*: "the marked mathematical character" on the one hand, and the commitment to "the explanation of the whole universe" on the other. On the one hand, what Aristotle had labelled *physics* offered the explicative principles, which could not be submitted to logical demonstrations. On the other hand, *astrology* took those principles for granted, and pursued "the mathematical development" of them. Tannery insisted on the image of Eudoxus as a mathematician who was also interested in "medicine, geography, literature, ethics, and law" but was not an innovator in the field of astronomic observations. When the cultural hegemony had been handed over from Athens to Alexandria, *astronomy-astrology* had undergone an important transformation: the main protagonist of that transformation, Hipparchus, was "a patient observer and an excellent computer." According to Tannery, he was not a talented mathematician: he had not invented trigonometry and epicycles. He could rely on mathematical models such as epicycles and eccentrics that had previously been introduced by Apollonius of Perga. He could also rely on Eratosthenes of Cyrene's "mathematical geometry" [Tannery 1893, pp. 26-31, 44, 56-9, 69, and 81].[14]

14 Tannery submitted ancient achievements to a careful analysis. He discussed the emergence of new mathematical models, the invention and circulation of instruments, and new techniques of computation. He also devoted many pages to the ancient historians of *astronomy-astrology*, and the tradition of historical researches that had emerged. He discussed Geminus' influential *Isagoge* or *Introduction to Astronomy*, Cleomenes' *On the Circular Motions of the Celestial Bodies*, Theon of Smyrna's *Astrology*, and Plinius' second book of *Natural history*. A plurality of historical reconstructions had emerged, and that plurality was a consequence of a plurality of world-views [Tannery 1893, pp. 82-3].

Tannery stressed that the hypothesis of the immobility of the Earth rested upon logical and physical reasons. The supposed difference in nature between the Earth and celestial bodies justified the choice of the Earth as a centre at rest: the expected effects of a possible rotation of the Earth confirmed that choice. He guarded against the attitude of regarding Aristarchus' heliocentric model in retrospective, with the benefit of hindsight. From the mechanical point of view, the heliocentric model would have represented a great step forward, but from the geometrical point of view, "that conception did not involve any real advantage." As Tannery rightly remarked, Greek *astronomy-astrology* had never gone beyond a geometric representation, and in some way, neither Tycho Brahe nor Copernicus had managed to go beyond this. He found that Copernicus' main achievement was the simplification of the hypotheses concerning epicycles and eccentrics, even though he had preserved "the ancient geometrical principles for the explanation of planetary motions." With regard to Brahe's system, he found that it was "the logical consequence of Ptolemy's astronomical hypotheses." In other words, Brahe's system was the geocentric system most consistent with the available observations in Ptolemy's time, in particular the observations of Mercury and Venus. Greek *astronomers-astrologists* (Aristarchus included), Brahe, and Copernicus also had in common the conception of the world as a finite sphere.[15] Tannery remarked that Giordano Bruno was the first scholar to have envisaged the infinity of the universe, and he was not an astronomer [Tannery 1893, pp. 99-103 and 120].

After having explored the traditions and the problems with reference to the motion of planets, the duration of the solar year, the tables for the Sun and the Moon, and lunar motions, Tannery concluded that "the mere observations of eclipses" would have led Greek *astronomers-astrologists* "to give up or at least seriously modify the hypothesis of epicycles." Although Hipparchus had suspected that some changes would have been necessary, "Ptolemy disregarded the scruples of his precursor," and he did not step back "even in front of a geometrical structure in blatant contradiction with observations." Ptolemy had underestimated what Tannery labelled "the capital vice of ancient astronomy," namely the unbridgeable gap between the theoretical scaffolding of epicycles and eccentrics and the observed "difference between minimum and maximum diameter of the Moon." Ptolemy had done worse than Hipparchus because he had not attempted to correct "a clearly perceived mistake": he had worsened

15 According to Tannery, the only exception was represented by Democritus and "the Hellenic-Babylonian Seleucus of Erythrea who supported the heliocentric system and the infinity of the world" [Tannery 1893, p. 101].

the mismatch by means of new geometric combinations. With regard to the average diameter of the Sun he had made similar mistakes, and the combination of the two mistakes had led to the impossibility of ring-shaped eclipses of the Sun. Tannery's synthetic evaluation of Ptolemy's contribution to astronomy was not positive. He credited Hipparchus with having already put forward "almost all that was valid" with regard to the Sun and the Moon: Ptolemy had confined himself to improving the theory of planets. Tannery ventured to claim that Ptolemy had made "science go backwards" [Tannery 1893, pp. 230-3, 242-3].[16]

With regard to the cosmological aspects of Greek *astronomy-astrology*, Tannery insisted on two main issues: the choice of a geocentric model, and the absence of a physical explanation. The two issues overlapped to a certain extent. Some time before Hipparchus, "mathematicians had given up the search for a mechanical explanation of celestial motions," a commitment that had been pursued by Eudoxus and probably Archimedes. Aristarchus' heliocentric hypothesis collided with religious biases, and it was not facilitated by the fact that he had lived some time later than Eudoxus and some time earlier than Hipparchus. The model of Eudoxus' concentric spheres was still hegemonic, and the theory of epicycles and eccentrics had not yet been put forward. Once more Tannery stated that the consistent development of Greek astronomy would have led to Brahe's mathematical model, but "at the very crucial point, Hipparchus deviated, and went back to the geocentric thesis." In his turn, Ptolemy "had blindly followed" Hipparchus. The historical reconstruction of what had presumably happened two thousand years before led Tannery to a remarkable conclusion that could be placed on the borderline between history and philosophy of science. The issue of geocentric and heliocentric models could be looked upon as "a meaningful instance of the importance of a priori (metaphysical) ideas in the development of science" [Tannery 1893, pp. 256-60].

16 Tannery pointed out that Ptolemy's theory of planets had not managed to account for the "the variations of distance" [Tannery 1893, pp. 243 and 245-6]. More recently, the historian of mathematics Lucio Russo put forward the thesis that what we call scientific method emerged in the fourth century B.C. in the context of Hellenistic civilization: afterwards science was defeated, and Western civilization declined. That ancient emergence, and the following defeat were forgotten, and the alleged scientific revolution in the seventeenth century was looked upon as a fresh start. The eighteenth-century *Enlightenment* emphasised the new historiographical view, and Voltaire violently disparaged any appreciation of ancient science. According to Russo, the existence of a long stage of decadence after the emergence of ancient science contradicted "the naïve and dangerous confidence in the continuous and automatic progress of mankind" [Russo 2013, pp. 17-8 and 450].

In the last pages of the book, Tannery focused on subsequent developments. After Ptolemy, the study of the sky in itself had lost ground: "judiciary astrology" had become even more important whereas astronomy had been reduced to "its humble servant." For many centuries from Ptolemy to the Renaissance, scholars had practised astronomy only because "it was necessary for astrology." For a long time, "the false science" had allowed scholars to preserve "the true one," and at the same time had allowed them to transmit fragments of pure mathematics. Even in the sixteenth century it had been astrology that had allowed "Kepler to earn his daily bread," and had assured the publication of Copernicus and Brahe's works. In brief, among Arabs, Byzantines, and Western people, the scientific body of knowledge of astronomy had been preserved and transmitted by means of another body of knowledge that was far more obscure and far less scientific in the modern sense. Science was preserved but no meaningful development took place in those centuries. From Tannery's point of view, this historical result was not surprising. It was consistent with one of his historiographical theses: science could flourish only "*when pursued as a value in itself*" [Tannery 1893, pp. 280-81].

4 Different Historical Perspectives

In the same year Berthelot published two volumes on alchemy in the Middle Ages. The volumes contained Syriac and Arabic manuscripts together with their translations into the French language; they were preceded by an *Introduction* that set out Berthelot's historical and scientific remarks. He honestly avowed that he had simply revised the experts' translations "from the technical point of view" in order to give "a scientific sense" to the texts. The manuscripts were the late outcome of a historical process begun "in the mid-fifth century" when Aristotle's researches and other texts concerning the liberal arts had been translated from Greek to Syriac. The exodus of Greek-speaking scholars towards Persia had given birth to intellectual communities where Greek scientific tradition was studied and commented. Alchemy and astrology belonged to that body of knowledge: they had been practised together with "medicine and mathematics, and by the same scholars." Syriac texts could be looked upon as a cultural bridge between Greek and Arabic civilisations. Together with Costantinople, the Abassid Caliphate had become a centre of attraction for the compilation and synthesis of the results of ancient science. The manuscripts Berthelot was to comment in the first of the two volumes came from the British Library and Cambridge University: they were copies of texts that had probably been written between "the seventh and the tenth century"

[Berthelot 1893a, pp. I-VII].

He stressed the empirical feature of that body of knowledge which included the melting of lead, iron, copper, and tin, the quenching of bronze, and the preparation of "sulphuric water" for colouring metals. The gold-colouring of metals, and "the actual transformation into gold" seemed persistently confused with each other by alchemists. The empirical relationship between the purity of tin and its melting temperature stood beside "the surprising conflation of Christian prayers and ancient Egyptian invocations." Scattered remarks on the usefulness of science and experimentation could even be found [Berthelot 1893a, pp. XI-XXXI, and XL-XLI].

In the second volume he focused on two Arabic manuscripts that came from Paris and Leyden. Latin translations of those manuscripts had circulated in the Middle Ages, but there were "great differences" between the original texts and translations. He questioned the attribution of some Latin treatises to Geber, namely Jābir ibn Haiyān, because those texts contained "new and original knowledge ... that was unknown to the Arabic author." He therefore claimed that "Geber's alleged Latin treatises" were apocryphal.[17] Berthelot stressed that the reliability of references and quotations in mediaeval manuscripts was quite low in general: frequently Pythagoras, Democritus, Plato, and Aristotle had been looked upon as outstanding alchemists. In the end, Berthelot hoped that the French translations could cast light on the problematic connections and genealogies that linked the body of alchemic knowledge of Egyptians and Greeks to Byzantine, Syriac, Arabic, and subsequent Latin traditions [Berthelot 1893b, pp. 5-7, 12, 16, 23, and 25-6].

---

17 In the 1920s, historians of chemistry and alchemy cast doubt on the reliability of Berthelot's historical reconstruction [Jenkins 2014, p. 2383; Brock 2009, p. 11]. Eric John Holmyard and John Riddick Partington, scientists and historians of chemistry, published brief papers in the journal *Nature*, wherein they pointed out some shortcomings of Berthelot's volumes on the history of alchemy. In 1922, after having stressed some erroneous transcriptions in Berthelot's texts, Holmyard focused on two issues: the questionable identity of the supposed Islamic alchemist who had been known with the Latin name of Geber, and the scant number of Arabic works of Jābir ibn Haiyān that Berthelot had analysed. Unlike Berthelot, who had associated Geber to a group of Latin forgers, Holmyard was inclined to assume that Jābir ibn Haiyān and Geber were actually the same person [Holmyard 1922, pp. 573-4]. The following year, in the same journal, Partington stressed how unsatisfactory Berthelot's historical reconstruction really was, and even suggested "a new start" in the researches on ancient alchemy. When he focused on the identity of Geber, he remarked that the theses on his identity remained substantially hypothetical, and hinted at the possibility of a set of Greek, Arabic, and Hebrew manuscripts as the sources of Geber's texts [Partington 1923, p. 220].

Berthelot's historical researches cannot be compared with Tannery's studies for at least two reasons. First, he did not master the original languages of the manuscripts, and therefore he could not deal with sensitive philological issues. Second, he had started from a Comtian historiographical perspective, and the confrontation with ancient alchemy had only managed to slightly modify his original attitude. He appreciated the amount of practical knowledge that he had found in those manuscripts: this was in tune with his image of modern chemistry as the most empirical of sciences, and with his general image of science as a practice that rested upon the observation of *facts*. As a consequence, his historical books can be considered a questionable contribution to the emergence of a professional history of science in French-speaking countries in the late nineteenth century. Nevertheless, their value should not be underestimated because they testify to the existence of a widespread cultural trend in that geographical environment at that time. A new sensitivity to the history of science also influenced scientists who were still in tune with quite a naïve positivistic stream.[18]

However, Tannery's style of research inspired a younger scholar, Gaston Milhaud, who had studied mathematics at the *École Normale Supérieure*, and had then taught mathematics in high schools and Universities. He therefore came to history and philosophy by way of mathematics, and eventually succeeded in becoming a Professor of Philosophy at the University of Montpellier.[19]

The book Milhaud published in 1893, *Leçons sur les origines de la science grecque*, was dedicated to Tannery "with respect and gratitude," and it was the

18 In 1924 Holmyard published a paper in the journal *Isis*, where a more extensive critical examination of Berthelot's volumes was put forward. The translations of Berthelot's colleagues, experts in the Arabic language, "were inaccurate from a scientific point of view," whereas Berthelot's subsequent intervention was uncritical from the linguistic point of view, because he had confined himself to checking the scientific soundness of the translations. In other words, the final outcome was the superposition of two subsequent shortcomings. As a consequence, Berthelot had not managed to clarify the long-standing problem of the questionable relationship between Arabic and Latin manuscripts, and more specifically the relationship between Jābir's Arabic texts and Geber's Latin texts. More in general, Berthelot had "no idea of the vast extent of Arabic chemical literature." However, at the end of his paper, Holmyard stressed that Berthelot's translations had been useful: putting them in perspective, they had been important "not in accomplishment, but in indicating what was to be accomplished" [Holmyard 1924, pp. 483-5, 487, 489, 495, and 499].

19 Milhaud set up "a programme of study in philosophy of science" at Montpellier University in the 1890s. A Chair of *History of Philosophy in its Relation to Science* was then created for him at the Sorbonne in 1909. He was one of the first scholars of Jewish origin to be appointed to a Chair in Paris [Chimisso 2008, pp. 25-6; Brenner and Gayon 2009b, p. 5].

outcome of some lessons he had held for students of sciences and humanities at Montpellier University. The volume did not present original researches, and unlike Tannery's books it did not offer a scholarly inquiry into history and philology. Nevertheless, Milhaud was consciously pursuing the aim of setting up a tradition of research that could rely on Tannery's innovative and authoritative researches. He started from the acknowledgment of a pure fact: from the institutional point of view, the history of science did not exist in France. The establishment of a well-grounded and professionalised field of research was a really demanding task: it required scientific competence, knowledge of general history, and familiarity with documents. Apart from specific competences, both scientific and philological, it required a marked philosophical sensitivity [un sens philosophique profond]. This sensitivity would have allowed scholars to go beyond the enumeration of "a huge number of scientific works in different ages" in order to focus on the history of ideas, methods, and theories. The history of sciences and the history of scientific method were components of a wider "history of the human mind [âme] in its different expressions." To a certain extent, science was a creation of the human mind [esprit], and therefore it could not be a complete set of objective truths but rather "a specific language" or "a specific implementation of human thought" [Milhaud 1893, pp. 3-5 and 8-9].[20]

In scientific practice he gave prominence to the creative power of the mind: scientific explanation was nothing other than "the search for constant relationships among indefinitely variable phenomena." The most important process was the reduction of a set of phenomena to a law: both scientific entities and laws were created by the human mind. According to Milhaud, the transition from Kepler's laws to Newton's law of gravitation represented a meaningful instance of that intellectual process where "new concepts emerged in the scientific language" even though no new empirical knowledge was available. This epistemological and historiographical perspective was in conflict with Comte's positivistic faith in the mighty pressure of *facts*. In reality, scientific progress was "a linguistic evolution," or in other terms, "a new explanation of the same phenomena." Following the above-mentioned instance, "Newton's language" had replaced "Kepler's language" in the scientific translation of the

20 In 1906, after Tannery's death, Milhaud warmly acknowledged the deep influence exerted by Tannery: "Quand je commençais à lire ses monographies sur Thalès, sur Anaximandre, sur Parménide, et sur la plupart des penseurs qui ont précédé Socrate, je me sentis captivé par le charme de ses études, au pont que dès ce jour mon désir eut été de prendre pour modèle et pour guide cet esprit si original, si ingénieux, et si minutieusement informé, qu'il s'agit de sciences, de philosophie, d'histoire ou de philologie" [Milhaud 1906, p. 4].

same planetary motions. The interpretation of the principles of dynamics as "truths which stemmed from observations" represented a historical and epistemological misunderstanding. Those principles were merely hypotheses; hypotheses and conjectures had always played an important role in the history of science [Milhaud 1893, pp. 11-3, 16-8, and 21-8].

After this historiographical and epistemological introduction, he focused on the dawn of Greek civilisation, and on the history of "the material conditions" that had fostered the emergence of the early philosophy. He indulged in a slightly idealised description of ancient Ionian civilisation, when Miletus, Samos, and Colophon emerged as intellectual centres of irradiation. Two elements were stressed by Milhaud: the existence of a network of communication between Ionian towns and other Mediterranean civilisations, and the relatively mild religious commitment of Ionian populations. Milhaud found that those elements amounted to suitable conditions for the emergence of what modern scholars called science and philosophy. He also credited ancient Ionians with cultural tolerance and confidence in the separation between science and religion. His historical reconstruction is perhaps too optimistic, but in the context of the late nineteenth century, it had the advantage of mitigating Comte's sharp distinction between the half-rational stage of ancient civilisations and the modern *positive* stage. With regard to the sources and the reliability of ancient historical reconstructions, he did not put forward original interpretations as he had honestly avowed from the outset. He could rely on Tannery's detailed analysis of primary and secondary sources "for mathematics," and Hermann Diels' analysis "for physics," namely the ancient natural philosophy. Milhaud's main aim was the establishment of the intellectual dignity and philosophical relevance of the history of science [Milhaud 1893, pp. 35, 42, and 50-5].

The third and fourth chapters of the book were devoted to the influence of "Egypt and Eastern countries" in the development of Greek science. He contrasted the views of Encyclopaedists, who had attributed "an advanced science" to ancient Eastern people: the level of that science had definitely been overestimated. At the same time he contrasted a more recent cultural trend that had influenced "all intellectual domains in the last fifty years," namely the concept of evolution. Scholars who made reference to that trend were not willing to see science as the outcome of a specific and historically determined process "that had taken place on Hellenic soil." After having briefly listed the different theses of some German and French historians, he went back to Tannery's thesis, which had already been put forward by Montucla the previous century: Eastern civilisations had confined themselves to developing "the material foundations" of all sciences. He found that recent discoveries like the

Rhind papyrus confirmed that thesis. The existence of ancient monuments certified that a high-level civilisation had originated there, but those monuments had required only "unsophisticated scientific knowledge." Abstract science, or in Milhaud's words "pure and unselfish science," had had a Greek origin [Milhaud 1893, pp. 69-70, 75-6, 86, 92, 114, and 120].[21]

In the second part of the book Milhaud divided Greek ancient science into "general physics," astronomy, and "pure mathematics." After having stressed that Aristotle could not be looked upon as a reliable source for "an unbiased history of ideas," he credited Thales, Anaximander, and Anaximenes with having been the first scholars to have pursued a scientific explanation of the universe. More specifically, he found in them a common commitment to reducing "the various and complex phenomena" to unity by means of a common principle. His discussion of Anaximander's απειρον does not deserve to be mentioned because it was based on Tannery's researches. On the contrary, Milhaud's interpretation of the whole of Milesian tradition is worth reporting. In that tradition he found "the idea of a unique homogeneous matter" that could give rise to all bodies and all perceived phenomena by means of transformations triggered by mere motions. In other words, in the Milesians he saw the roots of that mechanical world-view which allowed "a scientific explanation of the world." He also stressed that the advancements and stalemates in science had always been a consequence of the level of confidence in that world-view. Unfortunately, Aristotle had opposed the Milesians' approach: according to Milhaud, Aristotle's *qualities*, and "transitions from the potential to the actual" had led scholars astray. Nevertheless, the Milesian conceptual stream had managed to survive, and after many centuries had found "its most radical implementation" in Descartes's science and philosophy [Milhaud 1893, pp. 155, 160, 164, and 178-9].

Milhaud's historiographical thesis led him to the counterfactual prediction that the emergence of "mathematical physics would not have waited two thousand years" if Aristotle's philosophy had not gained the hegemony in subsequent centuries. The worst stumbling-block on the way to the development of modern science had been the lack of confidence in "the application of mathematics to the universe." It seems to me that Milhaud managed to grasp the

21 It is worth stressing that Milhaud's insistence on Greek "pure science" as opposed to the empirical character of Egyptian and Mesopotamian science was not in tune with one of the hallmarks of modern science, namely the close alliance between theoretical and empirical practices. The same remark might be made about Milhaud's rigid separation between "the search for truth" and the usefulness of "practical knowledge" [Milhaud 1893, pp. 141-2 and 152].

role played by Aristotle in the historical development of science, but the positive role played by "Anaximenes, Pythagoras, Democritus, and Plato" seems overestimated. The fact that he included those philosophers in the same set appears even less convincing. However, he did not fail to pursue a fair historical method: he did not exclude the idea that Aristotle's perspective could be revived by processes which could not be "reduced to quantity" or mechanical explanations. He did not even exclude that going back to Aristotle and his "substantial qualities" might find "a reasonable justification" in a future science. He also specified that his appreciation of the Milesians' views did not deal with the specific contents of their natural philosophy. He was not primarily interested in focusing on "false or true conceptions" but on perspectives that were "more … advantageous for the advancement of science." In competition with religion, the Milesians had affirmed the right of natural philosophy to "deal with cosmological problems" [Milhaud 1893, pp. 180-2].

With regard to "the specific role played by the Pythagorean and Eleatic schools," he found that the former could be credited with having speculated "on the abstract features of geometrical shapes," and having believed that "things could be *explained* by numbers" in a concrete and objective sense which was far from the modern perspective. Nevertheless, he found in Descartes "more than an analogy" with Pythagoras' mathematical attitude: Descartes had claimed that "all thing are *extended*" in the same way as Pythagoras had presumably claimed that "things are numbers." According to Milhaud, the Cartesian revolution was nothing else but the replacement of numbers with "*length* as the basic entity": it was a transition from "the abstract domain of numbers" to a geometrical-physical domain. On the track of Tannery's historical and conceptual reconstruction, he remarked that Zeno of Elea had challenged the Pythagorean view of bodies as a sum of discrete elements rather than the naïve conception of motion. The Eleatic school had contributed "to the positive development of science by divesting numbers of their absolute and metaphysical features," and by bringing them back to the scientific domain. Milhaud's thesis seems too strong, and his inquiry into the role played by the most ancient philosophers in the emergence of a scientific tradition seems too optimistic: even in Heraclitus' philosophy he found meaningful roots of that tradition. He found that, from the structural point of view, Heraclitus' fire was not so different from Thales' water, Anaximander's απειρον, and Anaximenes' air. In all them, Milhaud saw the search for a starting point, and "a chain of subsequent causes" [Milhaud 1893, pp. 190, 194-5, 202-4, 214, 218-9, and 223-4].

In the transition from the Milesians to Plato and Aristotle, and from them to Descartes and Leibniz, he saw a structural continuity that was much stronger than the existing meaningful differences. In that structural continuity, the difference between Milesians and Pythagoreans dissolved, as well as the differences between Pythagoreans and atomists. He found that Leucippus and Democritus could legitimately be considered as "precursors of Cartesian physics" even though their representation of the material world in terms of atoms and empty space was in opposition to Descartes' physical foundations. That difference vanished before the much more important commitment to explaining the universe by means of "an ideally geometrical and mechanical" physics. In Milhaud's historical reconstructions, general philosophical options or metatheoretical attitudes towards Nature became the most relevant issues [Milhaud 1893, pp. 220-1, 229, 255, and 264-5].[22] From the point of view of specific historical reconstructions of ancient cosmology, astronomy, and mathematics, Milhaud was indebted to Tannery, and this debt was explicitly acknowledged. Tannery was more sensitive to detailed historical and philological issues, and was more cautious about long-term processes, whereas Milhaud was more interested in interpretations and large-scale historical reconstructions. He dwelt on the borderline between history and philosophy of science, and he ventured to merge historiographical perspectives with epistemological remarks. His historical reconstructions appear less accurate but definitely more stimulating for readers interested in the development of cultural processes.

In the last pages of the book, Milhaud synthesised his main thesis: in the sixth and fifth centuries B.C., Greek scholars had defined the fundamental problems and had devised the first essential notions of modern science. At the same time, he guarded against a too naïve interpretation: "Claude Bernard or Pasteur's researches" could not be directly compared with "Empedocles or Aristotle's cogitations." There was a difference that dealt with empirical practices: ancient scholars had not managed "to fill the gap between observation and experiment." This lack of experimental practices in the modern sense had led some contemporary scholars to see an absolute incompatibility between Greek attitude [esprit] and modern methods, but the lack of experimental practices did not seem to Milhaud a good reason to assume incommensurability. From a more general point of view, Greek civilisation had seen "the

22 In the last part of the book, Milhaud once more remarked that the body of knowledge created by the "Pythagorean school" appeared to him as "enormously great." He found that "Pythagoras' genius" had only given "the first impulsion," and in the subsequent 150 years a wide community of scholars had intensively collaborated in order to develop that extraordinary legacy. In the end, the overestimation of Pythagorean achievements led him to state that "analytical geometry" was "implicit in Pythagorean works" [Milhaud 1893, pp. 299-300].

emergence of reason and freedom," in accordance with Renan's short passage which Milhaud quoted on the last page. In the end, he looked upon reason and freedom as the "essential features of our civilisation and modern science" [Milhaud 1893, pp. 301-2, 304, and 306].[23]

The following year Milhaud published a more philosophical book that focused on the complex relationship between logic and scientific practice, *Essai sur les conditions et les limites de la certitude logique*. He started from the alleged distance between subjective and objective elements in the construction of knowledge. In a strict sense, "the human mind could not exit from itself," and knowledge could not but be "essentially subjective." Every judgment could be associated with a variable proportion of subjective and objective, and concepts such as the concept of space were affected by that twofold nature. The main issue at stake was the compatibility between the outcome of "a reasonable number of scientific experiments" on the one hand, and linguistic, logical, and mathematical constraints on the other. Unfortunately, Milhaud's book is long-winded and not very clear, but he ultimately aimed to show how problematic the link between formal definitions and empirical content really was in scientific practice. He explored the problematic borderline between the formal laws of thought and the available empirical body of knowledge [Milhaud 1894, pp. 2, 4, 13, and 20].

Milhaud stressed the intrinsic tension between the formal language of thought and the empirical body of knowledge. Scientific theories offered "a language of unquestionable perfection" but that language belonged to "the domain of intelligible things" which unfolded beside "the observation of facts." According to Milhaud, the complex interaction among hypotheses, mathematical language, and experiments in scientific practice led to paradoxical effects: the more a hypothesis was "detached from perceived phenomena," the less could it be "jeopardised by observation." Even more paradoxically, the more it was *fictive* or *metaphysical*, the more it had "the chance to become a conclusive achievement of science" [Milhaud 1894, pp. 39, 41, 113-4, and 120].

The central part of the book was devoted to "the alleged conflict between freedom and the equations of mechanics" that had flourished in the

23 In the following years, Milhaud went on with advancing his historiographical thesis that "the foundation of rational science" could be traced back to the Greek cultural environment before the appearance of Plato and Aristotle's philosophical systems. According to Milhaud, Tannery had already shown that the first Greek philosophers had begun to speak "the language that we are still speaking." Apart from specific contents and specific theories, they had put forward the same problems, the same cultural reference frames, and the same patterns of explanation [Milhaud 1906, pp. 5-6].

1880s. He found that "the partisans of mechanical determinism" gave certain statements a concrete meaning that was not legitimate. They transferred very general and abstract statements to the real world: determinism rested upon the meta-theoretical belief that abstract theorems conformed to reality. The "general laws of mechanics" in themselves could not prove any meta-theoretical belief such as determinism. According to Milhaud, when some scholars put forward wide generalisations they were convinced of "abiding in the domain of science" whereas they were dealing with metaphysics. Physics was not allowed to assume "any equivalence between psychical and mechanical processes." When physics trespassed on the domain of metaphysics, they practised an extremely "dangerous kind of metaphysics." When someone thought that "the laws of modern science" entailed restrictions on human freedom, in reality those restrictions did not depend on scientific laws but on the "*a priori* belief" that nothing could escape determinism. However, he did not deem meta-theoretical statements or *a priori* beliefs as extraneous to science. The confidence in the uniformity and constancy of Nature was probably the most powerful metaphysical belief on which scientific practice was based: it was "its real raison d'être." The search for scientific laws stemmed from the meta-theoretical hypothesis that "they actually exist" [Milhaud 1894, pp. 125-6, 131, 133, 139, 142-3, and 146].

Finally, Milhaud came back to the issue that had been at stake since the beginning: the human mind had to "give up any pretension to logical certainty" in the domain of natural processes. The continuous interaction between *experiences* and *ideas* made scientific practice a dynamical process wherein provisional achievements were continuously discussed and updated. Scientific progress required an endless effort because the target could never be accomplished. Milhaud's *a priori* belief was that science would be pursued "as long as mankind survived" [Milhaud 1894, pp. 233 and 236].

In the late 1880s and early 1890s, we find three different histories of science that also correspond to subsequent chronological stages quite close to each other. At first we meet Berthelot's history of science: it was a typical positivistic history wherein both the march of scientific progress and the naivety of ancient science were emphasised. This historiographical reference frame was however mitigated by the acknowledgement of structural analogies between ancient and modern sciences. In particular, Berthelot found a common commitment to understanding the structure of matter in modern science and ancient alchemy. He also found that the focus on empirical practices was another common feature, as was the search for the existence of basic elements. The edition of some ancient alchemic texts was the joint effort of Berthelot, who assured the *soundness* of the content in chemical terms, and some philologists, who dealt with the literal translation.

Tannery's histories of ancient Greek mathematics, astronomy, and *science* in a broad sense (natural philosophy) offer a different intellectual landscape. First of all, they stemmed from a four-fold competence: scientific, philological, historical, and philosophical. Second, the historiographical reference frame included regressive stages and centuries of stagnation besides progressive trends: continuous developments could be followed or preceded by sudden transformations in the content and structure of knowledge. Third, the series of previous historical reconstructions and the analysis of their reliability represented an important component of Tannery's historical research: the critical analysis of the received views propagated over time was as important as the critical analysis of the existing primary sources. Fourth, he was also aware that the meaning of fundamental words and concepts such as mathematics, geometry, astronomy, and astrology had changed over time. Fifth, although the history of what was called ancient science could hardly be disentangled from the history of what was called ancient philosophy, he thought that a specific discipline, the history of science, should cover most of that body of knowledge because most of the content of ancient philosophy was in reality a specific kind of *science*.

The third stage was represented by Milhaud's histories of science, which could rely on the extensive researches already performed by Tannery. More specifically, he could rely on Tannery's historical-critical reconstructions and philological analyses of primary and secondary sources. Milhaud confined himself to synthesising all that sophisticated work, and embedded those histories in more explicit historiographical and epistemological frameworks. In contrast with the excess of empiricism he found in the Comtian tradition, he insisted on his concept of scientific practice as an act of mathematisation of natural phenomena and an act of intellectual and linguistic reinterpretation. He attempted to merge history of science and philosophy of science into each other in order to establish a new sophisticated discipline.

CHAPTER 8

From Theoretical Physics to Meta-theoretical Commitments

1 On the Borderline of Mathematical Physics

Berthelot, Tannery, and Milhaud had been trained in science, engineering or mathematics: their historical and philosophical interests stemmed from a scientific background. This is also true for the younger physicist Pierre Duhem, but like Berthelot and differently from Tannery and Milhaud, his professional career was as a scientist. His scientific training at the *École Normale Supérieure* was the starting point of a long series of original researches in theoretical physics.

Since the late 1880s Duhem had undertaken a demanding design of unification of theoretical physics that was based on two meta-theoretical pillars: the search for a common mathematical framework for mechanics, thermodynamics, and chemistry based on analytical mechanics, and the foundation of this generalised mechanics on the two principles of thermodynamics, which became fundamental principles for the whole body of knowledge of physical and chemical sciences. Although he labelled this generalised mechanics *Energetics*, we find a remarkable conceptual distance between Duhem and some upholders of Energetics like Georg Helm and Wilhelm Ostwald. The latter insisted on the principle of the conservation of energy as the sole foundation of physics whereas Duhem developed a sophisticated mathematical theory. In particular, Ostwald developed a physical world-view wherein the concept of matter had to be replaced by the concept of energy. Duhem's *energetism* was in tune with the *energetism* put forward by the Scottish engineer William Macquorn Rankine in the 1850s, where the concept of mechanical work was generalised in order to represent all kinds of physical and chemical actions.[1]

1 In order to appreciate the differences between Rankine and Ostwald's *Energetics*, see Rankine 1855, pp. 210-8 and 222, and Ostwald 1896, pp. 159-60. According to Anastasios Brenner, Ostwald's *energetism* represented a sort of disproportional answer to atomism [Brenner 1990, pp. 82 and 86]. It is worth mentioning that in the 1960s the scientist Donald G. Miller wrote that Duhem "belonged to the community of energetists, together with Ernst Mach, Georg Helm, and Wilhelm Ostwald" [Miller 1967, p. 447]. The warm relationship between Duhem

 DOI 10.1163/9789004315235_010

In 1886 Duhem had already published *Le potentiel thermodynamique et ses applications à la mécanique chimique et à la théorie des phénomènes électriques*, where thermodynamics offered the theoretical framework for the whole body of physical and chemical phenomena, and thermodynamic potentials offered the mathematical-physical toolbox for the description of those phenomena. At that stage, his theoretical design did not go far beyond what the American engineer Josiah Willard Gibbs, and the authoritative German physicist and physiologist Hermann von Helmholtz had put forward in the 1870s and the early 1880s. The content of the book corresponded to the doctoral dissertation Duhem had submitted to the *École Normale Supérieure* late in 1884, before the achievement of the *aggregation* in physics. This was an unusual procedure, but the faculty had allowed the talented student to present his dissertation, which was however rejected probably because of the new theoretical approach to thermodynamics, and because of the criticism it contained about Marcelin Berthelot's chemical theories. In 1888 he succeeded in obtaining his PhD after having defended his new dissertation, *L'aimantation par influence*, at the Paris Faculty of Science. At that time he was *maitre de conférences* in the Science Faculty of Lille University, and had published many papers on various subjects, electromagnetism, thermo-electricity, thermo-chemistry, and capillarity included. It is worth mentioning that the dissertation was presented in the class of mathematics despite its explicit physical content.[2]

In subsequent years, Duhem developed the structural analogy between mechanics and thermodynamics, and attempted to widen the scope of analytical mechanics in order to describe a wider set of physical and chemical phenomena beyond pure mechanics. In 1891 he published a paper in the official journal of the *École Normale Supérieure*, wherein he carefully put forward a historical reconstruction of the recent tradition of abstract thermodynamics. More specifically, he acknowledged the role played by the French engineer François Massieu, Gibbs, Helmholtz, and the German-speaking Russian physicist Arthur von Oettingen in the building up of a very general theory based

and Ostwald cannot be interpreted as an agreement on the meaning of *Energetics*. On their friendship, see Brouzeng 1981, vol. 2, pp. 226-8.

2 The word *thermodynamics* did not appear in the title of Duhem's second dissertation. Some historians have traced back Duhem's subsequent failure to be appointed to a Chair in Paris to his early criticism of Berthelot and Gabriel Lippmann's theories. As already remarked, Berthelot was perhaps the most authoritative scientist of the Third French Republic. For further details, see Jaki 1984, pp. 50-2, 78-9, and 437-9. For a complete bibliography of Duhem's scientific, historical and philosophical works, see Manville 1927, pp. 437-64, Jaki 1984, pp. 437-55, and Stoffel 1996a, pp. 24-129. For an essential chronology of Duhem's life, see Brouzeng 1987, pp. 161-5.

on thermodynamic potentials. In the paper he displayed what he called "the general equations of thermodynamics": the state of the system could be completely specified by giving a set of independent variables $\alpha, \beta, \ldots, \lambda$, and its temperature ϑ. It was a mechanical approach in the sense of abstract mechanics, where any microscopic mechanical model was explicitly banned. The theory aimed at the development of a common language for mechanics and thermodynamics. That common language required the widening of the mathematical structures and physical lexicon of mechanics [Duhem 1891, pp. 231-51].[3]

The following year he submitted a long paper with the very general title "Commentaires aux principes de la Thermodynamique" to the *Journal de mathématiques pures et appliquées*. It was the first part of a trilogy: the second and third parts were hosted by the journal in 1893 and 1894. The first passages of the first part clearly show one of the hallmarks of Duhem's scientific enterprise: an original combination of theoretical physics and historical and historiographical remarks. He set the history of thermodynamics in the wider context of the history of science, and the latter appeared as a periodical series of predictable stages. Scientific innovations were followed by a conservative trend, where recently established theories gave rise to many applications. At a given time, old and new problems led to the decline of consolidated theories, which was followed by the emergence of other innovations.

> Every science progresses by a series of oscillations.
>
> At certain times, the principles of science come under close scrutiny: hypotheses and specific restrictions are carefully analysed. Afterwards those principles seem firmly established, and the efforts of theoreticians are directed towards the deduction of the consequences: the number of applications increases, and experimental checks increase as well, and become more precise.
>
> The fact is that this development, at first easy and fast, becomes slower and more difficult. When the soil is unduly exploited, it becomes unproductive: some hindrances emerge, and cannot be overwhelmed by means of the established principles. Some contradictions cannot be solved, and some problems cannot be faced. At this stage, we must go back to the foundations of science: we must analyse their steadiness, and exactly estimate what they can bear without giving way. After having accomplished

3 For a detailed reconstruction of Duhem's thermodynamic theories, see Manville 1927, Brouzeng 1981, Bordoni 2012a, 2012b, 2012c, 2013, 2014a, 2014b, and 2015a.

> this task, we could build up the new consequences of the theory [Duhem 1892a, p. 269].

We find here the explicit acknowledgment of the intrinsic historicity and the dynamic nature of scientific practice. What Cournot and Bernard had courageously remarked in an adverse intellectual environment in the 1860s, and other philosophers such as Naville had restated in the 1880, received by Duhem a definite codification. We find in Duhem the awareness of the creative nature of scientific theories, the intrinsic necessity of a network of hypotheses alongside empirical and logical practices, and the plurality of possible conceptual scaffoldings to be associated with a given set of phenomena.

> Every physical theory rests upon a certain number of definitions and hypotheses that are arbitrary at least to a certain extent. Scientists are allowed to develop such a theory according to a logical order, even though we cannot claim that such an order is the only logical one. It would be an unjustified pretension, and we are aware of it. Thermodynamics might be developed in a way quite different from ours, even in a more satisfactory way. In the same way, we do not expect that no shortcomings might be found in our logical order [Duhem 1892a, p. 270].

In the 1892 *Commentaire*, Duhem was in search of the conditions under which the general equations of thermodynamics transformed into the traditional equations of mechanics. He found that the reduction could take place when heat fluxes and entropy variations vanished. His approach to thermodynamics had something in common with the point of view Poincaré had developed in the treatise he had published in the same year, *Thermodynamique, Cours de Physique Mathématique*. Poincaré agreed with Duhem on two fundamental issues: the essential role played by the two Principles of Thermodynamics "in all fields of natural philosophy," and the rejection of "the ambitious theories full of molecular hypotheses." With regard to the second issue, Poincaré stressed that microscopic mechanical models could not account for the second Principle. Since mechanics collided with *Clausius' theorem*, he would have reversed Maxwell and Boltzmann's approaches to thermodynamics in terms of molecular mechanics, and he would have built up "the whole structure of mathematical physics only on thermodynamics." He remarked that the exact computation of the internal energy of a body depended on the state of external bodies: to be precise, the conservation of energy in a given body called into play "the whole universe." A similar remark could be extended to the second Principle, although it was expressed by an inequality rather than an equality. An original combination of theoretical and historical remarks was also one of the hallmarks of Poincaré's treatise on thermodynamics: he shared with Duhem the sensitivity to the historical nature of scientific enterprise, and the awareness of the role played by *metaphysical* or meta-theoretical issues in that enterprise [Poincaré 1892, pp. V, XII-XIII, and XVIII].[4]

It is worth mentioning that in the late 1880s Poincaré had explored the complexity of physical systems and the shortcomings of classical mechanics. In 1891 he had stressed the conventional character of *geometrical axioms* against Kant, who looked upon them as *synthetic, à priori judgments*, and John Stuart Mill, who considered them as *experimental facts*. In the context of geometry, the search for truth had to be replaced by the search for "*the most convenient*" system of axioms [Poincaré 1891a, pp. 773-4; Brenner 2003, pp. 39 and 47]. In the same year, he had remarked that the physical and astronomic problem of three bodies still challenged mathematicians: the problem could not be solved by the known mathematical procedures. More specifically, the question of the stability of the solar system had remained an open question, even though, with customary understatement, he asserted that the problem had no actual importance from the physical point of view. Even simplified configurations, such as three planets moving in the same plane, the first being very massive, the second very small, and the third of negligible mass, could not be stable for all initial conditions [Poincaré 1891b, pp. 1 and 4-5].

Two years later Poincaré published some notes on the conceptual relationship between mechanics and thermodynamics. In a short paper sent to the *Revue de Métaphysique et de Morale*, he focused on the incompatibility between the theoretical foundations of mechanics and the most elementary experimental data. Mechanics required time reversibility: in other words, reversibility was "a necessary consequence of every mechanical hypothesis." Common experience was in conflict with that requirement: thermal conduction was a well-known instance of irreversibility. He found that every attempt to escape this contradiction was unconvincing, Helmholtz's hypothesis of "hidden motions" included. Poincaré's recent studies on the equations of mechanics had

4 On the attitude of Poincaré towards Duhem, see Poincaré 1892, p. XIX: "Twice I have been in disagreement with Duhem, and he could be surprised by the fact that I mention him only to contradict. I would be very sorry if he was inclined to believe in a malicious intention. I hope he does not suppose that I underestimate the services he has rendered to science. I have simply held that it was more useful to insist on the issues where his results deserved to be accomplished, rather than insist on issues which I would have merely repeated." Poincaré disagreed with Duhem on some entropy computations, and the rejection of Maxwell's theory of closed currents through dielectrics [Poincaré 1892, pp. 321-38, 366-83, and 390].

shown that "a closed system subjected to the laws of mechanics" could repeatedly approach its initial state over time. Two opposite world views were therefore at stake: a purely thermodynamic world-view entailed a sort of thermal death, wherein "all bodies will be found at rest at the same temperature," whereas a purely mechanical world-view could lead to a flow of heat from a cold body to a warm one, provided that scientists had "a little patience." As a consequence, the expectation that thermal irreversibility could stem from the laws of mechanics seemed pointless to Poincaré: he could not imagine a logical procedure where we found "reversibility at the outset" and "irreversibility at the end" [Poincaré 1893, pp. 534-7].[5]

2 Modern and Ancient Mechanics in Perspective

In the same year, Duhem had attempted to involve chemistry in his design of unification. In the book he published in 1893, *Introduction à la mécanique chimique*, he put forward a historical rather than a logical outline of the achievements of chemistry in the course of the century. In accordance with a deeply rooted meta-theoretical attitude, he found that the content of a physical law could have been better appreciated keeping readers in contact with both the efforts that had been required, and the mistakes that had been overcome. In the last chapter, he focused on experiments performed at high temperatures. Thermodynamics forbade certain transformations, and nobody had ever observed such forbidden transformations. On the contrary, there were transformations that were envisaged by the theory but did not happen. In Duhem's words, when "the system should be in equilibrium, it actually stays in equilibrium," but it can be found in equilibrium "even when, according to the theory, it should not." He labeled the first case as "true equilibrium," and the latter as "false equilibrium." When a mixture of hydrogen and oxygen, or hydrogen and chlorine, reached their *true* equilibrium, namely the transformation into water or muriatic acid, they released sufficient heat to trigger off an explosion. In Duhem's theoretical framework, an explosion was therefore a

5 Poincaré remarked that "the mechanical conception of the universe" had assumed two different forms: the mechanics of shocks and the mechanics of forces. In the first case, physicists imagined "atoms moving along a straight line, because of their inertia": the amount and direction of their velocity could not change unless "two atoms collide." In the second case, atoms were seen as subjected to a mutual "attraction (or repulsion), depending on their distance, and according to a certain law." Since he considered the first conception as a "particular case of the second," the distinction was overlooked in the course of the paper.

passage “from a state of false equilibrium to a state of true equilibrium,” and the remarkable amount of heat was the consequence of the transformation. A good theory would have accounted for sudden and disruptive events left unexplained by classical theories [Duhem 1893b, p. 176].[6]

At the end of his inquiry into the history of mechanics, thermodynamics and chemistry, Duhem drew an interesting conclusion about the role of scientific theories, wherein historical and meta-theoretical remarks were mutually interconnected. Scientific theories were necessary and fruitful, however provisional and incomplete they may be. This fact could explain why sometimes scientists had attempted to save a flawed theory when a better theory was not yet at hand.

> The history of physics shows us that a theory should not pride itself on being conclusive. We see that theories emerge and progress just to fall down. Nevertheless a theory, when it has been built up in search of truth, can never completely disappear. Among the debris we can find contents that might be employed in the building up of a better and more enduring theory. [...]
>
> It is unusual that the conflict with experiments leads science to get rid of a mistaken theory: the upholders of the theory will always try to justify the mismatch, and will search for a re-interpretation. Frequently, illogical behaviours are involved in these procedures: the self-esteem of every inventor, the persistent attachment to the received view, and the excessive deference to authority. Nevertheless, we must mainly take into account a natural leaning of the human mind, which would like to arrange the phenomena around some ideas. After having built up a theoretical system, the mind takes care of its preservation, in spite of the refutations imposed by facts, until a more complete theory emerges and offers a more satisfactory framework to experimental data [Duhem 1893b, p. 176].[7]

---

6 In the previous pages Duhem had described some processes giving rise to *false* equilibrium: “The decomposition of water absorbs heat. When we increase the temperature of a mixture of oxygen and hydrogen, and we let it grow gradually, we do not observe any chemical reaction until, at a temperature of about 500°C, by a violent explosion, part of the mixture will suddenly transform into steam” [Duhem 1893b, p. 155].

7 In 1902 Duhem was to publish a historical and critical account of the concept of mixture in the long-lasting tradition of philosophy, as well as in the recent tradition of chemistry. The book widened and deepened the researches Duhem had already published in 1892. The book has recently been translated into English and commented by Paul Needham [Duhem 1892c, Duhem 1902, Duhem 2002, and Needham 2002].

In 1894, in the third part of his *Commentaire*, in the second chapter, we find an astonishing reference to the Aristotelian meaning of the word *motion*: alongside the usual kinematical meaning, physical and chemical transformations in general could be looked upon as motions in a broader sense. From this point of view, the word *motion* was not opposed to the word *rest* but to the word *equilibrium*.[8] The generalization of the concept of passive resistances such as viscosity or friction bridged the gap between mechanics and thermodynamics, and allowed Duhem to give a mechanical interpretation of the second Principle of thermodynamics. In this sense thermodynamics became a generalization of mechanics, and the generalized mechanics could be looked upon as a general theory of transformations.

> In the present work, we have attempted to point out a third kind of relationship between Dynamics and Thermodynamics. We have transformed Dynamics into a specific instance of Thermodynamics, or better, under the name of Thermodynamics we have built up a science which encompasses every transformation of a body within common principles: changes of place, as well as changes of physical qualities [Duhem 1894a, p. 285].

Duhem transformed "dynamics into a specific instance of thermodynamics"; in other words, "under the name of thermodynamics" he had put forward a more general theory that encompassed within common principles "every transformation of a body." Both changes of place and changes of physical qualities found room in that generalized mechanics. The traditional science of motion became a specific instance of a more general science: it had to be understood that the change of position in space was not "a simpler modification than the change of temperature or any other physical quality." This generalization could bypass "the most dangerous stumbling block on the path of theoretical physics," namely "the search for a mechanical explanation of the universe." From the mathematical point of view, his design corresponded to a reduction of physics to the language of analytical mechanics, but from the theoretical point of view it was an anti-reductionist design that involved a generalisation of that language. In Duhem's "more general science" we find the coexistence of a mechanical approach, in the sense of a generalization

8 See Duhem 1894a, p. 222: "Nous prenons, dans ce Chapitre, le mot *mouvement* pour désigner non seulement un changement de position dans l'espace, mais encore un changement d'état quelconque, lors même qu'il ne serait accompagné d'aucun déplacement. [...] De la sorte, le mot *mouvement* s'oppose non pas au mot *repos*, mais au mot *équilibre*."

of Lagrange's mathematical physics, and the rejection of mechanical models and mechanical explications in the sense of the traditional mechanistic world-view [Duhem 1894a, pp. 284-5].[9]

In contemporary debates on thermo-chemistry and physical chemistry in general, two meta-theoretical issues were at stake: the role played by entropy and the reliability of mechanical models of matter. In 1894, in a paper published in the *Comptes Rendus de l'Académie des Sciences*, Berthelot defended his "experimental principle of maximum work," in which the word *work* could mean both *energy* and *heat*. The introduction of entropy led only to "a new utterance" for his principle: he found that the law expressed in terms of entropy had a more limited scope, and the results were "more obscure." Entropy was a physical quantity suitable for "people dealing with computations in the context of mathematical physics." Moreover, some chemical systems did not have "computable entropy." In brief, Berthelot firmly opposed Duhem's design of unification between thermo-chemistry and mathematical physics [Berthelot 1894, pp. 1378-9, 1382-5, and 1392].[10]

In 1895, in a paper sent to the *Revue générale des Sciences pures et appliquées*, Wilhelm Ostwald, then professor of physical chemistry at Leipzig University, sharply criticised scientists who relied on "the mechanics of atoms" as an intellectual *passe-partout* for the comprehension of the physical world. To this mechanical world-view, which Ostwald qualified as "physical materialism," he opposed a new theoretical approach he labelled *Energetics*. Mechanics could not explain the natural direction of natural processes, because it was time reversible. He thought that mechanical models could be easily dismissed in favour of a direct approach to experience, which would have allowed us "to see directly" the world, without "any picture, any symbol." Scientists had to

9 See Duhem 1894a, p. 285: "Il nous semble qu'une conclusion générale se dégage de cette étude: si la science des mouvements cesse d'être, dans l'ordre logique, la première des Sciences physiques, pour devenir seulement un cas particulier d'une science plus générale embrassant dans ses formules toutes les modifications des corps, la tentation sera moindre, pensons-nous, de ramener l'étude de tous les phénomènes physiques à l'étude du mouvement; on comprendra mieux que le changement de lieu dans l'espace n'est pas une modification plus simple que le changement de température ou de quelque autre qualité physique; on fuira dès lors plus volontiers ce qui a été jusqu'ici le plus dangereux écueil de la Physique théorique, la recherche d'une explication mécanique de l'Univers."

10 A historical and conceptual reconstruction of Duhem's opposition to Berthelot's "principle of maximum work," which involved the distinction between endothermic and exothermic reactions, and the separation between physical and *purely chemical* transformations, can be found in Brenner and Deltete 2004, pp. 204-7. On the criticism of Berthelot's law of maximum work outside France, see Needham 2002, p. xxiii.

confine themselves to quantitative relationships among "entities that could be handled and measured," and the most important of these entities was "the most general invariant, the energy," or better, any difference of energy. It seems a very naïve point of view because no physical theory can avoid some kind of symbols or representations. Nevertheless, this phenomenological attitude stood beside a more liberal conception of scientific development: although he had stressed "the advantages of the energetic theory over the mechanical theory," his energetics did not have to be looked upon as the final stage of science. In an unspecified future, Ostwald expected an even wider-scope theory, wherein energetics would have appeared as "a specific instance of more general relations" [Ostwald 1895, pp. 953-8].

When we compare Ostwald's with Duhem's energetics, we find a remarkable difference: the unifying power of a specific physical entity, namely energy, in the former, and the unifying power of very general mathematical structure in the latter. In 1894 and 1895, Duhem had already attempted to insert the theory of permanent deformations into his *Energetics* or generalised mechanics [Duhem 1894b, pp. 3-5]. The long essay he published in 1896, *Théorie thermodynamique de la viscosité, du frottement et des faux équilibres chimiques*, represented in some way the final stage of his theoretical, meta-theoretical and historical journey through the complex network of connections involving analytical mechanics, thermodynamics and chemistry. After having highlighted the structural analogy between chemical *false* equilibrium and mechanical friction, in the second part of the *Introduction*, he put forward the most general equations of motion. They contained five terms: generalised forces or *actions*, the derivatives of a thermodynamic potential, the traditional *inertial* terms of Lagrange's equations, terms corresponding to generalized viscosity, and terms corresponding to generalized friction. It applied this mathematical-physical structure to explosive chemical reactions, where generalized velocity corresponded to the velocity of reaction. When generalised viscosity vanished, velocity became infinite: this limiting case did not correspond to modern mechanics but to Aristotle's theory of motion [Duhem 1896a, pp. 8-9, 70-5, and 130-1].

The general equations contained both inertial and dissipative terms: when Duhem dropped dissipative terms, a reinterpretation of classical mechanics emerged, and when he dropped inertial terms, some mathematical and physical approximations led to a new mechanics for chemical processes. Therefore classical mechanics and *chemical* mechanics represented the opposite poles of Duhem's *Energetics*: the general equations offered a common mathematical framework for both physics and chemistry. From a different point of view, that pliable and general mathematical structure could include both modern and ancient physics: the new chemical mechanics appeared as a modern, mathematical implementation, and a radical reinterpretation of the natural philosophy of the Aristotelian tradition. In reality, in the context of Aristotle's *physics*, it was not strange that, in the absence of some kind of resistance, velocity became infinite. Obviously, a bold meta-theoretical reinterpretation was required in order to look upon Duhem's mathematical generalisation as a powerful design of unification between the ancient and modern theories of motions [Duhem 1896a, p. 205].[11]

Theoretical physics, history of physics, and meta-theoretical remarks were mutually interconnected in Duhem's actual praxis. In his search for a new generalized mechanics he had analysed the different stages in the history of mechanics: all of them had been fruitful and meaningful. At the time of Galileo, modern science had fought against the old physics of qualities, in order to supplant it: the complexity of the physical world had been set aside, and replaced by a simplified geometrical world. At the end of the nineteenth century, scientific progress allowed and required a new step forward, namely an abstract generalization which could account for that complexity. Duhem had endeavoured to retrieve dissipative effects within the boundaries of mathematical physics. A qualitative outline of the physics of dissipative effects could be found in the ancient Aristotelian natural philosophy. He was confident that the long-neglected complexity could suitably be included within the wide boundaries of his generalized mechanics-thermodynamics. He aimed at widening the scope of physics: the new physics could go beyond *local motion* in order to describe what Duhem labelled *motions of modification*. Maxwell and Boltzmann had started from microscopic local motions to attain the explanation of more complex processes like thermodynamic transformations, whereas Duhem had started from the mathematical laws of general transformations to arrive at local motion as a simplified, specific case.[12]

In the end, he hinted at the possibility that further developments and further generalisations of his generalised mechanics could account for some es-

11 As Monica Ugaglia pointed out some years ago, the Aristotelian theory of motion dealt originally with processes taking place through some kind of medium: it was not a *kinematic* theory in the modern sense, but rather a *hydrostatic* one. In the Aristotelian tradition after Johannes Philoponus, a "hybrid kinematic-hydrostatic system" emerged. According to Ugaglia, in the sixteenth and seventeenth centuries, Tartaglia, Benedetti and Galileo had to re-discover Aristotle's hydrostatic approach beneath that hybrid kinematics in order to overcome it [Ugaglia 2004, pp. 8-13].

12 Duhem's theoretical and meta-theoretical design was also explained in a book he published in 1903, *L'évolution de la mécanique* [Duhem 1903 (1992), pp. 199 and 218-9].

sential features of living beings. The fact that this kind of generalisation fell outside his explicit design of unification did not prevent him from imagining the positive and creative effect of generalised dissipations in the context of living structures.

> When we inquire into the properties of systems in which the work done by viscosity and friction might not be intrinsically negative, and uncompensated transformations might not be intrinsically positive, it is impossible not to be struck by the analogy between those properties and those of living tissues, both animal and vegetable. It is impossible not to notice how easily those processes could account for the majority of organic syntheses, which cannot be explained by ordinary chemical mechanics, and cannot be performed outside the living body, *under the thermal conditions that allow the living body to work* [Duhem 1896a, p. 206].

3 Duhem's First Philosophical Paper

In the meantime, in 1892, Duhem had published the first paper explicitly devoted to meta-theoretical issues or, to make reference to a recent tradition, to philosophy of science.[13] The paper, "Quelques réflexions au sujet des théories physiques," was the first of a series of papers he published in the 1890s in the journal *Revue des questions scientifiques*. The journal was published by the *Société scientifique de Bruxelles*, which was an association of Catholic scientists: its aim was the presentation, discussion and critical account of scientific theories, without having recourse to mathematical details but with particular attention to models, concepts, principles, and methodological issues.

From the outset, Duhem represented scientific enterprise as a three-stage task: from the knowledge of specific facts, the human mind derived some *experimental laws* by induction, and then created a scientific theory. The scattered set of facts dealt with the first level of pure *empiricism*, and the set of physical laws belonged to the level of the "*purely experimental science*," whereas the set of physical theories corresponded to "*theoretical science*." The objects of experimental laws were facts, and the objects of physical theories were experimental laws. The nature of theoretical physics was just the sub-

13 At that time, Duhem was 32, and was "maitre de conférences" at Lille University: in the same year, his wife died and he remained alone with a baby. For further biographical details, see Jaki 1984, pp. 97-9, and Brouzeng 1987, p. 54.

ject Duhem was to investigate, and the investigation would have focused on the tradition of French *mathematical physics*. Physical theories were a sort of "relief for memory": they synthesised the body of knowledge stored in experimental laws, and at the same time they offered a general mathematical framework. The transition from laws to theories corresponded to a passage between different fields of knowledge, both of them endowed with their specific languages. The mathematical engine of a theory performed a reinterpretation of the laws: the *nature* of the laws and the nature of their theoretical representations were definitely different [Duhem 1892b, pp. 139-40].

The transition from one level to another required a sort of conceptual shift: Duhem specified that even the best-known physical entities had no direct link with the corresponding experiences. Between the human experience of heat, which appeared "agreeable or disagreeable," and the mathematical representation of temperature there was a remarkable gap: differently from actual experiences, a given temperature could be added to another temperature, and multiplied or divided by a number. The correspondence between entities belonging to different levels entailed a sort of conceptual translation, and just as in every act of translation, a plurality of choices was available: there was no constraint, no necessity. The physical concept of temperature, for instance, had to satisfy two mathematical conditions: the same value had to be associated to equally warm bodies, and a greater value for a body A when A was warmer than B. According to Duhem, every physical entity endowed with these properties could be chosen as *temperature* [Duhem 1892b, pp. 143-4].

The hypotheses of a theory enjoyed the same freedom that was enjoyed by basic entities or definitions: the only constraint was the possibility of deriving "*experimentally verifiable consequences*" from them. If the set of consequences was wide-ranging, and in accordance with experience, the theory could be looked upon as good. Nevertheless, the choice of hypotheses should not be made at random: there should be an "ideal and perfect method." The last statement appears a little surprising because of Duhem's stress on the conceptual gap between laws and theories, and the plurality of theories corresponding to a given set of laws. At a first glance, such a method should consist in choosing hypotheses that were "the symbolic translation ... of some experimental laws belonging to the set to be represented," but Duhem found this unsatisfactory because a theory was something more sophisticated than a collection of laws, and should offer a conceptual content richer than the mere superposition of experimental laws. The most meaningful instance of this theoretical surplus was Newton's theory of universal gravitation, which was not a mere translation of Kepler's laws in a formal language of higher level: it also contained propositions which could not be derived by experience, but were the outcome of a theoretical elaboration [Duhem 1892b, pp. 145-8].[14]

As a result of this conceptual surplus, there was a conceptual gap between the "consequences of the theories" and the experimental laws which they should represent. This was a very sensitive issue, and Duhem was aware that he had "to insist on it": there was an inescapable gap between the abstract structures of a theory, and the laws derived from empirical practice.

> ... a good theory is not a theory giving rise to consequences never in disagreement with experience. If it was the real hallmark of a good theory, we would never find a good theory, and the creation of a good theory would be something outside human reach. A good theory is a theory that manages to represent a set of physical laws with satisfactory approximation [Duhem 1892b (1987), pp. 149-50].[15]

An important consequence followed: "the value of a theory" depended on "the set of laws to be summarised by this theory," and on "the degree of precision of the experimental methods" by which the laws were set up or applied. In other words, a good theory might become a bad theory if the boundaries of the field of application were enlarged or shifted, or the degree of experimental precision was improved. Another consequence followed: scientists could decide to replace a good theory with a better one, but the choice in favour of the latter would not mean the wrecking of the former. The logic of scientific theories was not of the kind *true or false*: the better theory could be derived from the previous good one by some kind of conceptual enlargement or re-arrangement. Only at this point did Duhem go back to the "ideal form" of a physical theory, which required something more than "the mere symbolic translation of an experimental law." This last solution would not be satisfactory because the theory would be "difficult to modify." On the contrary, when the hypotheses of the theory were far from the laws that they should explain, the theory was more general and pliable. Nevertheless, a heavy price was to be paid for this apparently paradoxical feature: a more general theory was also more precarious and subject to failure or confutation. The wider the scope of a theory, the

14 According to Duhem, the theoretical surplus of Newton's theory over Kepler's laws was two-fold: the reciprocity of the attraction, and the extension of the attraction to any couple of bodies.

15 These remarks probably astonished readers because modern science had emerged in opposition to the Aristotelian tradition. This was the philosophical wound that Duhem tried to heal.

greater the risk of default. According to Duhem, a theory was a complex entity: a good theory should be wide-ranging and pliable, but its strength was also its weakness [Duhem 1892b, pp. 150-1 and 153].

According to Duhem, the history of physics offered a bad example, or a "false ideal," of scientific theory, namely "the *mechanical theory*." In it, every physical entity had to be represented by means of "geometrical and mechanical elements of a given imaginary system." A specific instance of mechanical theory was the *mechanical theory* of light, where the properties of light had to be derived from the mechanical properties of aether. Different implementations of that model had been put forward over time: a continuous or a molecular aether, forces at a distance or contiguous actions between elements of aether, and so on. Duhem refused mechanical theories mainly because of the constraints they imposed on basic entities and hypotheses: in particular, non-mechanical entities had to be represented as "the combination, sometimes very complicated, of mere mechanical concepts." He made reference to basic concepts like temperature, which had actually received a complex re-interpretation in terms of microscopic motions. In any case, Duhem found that the historical trend was not in favour of mechanical theories: even the most complex mechanical theory could not manage to satisfactorily account for *Carnot's principle*, namely the second Principle of Thermodynamics. That failure had led to the transformation of "the *mechanical Theory of Heat*," into a physical theory, namely *Thermodynamics*, which he looked upon as "one of the most perfect *physical* theories" [Duhem 1892b, pp. 154-8].[16]

Duhem's specific distrust in every mechanical theory stemmed from a more general distrust in every attempt to explain the nature of the material world, which he labelled as "a metaphysical explanation of the universe." He thought that "the exact structure of the world" was unattainable, and every attempt to grasp it would have been doomed from the start: scientists who had attempted to pursue this target would have built up an extremely weak structure which would have quickly collapsed. He was aware that faith in scientific progress as the driving force of social and intellectual progress had exerted a strong pressure on scientists and their scientific practice in the last decades of the nineteenth century. Even cultured people demanded both "immediate appli-

16 In this context, "physical" means "non-mechanical." According to Duhem and a nineteenth-century tradition, physics and mechanics were looked upon as complementary research fields, rather than scientific sections linked by a hierarchical relationship wherein mechanics represented a subset of physics.

cations directed to satisfying material needs," and "explications ... directed to satisfying the ambition to understand all." The intellectual pressure was as strong as the social pressure, and he was also conscious that the search for "the nature of things" was deeply rooted in human nature, "from the most superstitious savage to the most curious philosopher." Nevertheless, the scientist had to resist this pressure, and had to oppose that expectation: a physical theory could offer nothing more than "a systematic coordination of laws," rather than "an explication of those laws" [Duhem 1892b, pp. 158-9].

Thermodynamics offered a meaningful instance of the struggle between the two theoretical attitudes. On the one hand were both the interpretation of temperature as a symbolisation of the notion of heat, and the principles of Thermodynamics as "generalisations of experimental laws." On the other, many elements were at stake: a huge number of microscopic bodies in stationary motion; the interpretation of temperature as their mean living force; a set of convenient assumptions on their number, dimensions, and motions; and an attempt to derive "the principle of equivalence between heat and work, not to say Carnot's principle." The second theoretical strategy represented a meaningful instance of a scientific practice aimed at unveiling "the nature and causes of physical laws," whereas the former confined itself to a symbolisation of those laws, and called into play physical entities which could not be reduced to geometry and mechanics. Duhem was strongly convinced that scientist should not be compelled to express basic concepts like temperature and quantity of heat in terms of space, time and mass [Duhem 1892b, pp. 160-2].

He acknowledged that Descartes, Newton, Huygens, Laplace, Poisson, Fresnel, and Cauchy's mechanical theories "had allowed science to progress greatly." How could "such a wholly mistaken idea on the role of physics" have led to that great success? His answer was both conceptual and historical: mechanical theories had represented the childhood of science, when its progress had been faster, but at the same time its role was "more roughly defined." The emergence of physics could really be associated with a sort of over-confidence in the power of mechanical models. However, for the time being, Duhem saw the "decline of mechanical theories," and the emergence of "purely physical theories." In this context, another issue was at stake: the existence of close bonds between physics and philosophy. When physicists considered a physical theory as "an explanation of the laws of nature," they were committed to a definite philosophical attitude, and they therefore leant towards a theory consistent with their philosophical belief. Symmetrically, when philosophers believed that "the nature of material phenomena" could be found in certain physical theories, they were inclined to draw their inspiration from those theories in order to develop their metaphysical systems. The close relationship between Descartes' physics and metaphysics was a meaningful instance of that close bond. In more recent times, Herbert Spencer's philosophy had been "heavily influenced by ideas borrowed from some thermodynamic theories." According to Duhem, only the awareness of the merely symbolic role of physical theories would have allowed scientists to become "independent of fashionable metaphysical systems," and would have led them "to give up imposing their influence on metaphysics" [Duhem 1892b, pp. 162-4].

Duhem quoted from and commented on two passages from the treatise which "the renowned analyst" Poincaré had published on optical theories in 1889. He agreed with Poincaré on the aim of physical theories: they could not "disclose the true nature of things," but only coordinate the physical laws which experience allowed us to understand. Nevertheless he disagreed with Poincaré on the meta-theoretical thesis that "different theories, associated with a given set of laws, could be look upon as equivalent." He believed in the possibility of judging the relative value of different theories, and insisted on three main features of a good theory: first of all the scope, and then the number and nature of hypotheses. The third feature was not easy to define because it dealt with qualitative elements: the hypotheses of the best theory had to be "the simplest and the most natural, and should offer the best translation of experimental data" [Duhem P. 1892b, pp. 165-6 and 169-70; Poincaré 1889, p. 11]. Since a precise explanation of that feature was far from easy, he confined himself to a specific instance: the comparison between Lamé and Cauchy's theories of double refraction. However, in the end, he optimistically thought that he had managed to put forward a *Pascalian* compromise between relativistic and dogmatic attitudes.[17]

In the last, brief section of the paper Duhem focused on the meaning and the *usefulness* of theoretical physics. It allowed physicists to go beyond the mere alliance between "experience and mathematical analysis"; it called for "systematic links" and some kind of speculation, in order to give sense and

17 See Duhem 1892b, p. 170: "Ainsi, en affirmant que la Physique Mathématique n'est pas l'explication du monde matériel, mais une simple représentation des lois découvertes par l'expérience, nous évitons l'obligation de déclarer vraie, pour chaque ordre de phénomènes, une théorie à l'exclusion de toute autre. Mais nous ne sommes pas condamnés pour cela à adopter toutes les théories, logiquement constituées, d'une même ensemble de lois: nous avons, pour choisir entre elles, des règles très sûres, qui, bien souvent, nous permettrons de préférer raisonnablement l'une d'entre elles à toutes les autres."

structure to the knowledge received from the experimental method. Theoretical physics allowed scientists to go beyond "the confused and inextricable accumulation" of laws derived by experience [Duhem 1892b (1987), p. 175]. Once more the complex nature of scientific theories was at stake: a network of logical and extra-logical skills was involved, and Duhem had only managed to hint at the main problems. Some specific issues were still waiting for a more satisfactory clarification.

4 Some Debates

Duhem's paper raised some debate. In 1893, the first issue of the *Revue de questions scientifiques* hosted a long paper, sent by the engineer Eugène Vicaire: the author immediately claimed that "Duhem's fundamental thesis" was false. The thesis under attack was what Duhem considered the aim of theoretical physics: the symbolic representation of physical laws, rather than their explanation in a metaphysical sense. Vicaire acknowledged that the thesis was widely shared by renowned scholars like the mathematician Poincaré and the German physicist Gustav Kirchhoff, and he found that it could be traced back to David Hume. Vicaire quoted with care from Poincaré's 1889 treatise on the mathematical theories of light, and he did not point out any difference between Duhem and Poincaré. He was committed to fight a philosophical battle against some ideas which he looked upon as "destructive of every science," and he did not hide his disappointment with regard to the "invasion of scepticism" in a journal which should have been extraneous to that philosophical trend. He was committed to putting forward a refutation because he was worried about the emergence of this dangerous scepticism: he had realised that "the danger was greater" than he had previously expected, and he found that "the necessity to oppose" that scepticism had become "more pressing" [Vicaire 1893, pp. 452-3].

Vicaire charged Duhem with not being able to distinguish between "applicative and explicative theories." In reality, what he termed *applicative theories* was nothing else but what Duhem had termed *laws*: the case he mentioned, the laws of reflexion and refraction of light, was clear in this regard. What he labelled as explicative theories was what Duhem had simply labelled theories, and the adjective *explicative* showed that Vicaire looked upon theories as actual explanations. In reality, he did not attempt to convince readers that physical theories were explanations of the physical world: he contented himself with mentioning a "common and traditional point of view," which had been "always correct." When he concluded that only explicative theories were "real scientific theories," this was neither a confutation of Duhem's point of view nor an effective line of reasoning, but a merely personal belief. The fact is that something more meaningful was at stake: "the essential merit of theories," if not their actual "raison d'être," dealt with the beauty and harmony they introduced in the web of knowledge. Not only was the contemplation of that beauty "the highest satisfaction of the mind [esprit]," but also the final aim of science. Beside this main aim there was the usefulness of theories, namely the possibility of "extending the scope" of laws, and even discovering new laws, or new fields of application. Vicaire extended the *merit* of theories to laws: they owned an "intrinsic beauty," because of "the order they let emerge from nature." The essential features of laws were "their practical usefulness, their intrinsic beauty, and the generation of theories." The third feature overturned the relationship between laws and theories that Duhem had put forward. The knowledge of the physical world proceeded "from phenomena to their relations, and from relations to causes." That laws could generate theories, and that theories could be assimilated to causes, was actually extraneous to what Duhem had written the year before [Vicaire 1893, pp. 453, 456, 459, and 461-3].

Vicaire faced many sensitive issues, and put forward some bold and interesting theses. He stated that "crude facts" and "the subtlest hypotheses" had the same nature, and a series of intermediate entities from the former to the latter could be imagined. An explanation of the physical world could legitimately be desired and pursued, even though it could never be accomplished. This desire was deeply rooted in human psychology: nobody could bear the effort of pursuing "a scientific research for more than five minutes" unless he was attracted by the unknown and by the possibility of grasping "some mysteries of nature." This natural momentum was stronger than "the prohibitions of a philosophy" that claimed to be *positive* but in reality was *negative* because it prevented scholars from following a natural ambition. It was at this point that Vicaire took into account similarities and differences between Poincaré and Duhem's meta-theoretical attitudes. According to Vicaire, not only had Poincaré stressed the plurality of the theoretical interpretations of a given class of phenomena, but he had also tolerated the existence of mutually inconsistent components of a theory. Vicaire acknowledged that Duhem had not dared so much, but attributed his milder attitude to a sort of fear. He blamed Duhem for having "let himself be intimidated" by Auguste Comte's school, in order to "gain his certificate of civic spirit." He went on with his unpleasant and psychologically-oriented remarks: the concern for "being compromised with metaphysics," had led Duhem to become similar to his adversaries. The accusation was unfair as to the style, and false as to the content, for Duhem was known for his independence of mind, and he had never tried to please anyone [Vicaire 1893, pp. 468, 472-4, 476, and 482].[18]

When Vicaire resumed his more rational line of reasoning, he stressed once more that "the search for natural truth and conformity to Nature, and the closely related concept of cause" were at stake in physical theories. The conformity, at least in part, to Nature was indeed the hallmark of a good theory, and it could not be attained by pure chance. Unfortunately, on this specific issue, Vicaire's remarks became quite puzzling: he acknowledged that, from the logical point of view, false assumptions could lead to true conclusions, but this fact did not imply the equivalence between true and false assumptions. He did not manage to grasp Duhem's reflection on the problematic link between the abstract, rational nature of deductive procedures, and the empirical nature of observation and experience. Both Poincaré and Duhem had realised that no automatic connection between the two domains could be found, and theoretical physics occupied the wide space between them. The plurality of theoretical interpretations dealt with the inescapable gap between reason and experience: their different personalities notwithstanding, Duhem and Poincaré were aware of that gap, and were aware of the complexity of scientific practice. In the last passage of the paper, Vicaire appealed to "old, perpetually true principles," and invited scientists to "know Nature, rather than handle symbols": it was a "noble ambition," and the value of "truth and science" resided just in this [Vicaire 1893, pp. 492-3 and 410].

In the same year, another engineer, George Lechalas, took part in the debate triggered by Duhem's paper. He published a short answer to Vicaire in the June-July 1893 issue of the neo-Thomist journal *Annales de philosophie chrétienne*, where Vicaire's paper had been re-published in the April-May issue. After having stated that he agreed with Vicaire on almost the whole content of his paper, he confined himself to pointing out what he considered "an inaccurate notion of mechanical representation." He stressed the difference between two different implementations of mechanics: there was a mechanics based on the concept of force, and a mechanics which was not based on it but did not reject "some hypotheses on the mutual actions among bodies." Lechalas' phenomenological attitude was not far from Duhem's: he stressed that when we perform experiments, we get in touch only with "motions, together with

18 Vicaire's reference to Poincaré was correct only in part: Poincaré's pluralism did not involve the presence of contradictions in the same theory. See Poincaré 1890, p. VIII: "Deux théories contradictoires peuvent en effet, pourvu qu'on ne les mêle pas, et qu'on n'y cherche pas le fond des choses, être toutes deux d'utiles instruments de recherches"

their velocities and accelerations." At the same time, unlike Duhem, Lechalas did not reject the concept of cause in general, but only "the anthropomorphic representation of causes" [Lechalas 1893a, pp. 278-80].

In 1893 Duhem published a paper in the *Revue des questions scientifiques*, where he reviewed a book which had been published by father Armand Leray, "philosopher, theologian, and scientist," some years before. Duhem appreciated the fact that Leary had separated the "rights of divine revelation" from "the rights of science": in particular, he had never based "a deduction leading to a scientific truth" on a revealed truth. Alongside the detailed description of Leray's scientific conceptions, and historical remarks on the nature of matter and actions at a distance, Duhem stressed two concepts: the different methods and aims of science and religion, and the different methods and aims of physics and metaphysics. He remarked that the first difference had clearly been pointed out by Leray, whereas the second had already been affirmed by Newton. Duhem credited Newton with having considered physics as the reduction "of a great number of experimental laws to a small number of theoretical principles," and metaphysics as the search for "the causes from which principles stem." Physics could really exist without metaphysics, although physics purified from metaphysics could only aspire to "an incomplete knowledge of the world" [Duhem 1893c (1987), pp. 41 and 76].[19]

It seems that the debate between Duhem and his critics was inevitably affected by a considerable intellectual gap. In other terms, in this debate we face the arduous exchange of views between scholars who put forward two kinds of historiographical and epistemological frameworks that I have already qualified as naïve and sophisticated. This sort of *impossible* exchange exposed Poincaré and Duhem' meta-theoretical remarks to misunderstandings that propagated throughout the twentieth century, and even afterwards. Nevertheless, here and there, some historians managed to grasp some essential features of their critical practice.

In the 1930s, the historian Margaret Eastwood interpreted Poincaré's awareness of the "breach between science and reality" as a specific implementation of the anti-dogmatic trend in the late nineteenth century, and his "idea of convenience" as a specific kind of criticism which was quite close to what she labelled "Pascal's *scepticism*". She also attributed to Poincaré the Pascalian love

19 With regard to the debate on the admissibility of actions at a distance, which was the subject of the second part of the paper, Duhem did not draw any conclusion: although "interesting from the point of view of metaphysics," any decision would "not affect physical theories" [*Ibidem*, p. 82].

of truth that should not "be confused with love of certainty" [Eastwood 1936, pp. 29-30, 37-9, 54-7, and 70]. In reality, the re-emergence of what might be qualified as Pascal's disenchantment can also be found in the researches of other scientists who lived in France in the second half of the nineteenth century, more specifically in the intellectual stream that flowed from Cournot to Duhem.[20]

Poincaré had cleverly benefited from a plurality of intellectual influences, the influences of some characters already analysed in the present book included. The historian Benrubi saw the influence of Kant, that he probably overestimated, and the influence of Boutroux, Bernard, and Cournot. Nevertheless Poincaré was a thinker of marked independence, and the word *influence* should be used even more cautiously when referred to him. However the stress on an objective closeness to Cournot seems to me appropriate. Not only had Poincaré been interested in probability both in mathematics and physics, but he had also shown a characteristically quiet, critical, and detached attitude towards scientific practice, an attitude that had always accompanied Cournot's writings.[21] The words Benrubi made use of for qualifying Poincaré's philosophical attitude, "Relativism and Pragmatism," had suitably been used for qualifying Cournot's attitude, and they definitely describe Poincaré more faithfully than the expression "Poincaré's Kantianism." Although I do not trust these labels, I am aware that philosophical debate has always involved synthetic classifications and the corresponding use of hypnotic words. Therefore, in this context, *pragmatism* seems a suitable word, definitely more suitable than *conventionalism*. It seems to me that Benrubi managed to grasp Poincaré's (let me say) rational pragmatism, when he specified that "Poincarè's Pragmatism is not to be taken in a Utilitarian and Nominalistic sense" [Benrubi 1926, pp. 96-9 and 101; Mazliak 2013, pp. 34-6].[22]

20 It is worth remarking that, from 1900 onwards, the interest in Pascal's intellectual enterprise continued to flourish: I confine myself to mentioning Boutroux,' 1900, *Pascal*, Léon Brunschvicg's 1904-14 edition of *Œuvres de Blaise Pascal* (fourteen vols., together with Pierre Boutroux from the first to the eleventh volume, and together with Félix Grazier from the fourth to the eleventh), and Fortunat Strowski's 1907-8, *Pascal et son temps* (three volumes).

21 In this context, it is worth noticing that in 1886 Poincaré had been appointed to the Chair of Calculus of Probability and Mathematical Physics at the Sorbonne, which he held until 1896, when he was replaced by Boussinesq. It was just in 1896 that he published the book *Le calcul des Probabilité*.

22 It seems to me that the word *conventionalism* is more suitable for other thinkers such as Édouard Le Roy. Benrubi rightly stressed that "Poincaré is very decidedly opposed to Le

Apart from some nuances, and apart from marked differences in personality, Duhem pursued a critical overview of scientific practice that was not so different from Poincaré's. As Brenner pointed out twenty years ago, the relationship between Poincaré and Duhem was studded with meaningful "absences and silences": it was a relationship that never became a public debate in the nineteenth century. Poincaré and Duhem's meta-theoretical theses have been looked upon as quite close to each other over time, and "have exerted a common influence" on the twentieth-century philosophical debate, logical positivism included. I find that the "important differences" some historians and philosophers of science have pointed out should not be overestimated, but this analysis falls outside the scope of this book [Brenner 1996, pp. 389-90].[23] A common attention to all the components of scientific practice (empirical, mathematical, logical, and philosophical), a common awareness of the shortcomings of the mechanistic view, a common awareness of the intrinsic historicity of scientific achievements, and a common intolerance of mythological received views on the history of science allow me to point out the existence of a meaningful convergence on fundamental issues. Ultimately, in Poincaré and Duhem I find a common critical attitude and a fruitful alliance between epistemological reflection and historiographical sensitivity that qualify the main characters of the present book.

Another issue deserves to be clarified. Duhem always mentioned the scientists who had contributed to the establishment of an abstract thermodynamics: Clausius, Massieu, Gibbs, Helmholtz, Oettingen, ... Duhem was acquainted with, and acknowledged the existence of, a specific scientific tradition: he even acknowledged the role played by Lagrange in the establishment of an abstract mechanics that had freed scientists from the yoke of microscopic mechanical models. He put forward a historical reconstruction of the

---

Roy's purely instrumental conceptions of scientific knowledge and to his Nominalism" [Benrubi 1926, p. 100]. What David Stump called "hypothetical method," in the sense of Poincaré's search for a middle way between "dogmatic claims" and "antirealist views," seems quite in tune with Benrubi's analysis [Stump 1989, p. 335]. Words such as *realism* and *antirealism* might be misleading when referred to Poincaré and Duhem.

23 According to Milena Ivanova, both Duhem and Poincaré "expressed a form of structuralism," namely structural realism, but they adopted different epistemological views with regard to "how knowledge of the structure of the world is reached" [Ivanova 2015, p. 88]. I have already pointed out the sterility of some philosophical labels when complex research programmes are involved, and when they are applied to historical contexts quite different from the context that has generated those labels. More specifically, can we find essential differences between what might be labelled as structural realism or pragmatism in the late nineteenth century?

research field he contributed to develop further, namely abstract thermodynamics. On the contrary, in the papers he published in the *Revue des questions scientifiques* from 1892 to 1896 we do not find explicit references to scientists and philosophers who had previously put forward similar remarks. We do not find the description of an existing, even though recent, critical tradition of research on the history and philosophy of science. This different attitude seems even more astonishing if we notice that the emergence of an abstract thermodynamics, and the emergence of new and sophisticated attitudes in history and philosophy of science took place in the same years, around the early 1860s.

Why these different behaviours? I can exclude that Duhem was not willing to credit Cournot, Bernard, and other scientists and philosophers with having given birth to a tradition that acknowledged the complexity of scientific practice, and pursued a fruitful alliance between new historiographies and epistemologies. Perhaps the reason might be found in the different landscapes of scientific and philosophical practices in the second half of the nineteenth century. Scientific trends could be more easily defined whereas philosophical trends were more various and fragmented, and more difficult to classify. There is probably another reason that deals with Duhem's professional profile: his interests in history and philosophy of science had mainly emerged from his researches in theoretical physics rather than from an autonomous philosophical research independent of any scientific practice. It is therefore true that Duhem was objectively in debt to some previous scientists and philosophers, but this debt did not correspond to a directly received influence. In reality, a specific philosophical influence, which acted beside his scientific researches, should not be excluded. More specifically, his scientific researches led him to the rediscovery and reinterpretation of the Aristotelian tradition: it was an intellectual pathway that led him from science to philosophy. On the other hand, we can find a definite philosophical influence that can be traced back to Pascal: as will be made clearer in the next chapter, there was a meaningful and explicitly acknowledged contiguity between Duhem and Pascal's epistemologies.[24]

I might say that Duhem found in Aristotle the awareness of the complexity of natural processes, and he found in Pascal the awareness of the complexity of scientific practice, in particular the intrinsically provisional, probabilistic, and incomplete nature of scientific practice. Traces of Aristotelian and Pas-

24 In this chapter and in the *Foreword* I have already mentioned the scholars who have acknowledged the role played by Pascal in the history of science and in Duhem's epistemology. I am particularly indebted to Jean-François Stoffel for having explicitly and repeatedly invited me to develop this historiographical pathway.

calian influences can certainly be found in the critical tradition that led from Cournot to Duhem, and has been mentioned in the previous chapters. Some hints and remarks deserve to be recollected.

In Bernard's texts we find the awareness of the specificity of life sciences in the context of science in general. We can also find some references to Leibniz and Pascal, even though the references to Pascal became more frequent and meaningful in the last, unpublished writings.[25] Bernard had aimed at the apparently paradoxical target of a philosophy of science without philosophy, in the sense of a philosophy of science that would get rid of cumbersome philosophical systems such as Comte's positivism.[26] In the last pages of his *Introduction à la physiologie expérimentale*, we find the praise of research, the charm of the unknown, and the awareness that truth is substantially unattainable even though fragments of truth can unexpectedly appear at the end of an exhausting research. Bernard had stressed that his attitude was not so far from Pascal's attitude that the talented mathematician and philosopher had expressed "in a paradoxical way" by saying that "we do not search for facts but we are in search of new researches." Bernard did not trust systematic philosophy in general, and normative epistemology in particular because he found that it stifled scientific élan, and actual everyday scientific practice [Bernard 1865, pp. 388-9 and 394].

Ten years later, the mathematician-philosopher Cournot had also focused on the new methodological and epistemological perspective opened by life sciences. The complexity of living systems, the link between the parts and the whole, and the existence of a network of processes that converged onto a common aim, had led Cournot to the appreciation of Aristotle's and Leibniz's natural philosophies. On the one end, the specific features of living structures were in tune with Aristotle's concept of *entelechy*. On the other hand, Leibniz' dynamism was more suitable for approaching complex systems rather than the simpler mechanical systems. According to Cournot, Leibniz' physics was more complicated than Newton's physics because Leibniz' natural philosophy had stemmed from biology rather than mechanics. Moreover, scientific knowledge had to be put forward in terms of probability rather than certainty. This *Pascalian* epistemology had allowed Cournot to pursue a third path between scientific dogmatism and anti-scientific scepticism [Cournot 1875, pp. 102-3,

25 I have already pointed out a bibliographic reference to Reino Virtanen for the mention of Leibniz and Pascal in Bernard's writings, and the influence of Pascal in Bernard's last writings [Virtanen 1960, pp. 32-42].

26 See Bernard 1865, p. 387: "Je pense que … le meilleur système philosophique consiste à ne pas en avoir."

348-9, and 360].[27] This epistemology could account for the complexity of the actual scientific practice: that complexity, together with the provisional and fallible nature of scientific theories was in striking opposition to the positivistic and scientist mood that had gained the hegemony in the 1870s.

As I have already remarked in the *Foreword*, in some literature Duhem has been looked upon as one of the starting points of French tradition in history and philosophy of science; in other texts, he has even been placed in the less definite set of ancestors of a tradition that was looked upon as a twentieth-century phenomenon. My thesis is different: I see Duhem, and Poincaré to a certain extent, as the accomplishment of a tradition that had emerged considerably earlier, in the 1860s. In different ways, and in different passages of the previous chapters, I have pointed out the awareness of the complexity of scientific practice and scientific tradition that was displayed in some historical and philosophical studies from Cournot to Naville.

I can only confine myself to reminding readers of some themes that have already been analysed in the previous chapters, more specifically the nature and structure of scientific theories. In 1865, Bernard had stressed the dynamic nature of scientific theories: the process of emergence, success, and partial or total failure of theories was the core of scientific progress. As a consequence, the fruitfulness of theories in the development of knowledge was more important than their content of truth: the evolution of the whole body of knowledge was more important than the steadiness of a particular stage [Bernard 1865, pp. 285 and 290].[28] In 1883, Naville had developed Bernard's themes, and had put forward the image of scientific enterprise as the stratification and interplay of different practices. The search for experimental laws interacted with the more sophisticated stage of theory-building, and in its turn that stage was deeply influenced by tacit and explicit principles or meta-theoretical options

27 See Cournot 1875, pp. 359-60: "Kant nie la valeur externe des idées qui sont le fondement de la philosophie naturelle, comme Pyrrhon niait en théorie l'existence des corps, sauf à y accommoder la pratique, ou comme Descartes lui-meme refusait d'y croire, à moins d'avoir Dieu et sa véracité pour garants. Kant aurait pu aussi s'autoriser de Pascal selon qui « la Nature confond les pyrrhoniens, et la raison (*lisez* le raisonnement) confond les dogmatiques »; tandis qu'à vrai dire la raison, d'accord avec la Nature, ne confond que ceux qui méprisent les principes nécessaires de toute critique."

28 See Bernard 1865, p. 300: "Il doit en conclure que les idées et les théories admises, dans l'état actuel de la science biologique, ne représentent que des vérités restreintes et précaires qui sont destinées à périr. Il doit conséquemment avoir fort peu de confiance dans la valeur réelle de ces théories, mais pourtant s'en servir comme d'instruments intellectuels nécessaires à l'évolution de la science et propres à lui faire découvrir des faits nouveaux."

which directed the actual research [Naville 1883, pp. 52-3]. The fluctuation of theories was the core of scientific progress: in the provisional and limited nature of theories Naville found the strength and the progressive momentum of science.[29]

Here we find the historiographical and epistemological outline that Duhem developed in a systematic way between 1892 and 1896. It is worth remarking that, in the same year of Naville's *La physique modern, études historiques et philosophiques*, the Austrian physicist Ernst Mach published the well-known book *Die Mechanik in ihrer Entwickelung Historisch-kritisch dargestellt*. It seems that neither this book nor the book Mach had published in 1872, *Die Geschichte und die Wurzel des Satzes von der Erhaltung der Arbeit*, exerted an explicit influence on Duhem. The subsequent establishment of a reciprocal sympathy and gratitude between Mach and Duhem should not prevent us from telling general commitments and specific theses apart. The two scholars had in common the interest in a critical and historical analysis of scientific theories, as a meaningful exchange of letters can testify. Nevertheless, important differences can be found in the role played by scientific theories: Mach acknowledged the pragmatic function of theories in the systematisation of the empirical body of knowledge, whereas Duhem attributed a creative power to theories in the establishment and advancement of scientific knowledge. Moreover Duhem gave prominence to speculative and deductive processes even in experimental practice.[30]

---

29 See Naville 1883, p. 55: "Les théories passent, la science demeure; [...] Mais l'histoire de la science, qui nous fait assister à la destruction des systèmes, nous montre qu'à un système détruit en succède un autre dont les conceptions sont plus solides et plus vastes. Elle nous montre que, sauf certains reculs passagers, il existe un progrès constant vers une plus haute intelligence de l'ordre universel,..."

30 Mach's *Mechanik* was translated into French only in 1904, when it had already been appreciated in the English environment. Mach's follower Friedrich Adler translated Duhem's *La théorie physique* (1906) into German in 1908, and Mach himself supplied an introduction. Philipp Frank translated Duhem's *L'évolution de la mécanique* (1903) in 1912. Before the First World War, Poincaré's books were also translated into German [Brenner 2003, pp. 6, 109-11, and 120; Howard 1990, pp. 364-5]. In 1985 the historian of science Roberto Maiocchi saw a striking opposition between Mach's inductivism and Duhem's deductivism: Duhem's stress on theory in scientific practice was to be misunderstood, and positivists considered Duhem a French follower of Mach. Maiocchi put forward his view of Duhem "against Mach" once more in 1990 [Maiocchi 1985, pp. 14, 297, 305, 348, and 355; Maiocchi 1990, pp. 387-8 and 398]. Although too sharp, Maiocchi's reconstruction helps us understand how meaningful the historiographical and philosophical differences between Duhem and Mach really were.

CHAPTER 9

Scientific Practice between Metaphysics and Experiments

1 The Debate on Science and Metaphysics

After some months, Duhem published another paper in the *Revue des questions scientifiques*, which was specifically devoted to the problematic link between physics and metaphysics, and was intended as an answer to his critics, in particular Vicaire. Duhem's starting point was the distinction between physics and *cosmology*. He reminded readers that physics encompassed "perception of facts, discovery of laws, and building up of theories." The label cosmology corresponded to "the search for the nature of material things as causes of physical phenomena," living matter included. In other words, cosmology was a subsystem of metaphysics. Since he was aware that his definitions could generate some misunderstandings when compared to the philosophical tradition, he specified that what he had labelled cosmology corresponded to peripatetic *physics*. What he had labelled metaphysics corresponded to the union of peripatetic physics (cosmology) and metaphysics. He specified that physics "in the modern sense" had no correspondence in peripatetic classification of knowledge: in ancient times, astronomy had been the body of knowledge and the actual practice that had more in common with modern physics [Duhem 1893d, pp. 55-6].[1]

According to Duhem, the distinction between physics and metaphysics did not depend on the subject matter under investigation, but "on the nature of our mind." Only "an angelic mind" could fill the gap between physics and metaphysics, and could attain "a direct insight into the nature of things." However metaphysics was not a superior kind of knowledge because metaphysical knowledge required physical knowledge as a premise. There was a profound asymmetry between physics and metaphysics: the knowledge of the physical world could legitimately lead to some metaphysical hypotheses "on the nature of material things," but those hypotheses could not stem univocally and automatically from physics or a specific field of physics. In Duhem's words, while the knowledge of causes entailed the knowledge of effects, a given effect "might stem from different causes." In the transition from physics to meta-

1 It seems to me that Duhem's *cosmology* might be translated into *natural philosophy*.

 | DOI 10.1163/9789004315235_011

physics, a plurality of metaphysical options were at stake: there was no necessary and unambiguous relationship between the two bodies of knowledge. Physics was based on its own method, which was independent of any metaphysics. It was the *experimental method*, which was something more than a purely empirical practice: it required a certain number of definitions and principles. Those principles appeared as "*self-evident, independently of any metaphysics*," even though they could become objects of investigation for metaphysics [Duhem 1893d, pp. 58-60 and 62-3].

In 1893, Duhem completed the three-level hierarchy he had outlined in his first paper on theoretical physics in 1892: starting from phenomena, we can devise some physical laws, then from laws we can frame theories, and finally from theories we can put forward a plurality of metaphysical options. The links between every level and the following were not ruled by necessity: there was a sort of independence between them. Looking at the four-level hierarchy from the point of view of theories, Duhem found that the scope of physics did not change when it reached the theoretical level: it only became "better as to the form, better ordered, simpler, and therefore more attractive." A theory could not modify the content of the physical laws that it linked together, and could not be in contradiction with a metaphysical truth: to accept or reject a physical theory on the basis of a "metaphysical truth" made no sense. In any case, a theory had nothing to do with truth: it could not be qualified as true or false, but "suitable or unsuitable, good or bad." The plurality of theoretical frameworks corresponding to a set of laws was consistent with this essential feature of theories. Duhem claimed that this thesis was "neither sceptic nor positivist" because it did not depend on specific philosophical commitments. The "destructive trend" of scepticism, and the positivist attitude to identify philosophical practice with scientific method, could be contrasted only by the precise delimitation and the "radical separation" between physics and metaphysics [Duhem 1893d, pp. 65-6 and 68-71].[2]

The second part of the paper was devoted to a historical reconstruction of the relationship between natural philosophy and metaphysics. Although his critics had insisted on tradition, Duhem claimed that his thesis was essentially in accordance with Aristotle and peripatetic tradition. He also mentioned Archimedes, Thomas Aquinas, and Copernicus, in order to show that the logical structure of deductive procedures was something different from the inquiry into the truth of hypotheses from which those procedures had

2 Once more, we find here a meta-theoretical analysis of scientific theories that had much in common with what Cournot, Bernard, and Naville had previously put forward.

started. The necessity of this separation had been acknowledged in the context of astronomy and applied mathematics until the emergence of modern science. Unfortunately some founding fathers of modern science had overturned that meta-theoretical attitude. He found that Bacon had tried to break the boundaries between the different fields of human knowledge; in a similar way Kepler and Galileo had followed the delusive belief that "physical theories can attain the true causes" of things. Once the borderline between the study of phenomena and their laws on the one hand, and the search of causes on the other had disappeared, scientists were allowed to "look upon physical theories as metaphysical explications." Duhem credited Descartes with having systematically blurred the boundaries between science and metaphysics. Descartes had had great influence, on Huygens in particular, even though he had not managed to extend his influence to Pascal and Newton. In any case, in the eighteenth and nineteenth centuries, the awareness of the problematic link between physics and metaphysics had "faded away progressively" [Duhem 1893d, pp. 71-3 and 79].[3]

However, alongside that widespread trend, Duhem found traces of a wise awareness. According to Duhem, some fragments of Laplace's *Exposition du système du monde*, and Ampere's *Théorie mathématique des phénomènes électrodynamiques* showed that physical theories could not be identified with the complete explanation of the natural world. He found that, even "in recent times," when people were so proud of the development of positive science, "the sound and wise peripatetic tradition" had not completely disappeared. At the same time, he acknowledged that a wrong point of view on science had not prevented science from experiencing a striking development. How could scientific progress have actually occurred in spite of a mistaken meta-theoretical attitude towards science? The question had already been discussed by Duhem the year before; he remarked that sometimes, unexpected lands had been discovered while looking for other countries. What scientists had found was independent of what they had searched for: even this was a meaningful instance of the mutual independence between scientific practice and meta-theoretical commitment [Duhem 1893d, pp. 80-2].[4]

3 At least from Aristotle's *Posterior Analytics* onwards, scholars were aware of the mutual independence between the correctness of deductive procedures starting from certain hypotheses, and the truth or reality of those hypotheses. Actually, some founding fathers of modern science believed that the gap could easily be filled.

4 This line of reasoning was obviously a double-edged sword: if every metaphysical or meta-theoretical commitment could substantially be fruitful, the advantage of a clear separation between scientific practice and a realistic philosophy vanished.

Duhem's paper raised some debate in the philosophical environment. In an article published in the *Annales de philosophie chrétienne*, the philosopher Edmond Domet de Vorges, remarked that Duhem had faced "a question of general methodology and high philosophy": therefore philosophers "should be allowed to have their say."[5] In particular, he regretted that metaphysics and the philosophy of Saint Thomas were involved in a dangerous trend. He did not disagree with Duhem on the distinction between physics and metaphysics, but on the use of the word *cause*, which he considered extremely ambiguous. According to Domet de Vorges, Duhem had not grasped the difference between physical and metaphysical causes. In physics the concept of cause had at least two different meanings: the meaning of an immediate connection, as in the case of heat which causes the dilatation of the thermometer, and the "more profound" meaning of hidden actions underlying apparently irregular phenomena, as in the case of the paths of planets in the sky. Science could not "content itself with appearances": it was in search of causes, and it was a legitimate search. The success of Kant's conceptions had led to imagining "a sort of subjective science," namely rational and computational practices without any objective value, and therefore without scientific value. The critical point, and the reason for the disagreement between Domet de Vorges and Duhem, was the gap between human reason and the material world. Domet rightly grasped the philosophical issue at stake: the rejection of the explicative power of scientific theories entailed the existence "of an uncharted land between physics and metaphysics." He claimed that Duhem would definitely have agreed with him on this [Domet de Vorges 1893, pp. 137-141.].[6]

The gap between science and metaphysics was exactly the gap which could not be easily bridged according to the Aristotelian tradition, and according to the more recent Kantian tradition. Domet de Vorges was in tune with a different and still common meta-theoretical attitude in the scientific community, some puzzling specifications included: he stated that "scientific hypotheses are physical explanations," provided that "scientists, according to Duhem, do not attribute a metaphysical nature to hypotheses." The fact is that, according to Duhem, the identification of physical hypotheses with explanations transformed them into metaphysical statements: there was an objective philo-

5 Domet des Vorges was the honorary president of the *Société de Saint-Thomas d'Aquin* in Paris [Stoffel 2007b, p. 294, fn. 110].

6 As I have already remarked in the case of Duhem 1893d, it seems to me that Domet's identification of "objective value" with "scientific value" was not only in opposition to Duhem and Kant, as he himself acknowledged, but also to Aristotle's views.

sophical mismatch between the two points of view. In the end, the philosopher Domet de Vorges agreed with the engineer Vicaire on having recourse to deeply rooted habits: it was "accepted by everybody" that hypotheses were "the pathway towards truth." Unfortunately, the new trend in science, which was upheld by Duhem and "a large number of supporters" and was taught at the Sorbonne, had attained great success without any evident reason. In reality, Domet saw two reasons, the former being "outside actual scientific practice," and the latter inside. On the one hand he saw the influence of Kantian philosophy, which had excluded any entity that could not be seen or seized: the *phenomenon* was attainable, but the *noumenous* was not. On the other hand he saw and regretted "the overwhelming presence of mathematics in physics teaching." He expected that "the misuse of computation" would have doomed modern physics to failure in the same way as "the misuse of logic" had brought discredit on mediaeval philosophy [Domet de Vorges 1893, pp. 141 and 144-7].[7]

With regard to the second reason, Domet actually hit the mark, since the second half of the nineteenth century had seen the emergence of complex mathematical theories in mechanics, thermodynamics, and electromagnetism. On the contrary, the first reason made reference to a philosophical trend that had emerged a century before, and could not account for more recent cultural processes. Moreover Domet underestimated the meaning and scope of theoretical physics, and the emergence of new, sophisticated theories on electrodynamic and thermodynamic phenomena. It seems that he could not appreciate the increasing mathematisation of theories in the context of a new fruitful alliance between the tradition of mathematical physics and the tradition of natural philosophy. However, when Domet attempted to frame a conclusion, he focused on the keystone of the whole philosophical debate, and remarked that "the *esprit* of the ancient philosophy" was essentially objectivist. This was the real issue at stake: the interpretation of Aristotelian tradition in realistic terms. The philosopher Domet could not accept that a physicist inquired into that philosophical tradition, and gave prominence to one of the hallmarks of that tradition: the mutual independence between the correctness of rational procedures and the truth of the hypotheses on which those procedures were based. The neo-Thomistic trend and apologetic com-

---

7 Domet de Vorges conceded that hypotheses might not be "directly verifiable," and he made reference to "facts that not only might be hidden at present, but also intrinsically unattainable." At the same time, hypotheses might reach a "high degree of certainty," and the whole science might be based on them.

mitments led to a naive realism that Duhem could not share. The last passage of Domet's paper could be looked upon as a ban on any further attempt to put in danger the realistic received view: Duhem could not find any support "in traditional philosophy" [Domet de Vorges 1893, p. 151].

Many Catholic scholars were following the neo-Thomist trend; Duhem was Catholic but he was not neo-Thomist. Although some historians have attempted to enrol Duhem in that philosophical family, this thesis can hardly be upheld: Duhem's problematic realism and the sharp separation among scientific practice, philosophical commitment, and religious belief were not in tune with the more confident realism and mild apologetic temptations of some neo-Thomists. Domet had realised that Duhem did not belong to his company: he did not appreciated Duhem's critical attitude, and moreover he did not manage to grasp the sense and usefulness of Duhem's sophisticated distinctions. He therefore decided that Duhem belonged to the positivist company, namely the enemy.

The relationship between Duhem and neo-Thomists has been widely debated by historians and philosophers, and we can find a wide range of interpretations.[8] The legend of Duhem as "a follower of Thomas Aquinas" [Benrubi 1926, p. 103] was propagated throughout the twentieth century. It is worth stressing that not all Catholic philosophers were neo-Thomists: the influential philosopher and Duhem's close friend Maurice Blondel opposed Scholastic philosophy, and considered Pascal "his foremost spiritual ancestor" [Stoffel 2008, pp. 111-2; Benrubi 1926, pp. 204 and 206]. I agree with Stoffel on the prominence of Pascal among Duhem's philosophical and theological reference frames, and with Helge Kragh on the fact that Duhem had to face the hostility of scientists and anti-clericals, as well as the hostility of many Catholic scholars.[9]

8 On the one hand, we find the physicist, historian, and Catholic priest Stanley Jaki, who in 1984 lent towards a neo-Thomist Duhem. On the other hand, we find Niall Martin, who in 1991 opposed the *Pascalian* Duhem to neo-Thomist philosophical and theological attitudes. In 2000, Martin Hilbert expressed his disagreement with Martin on the detachment of Duhem from neo-Thomists on the grounds of Duhem's "penchant for Pascal." He mentioned some "neo-Thomists with an admitted admiration for Pascal," and in general Duhem's collaboration with Catholic journals and activities. Although Hilbert remarked that Duhem deserved "a large share of the credit" for having helped "Thomist philosophers come to grips with modern physics" I find that this credit is definitely something different from an *entente cordiale* [Hilbert 2000, pp. ii, 3, 6, 350-1, and 358-60].

9 The latter "suspected him of philosophical scepticism and *fideism*, the heretical belief that faith rests on faith and nothing else" [Kragh 2008b, p. 391].

As a further instance of the Catholic attitude towards Duhem, the same year, in a brief note, Lechalas also avowed that he would have confined himself to demonstrating "serious evidence of Duhem's intimate leaning towards positivism."[10] What was really at stake in Lechalas' *Note*? The unbridgeable gap between his naïve realism and Duhem's more sophisticated realism, I suppose. The majority of contemporary readers did not seem so intellectually equipped as to follow Duhem in his historiographical and epistemological remarks: the intellectual gap between Lechalas and Duhem represented a meaningful instance. The majority of Catholic scholars preferred to rely on a strict realism, which appeared more in tune not only with the apologetic aims of radical Catholics but also with the milder neo-Thomist commitment. That the catholic Duhem did not share the same attitude appeared incomprehensible and quite unsettling.

It is worth remarking that in the last decades of the nineteenth century, a wide debate on the relationship between science and theology took place, and the second Principle of Thermodynamics played an important role in it. Duhem did not appreciate the cosmological interpretations of the two Principles of Thermodynamics, and theological arguments based on them. He was a firm believer and at the same time "an independent mind": he disliked transforming scientific contents into apologetic arguments, and always insisted on "a clear separation between science and faith" [Kragh H. 2008a, pp. 113-7].[11]

---

10 After having compared Duhem's treatise on acoustics with that on optics, he realised that the latter did not mention any elastic medium, whereas the former relied on "the usual conceptions on the cause of sound." How could Duhem "reject, *in principle*, the explicative power of aethereal oscillations" together with what he had labelled "*mechanical theory* of light"? At the same time, Lechalas acknowledged that, in the case of "sonorous perceptions," the elastic medium really interacted "with our organs," differently from the case of optical perceptions. In the end he acknowledged that the difference between the two theoretical representations corresponded to "a real difference" dealing with real phenomena. The whole line of reasoning sounds quite strange: at first it seemed that he did not succeed in comprehending the difference between the status of ordinary matter and aether, but then he pointed out that the supposed contradiction was not so [Lechalas 1893b, pp. 312-14]. However, with regard to the specific case of optics and acoustics, Lechalas missed the point: according to Duhem, neither oscillation in the air nor aethereal oscillations could be looked upon as actual explications of the natural world.

11 See also Kragh 2008a, pp. 116-7: "According to Duhem, the controversy between Catholic thought and modern science was essentially a misunderstanding based on a failure to appreciate the separate domains of the two fields." Kragh also noticed that Duhem's conceptions "made him a target for some Catholics, who suspected him of philosophical scepticism" [*Ibidem*, p. 117].

2 The Problematic Concept of "Natural Classification"

In the same year, Duhem published another paper in the *Revue des questions scientifiques*, where he collected some remarks triggered by the translation of William Thomson's *Popular Lectures and Addresses* into French. He started from the astonishment of French scholars faced with the specific English way of interpreting and practising physics. That specificity consisted of an imaginative faculty which allowed English scholars to envisage a complex net of concrete representations. In the case of electromagnetic theories, where French and German physicists conceived a bundle of lines of force, the English physicists imagined a bundle of elastic threads that could decrease their length and increase their section at the same time. He had found in Oliver Lodge the most radical implementation of that attitude.[12] In general terms, that faculty represented "an extraordinary ability to grasp the concrete, and an extreme difficulty in appreciating the abstract." After having quoted a well-known passage from W. Thomson's *Baltimore Lectures*, where Thomson stated that he could not be satisfied with any physical explanation that did not correspond to mechanical models, Duhem remarked that "the English school" relied on an extremely specific kind of "reduction of matter to a pure mechanism." He could not accept that reduction, and that kind of mechanism [Duhem 1893e (1987), pp. 113-117 (footnotes at p. 117 included), and 119].[13]

This was a very important issue: English *mechanism* had its roots in Descartes' *mechanism*, but different kinds of *mechanism* could be consistently conceived. An abstract kind of *mechanism* was not far from Duhem's horizon: in building up his generalised mechanics, he relied on a very general and wide-ranging mechanical approach, on the track of Lagrange's analytical mechanics. He had realised that specific mechanical models did not allow him to describe the complexity of the physical world, but a more abstract mechanical approach really could. Duhem's *mechanism* was a structural *mechanism*: thermodynamic processes and chemical reactions could be represented by the generalisation of analytical mechanics rather than by microscopic models of interacting particles. We are considering here two different traditions of Mechanics: wide-ranging mathematical structures on the one hand, and specific mechanical models on the other. Another sensitive issue was the need for a

12 For Lodge's approach to electromagnetism see Lodge 1883, pp. 328-30, Lodge 1885, pp. 486-7, and Bordoni 2008, pp. 139-44.

13 He remarked that English genius "had generated Shakespeare, but never a metaphysician" [Duhem 1893e (1987), p. 116].

compact structure and logical consistence of a physical theory: that necessity was one of the hallmarks of the "French school." Duhem considered himself a follower of that illustrious tradition which included both Lagrange's abstract line of reasoning and Laplace's different approach, which was not insensitive to the fascination of mechanical models. Even "a geometer of Laplace and Cauchy's school" would not have accepted the proliferation of contradictory models typical of the *English school* [Duhem 1893e (1987), pp. 128-9].

Duhem could not accept the superposition of different mechanical models, or a series of partial models that could be "developed independently of each other." Browsing through Thomson's book, he had found a gas represented as a set of tiny bullets, "endowed with unthinkable velocities," which collided with each other "during their crazy race." But he had also found material molecules represented as a structure of concentric shells connected by springs, or a gyrostatic system composed of aethereal vortices. He acknowledged that his rejection of theories as mere collections of scattered models, which were consistent in themselves but mutually inconsistent, could appear in contradiction with his mistrust of metaphysical explanations. If he did not rely on the explanatory power of theories, but only on their symbolic representative power, why should he have demanded a strictly logical foundation of such theories? This was a subtle objection indeed, and Duhem only managed to offer a very general answer: he was pursuing a third path between dogmatism and scepticism. He could not agree with scientists who "firmly attributed an ontological value to physical theories," and at the same time he could not agree with those who looked upon Laplace, Ampère, Thomson and Maxwell's methods as equivalent. From the purely logical point of view, a physicist could not be prevented from representing "different sets of laws, and even a single set, by several incompatible theories." It was just logic which granted the physicist this right. Nevertheless, there were extra-logical reasons that did not recommend the proliferation of incompatible theories [Duhem 1893e (1987), pp. 132-3].[14]

Extra-logical reasons could be synthesised in the belief that perfection should be pursued in scientific practice. The physicist aimed at a "*natural classification* of laws," where the demanding adjective *natural* echoed suitability but also perfection, in the sense of a "perfect and ideal theory" or a "complete and appropriate metaphysical explanation of the nature of material things."

---

14 Duhem remarked that the English way of practising theoretical physics was consistent with English law, which consisted of a countless collection and superposition of both "laws and habits" [*Ibidem*, p. 131].

Obviously, an ideal theory was not a real entity: it could be attained neither in practice nor in principle because of the unbridgeable gap between physics and metaphysics. It was definitely outside the scope of the human mind: it was like a visible but unattainable horizon. Actual physical theories, put forward by real physicists, had to "strive for perfection," even though perfection could not be attained. Duhem was facing here a component of scientific practice that did not deal with experimental or logical-mathematical practices. It dealt with a network of theoretical and meta-theoretical commitments that resisted any definite specification. It would have been better to explicitly acknowledge the existence of that intrinsic complexity, but Duhem preferred to engage himself in a further philosophical analysis. In the end, he ventured to write that a natural theory "would class the physical laws in a way which would reflect the metaphysical relationships among the essences from which the laws emanate." Essences were obviously no less unattainable than ideal theories, and this definition made theoretical practice even more mysterious [Duhem 1893e (1987), pp. 136-7].

It is worth remarking that the concept of natural classification had already been introduced by Comte even though the latter had made use of it in the meta-theoretical rather than theoretical context. According to Duhem, a physical theory became a natural classification when it managed to grasp the deepest links among phenomena. According to Comte, a natural classification was a relationship among different bodies of knowledge that was in tune with actual genealogies. A natural classification had to take into account both the historical development of knowledge and the network of logical links among the different research fields [Comte 1830, pp. 60-1, 76-8, and 86]. In 1861, Cournot had focused on classifications in the domain of life sciences. Although every classification was provisional and incomplete, some classifications could grasp some essential features of reality and the actual relationships among different entities. In this case, the classification could deserve the qualification of *natural classification*. However, the suitability of a classification could not be demonstrated but only appreciated by a synthetic overview: the judgement could not rely on purely logical or empirical procedures. Pragmatic or "intuitive assessments" were at stake: the wide domain of human faculties between logical and empirical practices was involved. As a consequence, even a natural classification was intrinsically provisional and incomplete, but it could also be reliable and effective [Cournot 1861, pp. 423 and 425-6]. On the contrary, in 1874 Boutroux had expressed his dissatisfaction with the concept of natural classification itself. This distrust was based on his radical anti-reductionism, and on the perception of an unbridgeable gap between the domain of rational procedures and the domain of empirical practices. The complexity of nature defied any attempt to match facts and theories: every classification could not but be artificial.[15]

Duhem did not manage to define his third way between dogmatism and scepticism, and the concept of natural classification, in a satisfactory way. The fact is that the third way could not be defined in formal terms: if he had explicitly acknowledged this impossibility, he would have avoided subsequent fruitless debates.[16] The extremely delicate balance that emerges from the following passage shows us how difficult it was to undertake that third way.

> Although our theories are obviously imperfect, they should aim at perfection. They are simply classifications that links laws to each other, and in no way might they grasp the relations among the essences; nevertheless, we should manage to set up a network of relations that give rise to actual analogies rather than contingent similarities. Theories should approach the actual relations among the essences: in other words they should aim at a *natural* classification rather than an *artificial* one [Duhem 1893e (1987), p. 137].

In the last sections of the paper, Duhem was committed to pursuing an *entente cordiale* between abstract and imaginative attitudes in scientific practice. At first, he criticised the new generation of British physicists, who had crossed the boundaries that neither W. Thomson nor Maxwell had overstepped. They dared to inquire into "strange and disconcerting" subject matters with the same self-confidence as in their researches on optics and electricity. In particular, Duhem criticised William Crooks, Lodge, and Peter Guthrie Tait because of the researches they had undertaken on the "transmission of thought at a distance, spiritualism, and magic."[17] At the same time, he acknowledged that

15 See Boutroux 1874, p. 46: "Fort de l'idée de genres et des lois, l'esprit humain espérait remplacer les classifications artificielles par des classifications naturelles. Mais, avec le progrès de l'observation, telle classification, que l'on croyait naturelle, apparaît à son tour comme artificielle; et l'on se demande s'il ne conviendrait pas de substituer pas à toute systématisation rationnelle le dessin pur et simple d'un arbre généalogique."

16 According to Sindhuja Bhakthavatsalam, beyond the different representations of Duhem as an instrumentalist and a realist, Duhem's concept of natural classification helps us understand "the pragmatic rationality of a physicist," and the necessity of "a pragmatic rationality requirement" [Bhakthavatsalam 2015, pp. 11 and 21]. I substantially agree.

17 In the last decades of the nineteenth century, when the process of specialisation and professionalisation of scientific practice was accomplished, a marked interest in researches at the borderline between physical and psychic phenomena flourished in the British in-

their attitudes "encouraged invention in the highest degree." Although those researches could become dangerous for science, Duhem noticed that "the inventions which had blossomed on the Continent had not been as numerous and audacious as those in England and America." The partial re-evaluation of the imaginative side of scientific enterprise was consistent with the appreciation of the extra-logical elements that concurred in the emergence of that complex scientific practice that he labelled theoretical physics. Theoretical physics was not completely "subject to the inflexible laws of logic": the choice of hypotheses was an instance of extra-logical process, wherein personal attitudes, historical heritage, and other influences were at stake. A physical theory was therefore the result of a historical process, and showed the hallmark of a specific place and time. According to Duhem, extra-logical influences affected "the hypothetical part of the theory": that component might be discarded after some time, whereas the logical or formal structure of the theory was apt to survive. Faults and mistakes were influenced by "the environment and race, and by physical barriers and political borderlines," whereas the structural features of a physical theory were, in some way, perpetual creations of the human mind [Duhem 1893e (1987), pp. 139-40 and 145-6].[18]

3 Experimental and Theoretical Practices

After having examined the aims and features of theoretical physics, in 1894 Duhem inquired into experimental practice. The fifty pages of the paper "Quelques réflexions au sujet de la physique expérimentale" were ordered into a list of fourteen theses, divided into two parts: ten theses appeared under the question "what is an experience in physics?," and four under the subsequent question "what is a physical law?". If the second part resumed and

tellectual environment [Thomson J.J. 1936, pp. 159-63, 298-9, and 383; Oppenheim 1985, pp. 3, 199-202, 334-5, and 393; Bordoni 2008, pp. 266-70].

18 Duhem found that W. Thomson and Hermann von Helmholtz were the champions of the two different attitudes: imagination and specific models on the one hand, and abstraction and generalisation on the other. See Duhem 1893e (1987), pp. 141-3, in particular, p. 143: "… partant de la physique, Helmholtz remonte par l'analyse, de principe en principe, jusqu'à rencontrer la métaphysique; Thomson descend, de conséquence en conséquence, jusqu'aux applications industrielles; le premier est un des plus profonds philosophes de notre siècle; le second en est un des ingénieurs les plus inventifs." It is worth stressing that not only was Thomson an imaginative theoretician, he was also a talented engineer. On Thomson's role in the realisation of a submarine cable for transatlantic telegraphic connection, see Smith and Wise 1989, pp. 661-83.

completed some remarks he had already made in previous papers, the first part was substantially new. It pivoted on three fundamental theses: first, a physical experiment did not consist in a purely empirical process; second, it could not be as powerful as to lead to the refutation of a single hypothesis; third, it was less reliable, even though more precise, than ordinary experience. He was aware that those theses would have astonished the readers of the *Revue des questions scientifiques*, and would have scandalised some minds "concerned about scientific rigour": he knew that, from Francis Bacon onwards, the rhetoric which accompanied scientific practice had led to more comfortable conclusions [Duhem 1894c, pp. 147, 151, and 181].[19]

With regard to the first fundamental thesis, Duhem invited readers to enter a laboratory full of electromagnetic devices and ask a physicist what he was doing: he might answer that he was measuring an electric resistance or the oscillations of a bar which carries a little mirror. In other words, his answer did not deal with pure facts or observation of facts, but physical concepts and abstractions: electric resistance, temperature, and gas pressure were not facts but rational representations of facts. In reality, any actual experimental practice consisted of two parts: at first the observation of certain phenomena, and afterwards the active interaction with those phenomena, and "the interpretation of the observed facts." In other words, the performance of a physical experiment required both the knowledge of physical concepts and accepted theories, and "the ability to apply them." From the structural point of view, experimental practice was not so different from theoretical practice. The former involved the interaction between two different components: empirical practices in a proper sense, and basic theoretical knowledge. The latter involved the interaction between the formal structures of logic and mathematics, and a complex network of speculations [Duhem 1894c, pp. 148-9].

In the context of experimental practice, the new key concept pointed out by Duhem was *interpretation*.

> A physical experiment is a definite observation of a set of phenomena accompanied by an INTERPRETATION of such phenomena. The interpretation replaces the actually performed observations with abstract representations that correspond to the former in accordance with the physical theories accepted by the observer [Duhem 1894c, p. 150].

19 In this paper, Duhem mentioned Claude Bernard's *Introduction à l'étude de la médicine expérimentale*.

Duhem acknowledged that a purely empirical practice made sense in the early stages of the history of science, when theoretical frameworks were missing or only roughly outlined, but it made no sense in the case of advanced sciences, where "mathematical structures" played a fundamental role. When a science progressed, the role played by theory in the interpretation of experimental facts increased progressively. In the late nineteenth century, the pretension to having written a *purely experimental* paper on physics was not different from the pretension to having expressed "an idea without making use of any sign, neither spoken nor written." These remarks led Duhem to the second fundamental thesis, namely "*a physical experiment can never condemn an isolated hypothesis, but only a theoretical system.*" When a physicist performed an experiment in order to check the consistency of a given law or statement, he could not confine himself to the statement under investigation. The experiment involved a wider body of knowledge that dealt with basic assumptions, definitions, concepts, and other fundamental laws. The physicist had to make use of a wider set of theories, which could be assumed "without questioning them." If an expected prediction did not take place, the scientific community was allowed to conclude that there was "something wrong" in the complex net of theories, but it would not be able to identify exactly the mismatch [Duhem 1894c, pp. 153-5, and 157].[20]

In brief, the complexity of scientific practice forbade scientists to perform logical procedures of decision on single statements: however paradoxical it might be, the *modus tollens* procedure could only be applied to the body of knowledge of science as a whole.

> In brief, a physicist cannot submit an isolated hypothesis to experimental check, but only a set of hypotheses. If an experiment were in disagreement with theoretical expectations, it would only show that one hypothesis at least should be modified, but it cannot single out the mistaken ones [Duhem 1894c, p. 160].

Physics could not be looked upon as a machine: it could not be easily disassembled in order to "check its components separately." Physics was rather a living system wherein when "every part operates, even the most distant parts come into play." This could be qualified as methodological or meta-theoretical

20 The second fundamental thesis has subsequently become known to philosophers as Duhem's holistic thesis. Under the label "Duhem-Quine thesis," it has been widely discussed and criticised. See the following *Afterword*.

holism. It is worth stressing that the awareness of the complexity of the natural world, the rejection of a reductionist approach to it, and the design of a generalised thermodynamics preceded Duhem's meta-theoretical holism. This does not mean that Duhem's holism directly stemmed from his physical theories, but we can take note of the existence of two subsequent steps in Duhem's intellectual pathway, the first being theoretical and the second philosophical and methodological or more generally meta-theoretical. At first he had attempted to go beyond purely mechanical models of the natural world and put forward mathematical theories that included irreversible processes. Then he attempted to go beyond simplified accounts of both theoretical and experimental physics. In both cases, a rediscovery and reinterpretation of Aristotelian tradition was at stake. He arrived at a representation of scientific practice as a living process, where each part concurred in the realisation of the whole, and at the same time was influenced by the whole. Skipping from the theoretical to the meta-theoretical level, not only has Duhem shown that we cannot act on the natural world as if it were clockwork, but also that we cannot look upon the body of scientific knowledge as if it were mechanically assembled.

> ... when a clockmaker receives a clock that does not work, he/she disassembles all wheels and gears, and checks them individually until he/she finds the one broken or deformed. On the contrary, when a physician examines a sick person, he/she cannot dissect the patient in order to put forward a diagnosis: he/she can only guess the seat of the illness by observing the effects on the whole human body. The physicist who attempts to rearrange a deficient theory resembles the latter rather than the former [Duhem 1894c, pp. 160-1].

According to Duhem, geometry grew by accumulation of new theorems, which could be looked upon as "demonstrated once forever," whereas physics could only offer "a symbolic picture," which had to be continuously retouched, in order to widen its scope, and improve its unifying power. In geometry, there was "no place for a third alternative between two contradictory statements," whereas in physics the actual demonstration of a statement could never be attained. The reduction to absurd, which was typical of geometry, assured that a statement was true when its contradiction led to false consequences. In physics such a reduction could find no place, and therefore "*the experimentum crucis*" was impossible [Duhem 1894c, pp. 161, 163, and 165]. This impossibility led Duhem to his third fundamental thesis that dealt with the difference between ordinary experience and physical experiments. He remarked that when honest people reported a bare fact, that fact could be considered true and certain. On the contrary, a physical experiment was much more that the perception of a bare fact, and therefore its degree of certainty was much less. Even the communication of a physical experiment was much more complex than the communication of a bare fact. An experiment performed by a physicist who shared our "interpretation of phenomena," and relied on the same set of accepted theories, was not difficult to understand. On the contrary, an experiment performed by a physicist who did not share the same body of knowledge, or which had been performed in different periods of history, could appear nonsensical. In that case, things went as if the two scientists spoke "foreign languages": the experiment of the one might appear meaningless to the other [Duhem 1894c, pp. 175-7].

The disentanglement of the complex relationship between theory and experiment required a specific sensitivity or some kind of flair that dealt with "the *esprit de finesse* rather than the *esprit de géometrie*." The uncertainty that affected experiments was intrinsically linked to the provisional nature of physical laws: a law represented a set of phenomena with an approximation that "physicists could consider satisfactory at present, but might not be accepted afterwards." This provisional nature of laws could not be mended because their content of truth was historically determined. It would have been too naïve to look upon a law as true today and false tomorrow: it was necessarily "neither true nor false" at any time [Duhem 1894c, pp. 179 and 188]. In Duhem's epistemological framework, physical laws were the link between experiments and theories: their provisional condition stemmed from the intrinsic uncertainty of experimental results but also from a sort of *linguistic* inadequacy of concepts and symbols. Physical laws needed to be updated, revised, or replaced. According to Duhem, something like a final truth could not be attained, and Pascal had managed to express this concept in an effective and imaginative way. He had compared truth to a tip that was "too subtle to be handled with care by our blunt tools." When scientists attempted to direct the tip towards a definite point, the tools exerted a blunt pressure, crushed the tip, and the approximate contact was "more onto the false than onto the true" [Duhem 1894c, pp. 190 and 195].[21] In no way could scientists rely on automatic procedures for

21 See Pascal 1897 (1976), p. 76: "La justice et la vérité sont deux points si subtiles, que nos instruments sont trop mousses pour y toucher exactement. S'ils y arrivent, ils en écachent la pointe, et appuient tout autour, plus sur le faux que sur le vrai." Duhem also offered a nice metaphor for the usefulness of precision, the illusion of formal certainty, and the complex relationship between precision and certainty. A botanist was in search of a rare

the attainment of perfect knowledge: the actual procedures were affected by approximation and uncertainty.

In the same year Duhem published a paper on the history of optics, "Les théories de l'optique," in the *Revue des deux mondes*, and the dynamic and provisional nature of scientific theory was at stake once more. His history was something more than a mere collection of meaningful facts: from the outset he put forward a definite historiographical framework. He looked upon the landscape of physical sciences as a twofold entity. On the one side were systematic bodies of knowledge, which had been considered as such even in ancient times and in the Middle Ages: among them he mentioned astronomy, hydrostatics, and the general principles of statics. On the other side he saw only scattered experiences, often "mutually inconsistent and roughly performed": scholars interested in the history of natural philosophy could find neither "steady evolution" nor a systematic network of logical links. According to Duhem, a reliable history of science required the existence of a body of knowledge endowed with logical and progressive development. This feature could not be found in the researches on electricity, magnetism, heat, and other subjects which natural philosophers and practitioners had undertaken over a long time span, from the ancient Egyptian civilisation to the threshold of the European Renaissance. He specified that he was writing "a history of physics for the benefit of physicists," and for this reason he was to confine himself to "the modern tradition," which had emerged at the end of Renaissance [Duhem 1894d, pp. 94 and 118].[22]

The starting point of his history of physical sciences was Descartes' mathematics and natural philosophy; he then introduced readers to Fermat, Huy-

---

tree and asked two countrymen for information: the first simply said that the tree was actually in that wood, whereas the second specified what track she/he had to follow, and once there, how many steps in a specific direction were required. The botanist managed to find the tree, but after a slightly different number of steps. The content of the first piece of information "was true, and the second false," even though the latter was definitely more useful and precise [Duhem 1894c, p. 195].

22 In 1872, Cournot had already pointed out the existence of two different paths in the history of science: on the one hand, the history of systematic sciences such as the four sciences of the "*quadrivium*," and on the other, the history of scattered empirical practices, "extraneous to official lecturing." Systematic sciences, which could be traced back to the age of Greek civilisation, had experienced a striking revolution in the sixteenth and seventeenth centuries. The scattered bodies of knowledge, which included "theories of heat, magnetism, and electricity," had preserved their character of semi-speculative and semi-empirical sciences throughout all the stages of the so-called scientific revolution [Cournot 1872, pp. 50-65 and 292-4]].

gens, and Newton's theoretical models, and finally described the models of optical and electromagnetic aethers. The last pages of the paper were devoted to Maxwell's electromagnetic theory of light, which Duhem criticized in some detail. He made use of this theory as a suitable case study in order to discuss his meta-theoretical remarks on the features of a physical theory. He focused on Maxwell's key-concept of *displacement current*, namely the variations of polarization in dielectrics, which showed a formal analogy with ordinary electric currents in conductors. Two main issues were at stake: the polarization of aether as a specific instance of polarization in dielectrics, and the nature of the analogy between electric currents in conductors and dielectrics. Duhem remarked that no experience, but only "an incomplete analogy," had led Maxwell to the new general concept of electric current: in some way, "the electrodynamics of displacement currents" was a skillfully tailored, purely theoretical artefact. The theory could only rely on one experimental datum: the equality between the velocity of propagation of displacement currents and the velocity of light in the same medium. There was "a logical chasm" between a numerical analogy, which might even be contingent, and the hypothesis that light consisted of rapidly oscillating displacement currents [Duhem 1894d, pp. 118-20].[23]

It is worth mentioning that in 1894 the foundations of Maxwell's theory were still questioned even though the theory had recently received a meaningful corroboration through Hertz's experiments. The sharpness of Duhem's criticism mirrored the gap between his meta-theoretical attitude and the *imaginative* attitude of British physicists he had already discussed in 1893. He wrote as if he were the watchman of French scientific tradition, and in particular the spokesman of a community of scholars who "loved clarity and were concerned about exactness." Maxwell's theory induced "a painful surprise" in minds that had been trained in the tradition of French mathematical physics. According to Duhem, "the murky and puzzling principles on which that theory was based" could not fit in with the research style of French masters from Laplace to Cauchy. At the same time, he was aware that, although at the beginning Maxwell's theory had been regarded as "a paradoxical and ingenious overview," subsequently it had gained much more consideration in most of the scientific community. Duhem considered himself as a follower of a tradition where every physical theory had to satisfy at least three basic features. First, "not even the slightest presence of contradiction" could be tolerated. Second, the differ-

23 He found that Maxwell was not so different from Fresnel as to the ability "to devise rather than justify his inventions" [Duhem 1894d, p. 120].

ent parts of a theory had to be logically connected. Third, the number of independent hypotheses should be minimum. He found that recent and successful meta-theoretical views had driven scientific practice in the opposite direction: heuristic power and fruitful applications had become the most appreciated targets of physical theories [Duhem 1894d, pp. 119-20].[24]

Duhem acknowledged that Maxwell's electromagnetic theory offered some advantages over the previous elastic theory, but he found that a better logical consistency could be attained by Helmholtz's theory, which was credited with having "a wider scope," and did not rely on specific hypotheses on the structure of aether. However, the replacement of Fresnel's elastic theory by the electromagnetic theory represented a progress with regard to the tradition of mechanical models, because dielectric polarization could be looked upon as a new "primary quality" which did not require to be reduced to matter and motion. In reality, the success of the electromagnetic theory could not give a definite answer to the question whether a mechanical world-view was the best representation of the physical world. Maxwell himself, and more persistently William Thomson, had attempted to reduce the whole set of electric, magnetic and optical phenomena to matter in motion. Nevertheless, in order to perform that reduction, they had paid a very high price: they had devised "a strange and complex" structure for aether, which on the other hand, had to be looked upon as the simplest dielectric [Duhem 1894d, pp. 120-1].[25]

24 The following passage shows Duhem's sharp criticism: "Mais, ces esprits-là se font rares aujourd'hui; leurs exigences semblent exagérées à beaucoup de physiciens; plusieurs même les trouvent un peu ridicules, et, avant la précision et la logique, qui ne satisfont que la raison, font passer la généralité des aperçus et l'imprévu des rapprochements, qui séduisent l'imagination; aussi fait-on grâce à la théorie électromagnétique de l'obscurité de ses origines; on lui demande seulement d'être féconde en applications" [*Ibidem*, p. 120].

25 In reality, as far as it is developed in his *Treatise* (1873), Maxwell's theory was a twofold mechanical theory, in accordance with the two meanings that could be associated with the adjective *mechanical*: specific mechanical models, on the one hand, and the mathematical structures of abstract mechanics, on the other. In Maxwell's *Treatise*, we find a sort of superposition between the two approaches. In 1893 Oliver Heaviside showed that Maxwell's electromagnetic equations, which he had synthesised in the vector language, could not fit in with any mechanical model of aether [Heaviside 1893, pp. 128-31; Buchwald 1985b, pp. 288, 294, and 234; Buchwald 1985c, p. 236; Bordoni 2008, pp. 163-5]. Hertz's well-known experiments on the propagation of electromagnetic perturbations were also performed with the hope of detecting empirically the difference between Maxwell and Helmholtz's theories [Hertz 1892 (1962), p. 20, Doncel 1991, pp. 1 and 6; Darrigol 1993, p. 233; Buchwald 1994, p. xiii].

Duhem's meta-theoretical commitment was both anti-mechanical and anti-reductionist. Maxwell's and Thomson's aethereal vortices appeared to him not so different from Descartes' vortices, which had been superseded by Newton's mechanics. Once more Duhem quoted Pascal's irony: the talented mathematician had mocked any reductionist methodology making use of the best known symbol of determinism and reductionism in the seventeenth century, namely the mechanical clock. As Pascal had remarked, we could envisage rough mechanical models for every phenomenon. We might even disassemble a body into its basic elements, in terms of "geometry and motion," but how could we identify exactly those elements, and how could we "build up the whole structure" starting from them? [Duhem 1894d, pp. 121-2].[26]

4 The Systematic Link between Historiography and Epistemology

Duhem outlined a complex historiography where both linear progress and cyclical processes were at stake. He started from "the leading hypothesis" of a successful theory that might be considered as such by a given generation of scientists. It might have been considered as "an evident mistake by the previous generation," or "an evidence of the ignorance of their forefathers" by a subsequent generation. According to Duhem, the history of optics was a meaningful instance of that oscillating trend: seventeenth-century scholars had contemptuously rejected the model of emission, eighteenth-century natural philosophers had relied on this model, and had despised the wave model, and then nineteenth-century physicists retrieved the latter, and found surprising the confidence in the former as a serious theory. He depicted physical theories as dynamic entities which experienced a predictable chain of events that could be easily outlined: they emerged, then they multiplied their successes "accounting for disregarded or poorly-understood phenomena," but at that stage they repeated a common mistake. The hypotheses on which the theory rested were looked upon as "absolute certainties," and the representation of the external world offered by the theory was transformed into "a faithful description of the world structure." But some difficulties soon emerged, and the weight of the failures led to the collapse of the theory: in the end, scientists hastened "to sweep away the debris in order to make room for another

26 Pascal's passage was drawn from one of his *Pensées*, which was explicitly addressed to Descartes: "Il faut dire en gros: cela se fait par figure et mouvement, car elle est vrai. Mais de dire quels, et composer la machine, cela est ridicule. Car cela est inutile et incertain et pénible." [Pascal 1897 (1976), p. 72; see also Pascal 1951, vol. 1, p. 66].

theory." In brief, Duhem outlined a historiographical framework where theories emerged, were successfully upheld, then suffered a dogmatic drift, were afterwards overwhelmed by subsequent defeats, and eventually were replaced by new theories [Duhem 1894d, p. 122].

He was aware that his representation of physical theories as provisional creations of the human mind might lead to a sceptical attitude and an underestimation of science. He was aware that both scholars and common citizens could question "the fruitfulness of the efforts directed to built up and then destroy all those structures": in particular they might wonder whether the scholarly élite had manage to attain actual progress in the knowledge of the physical world. This state of uncertainty might lead to two meta-theoretical attitudes that Duhem could not accept: the idea that "the secrets of nature are unintelligible," and "the confidence in mere experiences" at the expense of theoretical practice. Duhem acknowledged that "the breeze of scepticism" was blowing through French intellectual environment, but he invited his readers "not to let that wind shake them." He hoped that an attentive reader could single out traces of a slow progress underneath the theoretical oscillations [Duhem 1894d, pp. 122-3].

This was the keystone of Duhem's historiographical outline: the periodical series of successes and failures in scientific practice hid an actual and higher-level progress. There was a positive heritage in the history of science, and it could be found even in outmoded or totally disappeared theories. First of all, a theory might disappear, but physical laws might survive. Descartes' optical theories were definitely outdated, but the law of light refraction was still valid, and it had continuously been re-interpreted by new theories. Huygens' law of refraction and "Newton's law on the series of colours in thin layers" had shared the same fate: they had become part of the contemporary body of knowledge. However, physical laws were not the only positive heritage of dead physical theories: even mathematical language and the logical connections that gave a consistent and unitary structure to a theory could survive. The mathematical structures of physics were born together with Descartes' "mechanical hypotheses": mechanical hypotheses had disappeared, but the mathematical structures [la physique mathématique] still survived [Duhem 1894d, pp. 123-4].[27]

27 What I have labelled "theoretical physics" or "theoretical practice" in the *Introduction* was labelled *mathematical physics* by Duhem in the above-mentioned passage. At the same time, Duhem made use of the same label for the formal structure of physical theories. As I have attempted to explain, late nineteenth-century theoretical practice was something more sophisticated than the tradition of French mathematical physics, and Duhem was one of the champions of that new practice.

In brief, the emergence, development, dogmatization, crisis, and fall of every meaningful physical theory left behind a permanent and valuable heritage: empirical laws and formal structures. In this context Duhem stressed the extra-logical concept of the *fruitfulness* of a physical theory. Although philosophers and scientists have traditionally focused on the concepts of truth or falsity, he claimed that the value of scientific theories could be found beyond their supposed truth, because their truth was the outcome of a historical process, and therefore it was a provisional value. On the contrary, the fruitfulness of a theory was a permanent value. The *scaffolding* of a theory, namely the specific models and hypothesis on which it was based, represented the provisional component of a theory. These components could be separated from the hard core of the theory, which consisted of permanent mathematical entities. Although Huygens' hydrodynamic models were outdated, they had generated the notion of wave surface; although Newton's luminiferous particles had disappeared, they had generated the concept of "very short periods which correspond to colours"; although scientists could not accept Young's analogy between the aether of a light ray and a column of vibrating air, that analogy had allowed them "to associate a direction with the quantity representing the phenomena of light"; although Fresnel's aether and its motions had disappeared, physicists still relied on the formal analogy between the equation of light vibrations and transverse vibration in elastic solids [Duhem 1894d, pp. 124-5].

At the end of the paper, Duhem synthesized his historiographical view, where the superposition of two historical processes was at stake: the short-term emergence, development, crisis, and finally replacement of specific hypotheses and models, and the long-term progress of mathematical structures, key concepts, and laws. On the one hand, he saw the series of theories that rose "only to be overthrown," and the hypotheses "that a given century contemplates as the secret machinery of the Universe" but then "the following century will shatter like a child's toy." On the other hand, beneath that apparently idle process, he saw "the slow but constant progress of mathematical physics." A long-lasting and persistent stream of progress flowed underneath the transformations that affected the history of science [Duhem 1894d, p. 125].[28] The last passage of the paper was extraordinary lyric: Duhem had found a suitable

28 The original passage deserves to be quoted: "Ainsi, sous les théories qui ne s'élèvent que pour être abattues; sous les hypothèses qu'un siècle contemple comme le mécanisme secret et le ressort caché de l'Univers, et que le siècle suivant brise comme des jouets d'enfant, se poursuit le progrès lent, mais incessant, de la physique mathématique."

metaphor, which could poetically express his overview of the dynamic complexity of scientific practice, and scientific progress.

> When waves go towards the beach, a water layer ripples and swarms into the dry sand before retreating from the beach giving up its conquest. Waves fade away and let the sand dry up before new waves come one after the other. This superposition of waves that rise and then collapse seems a shallow effort of the sea, an idle combination of foam and noise. Nevertheless, two hours later, the beach that had been trodden by our footsteps is now sleeping under deep water: during the relentless oscillations of water back and forth, the Ocean tide has really gone up [Duhem 1894d, p. 125].[29]

In 1896 Duhem accomplished his scientific design of a wide-ranging thermodynamics, which was, at the same time, a generalised mechanics. In particular he generalised Lagrangian equations of motion in order to describe irreversible processes and chemical reactions. Then he attempted to look at his theory from the outside, in the context of the history of physics. He outlined a history of physical theories from the seventeenth century to his day, and once more he sent the corresponding paper "L'évolution des théories physiques du XVIIe siècle jusqu'à nos jours" to the *Revue des questions scientifiques*. He started from a widespread historiographical thesis: modern physics had emerged as a reaction to Scholastic philosophy, but he immediately claimed that the origin and the evolution of modern science could not be understood without the contribution of that philosophical heritage. The Aristotelian tradition had produced countless commentaries, in which Aristotle's philosophy had been "explained, developed, and sometimes transformed." A sophisticated body of knowledge had been accumulated and re-interpreted, and the possibility of bridging the gulf between ancient natural philosophy and modern physics did really exist. The modern geometrisation of natural philosophy was suitable for a specific class of transformations, namely the change of shape and spatial position or *local motion*. Nevertheless physics dealt with a more

29 Stoffel pointed out the striking analogy between Duhem's passage and one of Pascal's *Pensées* on cyclic, historical processes [Pascal 1951, p. 417; Stoffel 2007b, pp. 292-3]. I find an even more striking analogy with Naville's passage on the slow, scientific progress underlying the appearance and disappearance of theories. I have already quoted the passage in the sixth chapter: "Les théories passent, la science demeure: ... à un système détruit en succède un autre dont les conceptions sont plus solides et plus vastes" [Naville 1883, p. 55].

general class of transformations, or motions in the Aristotelian sense, which encompassed every kind of transformations and qualities of bodies. Among this wide class of transformations Duhem listed a body that became warmer or colder, a source of light that became more or less bright, but also a solid that became fluid and a liquid that vaporised, any kind of electrisation, and eventually chemical transformations [Duhem 1896b, pp. 463-7].[30]

According to Duhem, some specific contents of Aristotelian natural philosophy were in contrast with the corresponding contents of modern theories, but a surprising structural analogy linked that tradition to the modern theories of physical and chemical transformations.[31] The declining Scholastics had contented themselves with commentaries wherein "dull repetitions and distortions of Aristotle's views" had been put forward. Even worse, those scholars had given up pursuing an autonomous study of Nature. At the dawn of the modern age, the situation had generated a marked distaste in scholars who held in high esteem rigour and clarity: unfortunately they had identified "the great work of Aristotle and masters like Thomas Aquinas" with the idle and trivial exercises of their last followers. In that context, it had been easy for Francis Bacon to put forward a new kind of logic, and a naïve empiricism that had oversimplified the connections between facts and interpretations. From the seventeenth century onwards, some scholars had hoped that his *Novum Organum* could replace Aristotle's *Organon*. From a more rationalistic point of view, Descartes had put forward a new natural philosophy where qualities had been banned completely by the study of material phenomena, and had been replaced by geometrical and kinematical representations. This was the first historiographical thesis that emerged from Duhem's historical reconstruction: the founding fathers of modern science had dismissed a fruitful and subtle

30 For his general theory of physical and chemical transformations, and in particular his generalised *equations of motions*, see Duhem 1896a, pp. 70-4 and 89-107. For the transformations experienced by Aristotle's natural philosophy in the first century of the Christian era, and in particular Johannes Philoponus' re-interpretation, see Ugaglia 2004, pp. 8-9, 90, 113, 130, 133, and 135-6.

31 Pierre Boutroux emphasized the role played by Tannery in the discovery of mediaeval mathematicians and natural philosophers, and claimed that "Tannery had opened the way to Duhem." At the same time, he stressed the distance between Tannery's historiographical discontinuism and Duhem's continuism [Boutroux P. 1938, pp. 696 and 701]. I find that the representation of Tannery as a discontinuist and Duhem as a continuist [Clevelin 2006, pp. 7-8, 12, and 15] should be smoothed. As I have already pointed out, Tannery saw a meaningful continuity between Copernicus and Brahe's *geometrical* theories, and ancient astronomical representations, in the same way as Duhem stressed both elements of continuity and discontinuity in Galileo's new science.

philosophy together with their late, fruitless outcomes [Duhem 1896b, pp. 468 and 470-1].[32]

Alongside a new scientific practice, a new philosophy had emerged: the natural world could easily be dismantled like a mechanical clock, and "the mechanism of Nature" could easily be explained as the mechanism of a mill. Physics had been reduced to geometry, and every thing could be explained by means of extension and motion. The situation became not so different from the declining stage of the Scholastics: Descartes' philosophy had given rise to "a crowd of ignorant followers" who had invented the most odd and complicated machinery to account for phenomena that they "had not condescended to study." Vortices of subtle matter and ribbed microscopic particles were such pliable models that they could explain everything. With regard to "the ridiculous devotion" towards their master, Descartes' followers had behaved no better than the last decadent Scholastic philosophers. Once more Duhem resorted to Pascal, who had devoted sharp remarks to mechanism and mechanical reductionism. Although everything might be analysed in terms of extension and motion, how could we reverse the process, and build up the *machine* that corresponded to the complex structures under investigation? Pascal had labelled such a pretension as "ridiculous, useless, trying, and questionable"; moreover, even if it had been possible, the pursuit of that natural philosophy would not have deserved "an hour of pain" [Duhem 1896b, pp. 474-6; Pascal 1897 (1976), p. 72].

Duhem reminded readers that even Leibniz had compared some of Descartes' followers to Scholastic philosophers who replaced the study of nature with commentaries to Aristotle's *Physics*. Duhem also remarked that Cartesians' "concern about occult causes" or qualities should have been directed at themselves. Their "vortices of subtle matter, elusive as they were" were almighty mechanical structures that could account for everything. Besides extension and motion, Leibniz had ventured to introduce the notion of force, which allowed matter to act and resist other forces. He had appreciated the solidity and sharpness of Scholastic philosophers and theologians, and had expected that "Descartes' fancy tale of physics" would have been forgotten soon. According to Duhem, the foundations of Newton's physics were not so different from Leibniz's, even though Newton had relied on "experimental

32 Duhem credited Galileo with having been the actual founder of modern science: he had shown "how experiments had to be performed, how the results had to be interpreted," and how a mathematical law had to be formulated [*Ibidem*, p. 470]. With regard to Duhem's interpretation of the role played by Bacon's empiricism and Descartes' rationalism in the foundations of modern science, it is worth mentioning the similar remarks Cournot had put forward in 1872 [Cournot 1872, pp. 300-5 and 309-14].

induction" whereas Leibniz had started from "metaphysical intuition." Their physics was based on the same entities, namely matter, motion, and force. Nevertheless Newton's reference to "experimental method" was in tune with the new philosophy, while Leibniz's design had not managed to become a fashionable trend [Duhem 1896b, pp. 476, 478-80, 482, and 484].

At the turn of the nineteenth century, Newton's approach had been extended to the emerging new chemistry and the theory of elasticity, which was based on "the hypothesis of molecular attraction." At first Tobias Mayer and Charles Augustin Coulomb, and then Poisson, had applied a Newtonian approach to electric and magnetic actions. Poisson had "announced the advent of *physical Mechanics*," in which all kinds of interactions were reduced to molecular attractions, and Laplace had claimed that those microscopic forces ruled all phenomena on Earth, in the same way as gravitation ruled phenomena in the skies. Nevertheless, in the first half of the nineteenth century when Laplace's design of unification had appeared successfully accomplished, Fourier's new science of heat had violently shaken that design. Afterwards the emerging kinetic theory of gases, where the ultimate microscopic parts of a body were imagined in fast irregular motion, had led physics back to Cartesian theoretical models. Moreover, the mathematical stability of *vortex rings* in a perfect fluid had led Thomson, Tait, Maxwell and Lodge to imagine vortices as models of real *atoms* [Duhem 1896b, pp. 485-7 and 489-91].[33]

Nevertheless, "the complication and eccentricity" of the vortex machinery had dissatisfied physicists, and this was not so different from what had happened to "original Cartesianism" two centuries before.[34] In the meantime, the second law of thermodynamics had appeared as a watershed in the history of

33 Poisson's *physical Mechanics* was in opposition to Lagrange's *Analytical Mechanics*, in which the specific features of interactions were not taken into account. Duhem specified that "the British physicists had developed the vortex theory" in accordance with the *imaginative* character he had already pointed out in the paper he had published three years before [Duhem 1896b, pp. 490].

34 Duhem remarked that Helmholtz had not been attracted by vortex theories [*Ibidem* p. 228]. In reality, as the historian John Theodore Merz had remarked more than a century ago, the vortex-atom theory had marked "an epoch in the history of thought," and had slowly faded away because it "explained too much — and therefore too little," according to the historian of science Helge Kragh [Merz 1912, pp. 62 and 64-6; Kragh 2002, pp. 34, 71, 88-9, 92 and 95]. For a wider analysis of that stage in the history of British physics, see also Thomson W. 1860 (1872), pp. 224; Giusti Doran 1975, pp. 140, 142, and 189; Bordoni 2008, pp. 25-8.

physics. This is the second historiographical thesis to emerge from Duhem's paper: new physical concepts had emerged, and it was questionable whether they were consistent with traditional mechanics. He mentioned "the geometer Poincaré" and the engineer Rankine as upholders of a suitable approach to thermodynamics. According to Duhem, the former had pointed out that neither Leibniz, Newton, and Boscovich's dynamism, nor the Cartesians' pure mechanism were consistent with thermodynamics, whereas the latter had previously pointed out the fruitlessness of Cartesian rejection of any quality. The mathematisation of qualities could not be looked upon as an *explanation* in traditional terms, because it could not be reduced to mechanical models [Duhem 1896b, pp. 493-4 and 496-7].

Duhem found that his generalised mechanics or *Energetics* had realised a new alliance among the tradition of Aristotle's natural philosophy, the more recent tradition of analytical mechanics, and the even more recent developments of thermodynamics. It is worth stressing that Duhem's generalised mechanics was as much a new perspective in physics as a new field of physics. It had been Rankine who had labelled *Energetics* that new perspective, because he had been the first to understand the new role of thermodynamics. The new mathematical approach was consistent with a modern re-interpretation of Aristotle's natural philosophy, and at the same time it had been realised thanks to "the relentless efforts of experimenters and mathematicians during the last three centuries." The last passages of Duhem's paper are quite emphatic, and a patent exaggeration emerged: he saw his generalised mechanics or *Energetics* as the accomplishment of the whole history of physics.[35] After having impatiently abandoned the Scholastic tradition, the human mind "had spent three centuries... paving the way to the authentic knowledge of the material world." He dared to claim that science had attained a target that had been foreseen by the Supreme Being who ruled "all those fluctuations" [Duhem 1896b, pp. 498-9]. The excessive confidence in his theoretical physics, which was credited with being the accomplishment of the whole history of physics, was definitely in contradiction with his representation of science as an intrinsically provisional attainment. Duhem's enthusiasm was tainted by

---

35 Duhem ventured to write that *Energetics* was the physics of Aristotle, but also the physics of Descartes, and a revised implementation of "the *universal mathematics*, which had been dreamed by the great philosopher of the XVII century." The convergence of different traditions allowed Duhem to claim emphatically that his theory was even the physics of Kepler, Galileo, Pascal, Newton, Euler, Lagrange, Poisson, Green, Gauss, Robert Mayer, Sadi Carnot, Joule, W. Thomson, Clausius and Helmholtz [Duhem 1896b, pp. 498-9].

dogmatism, and this dogmatism was in contradiction with his sophisticated philosophy of science.[36]

As a philosophically competent scientist, Duhem was aware of the problematic links among the concepts of reality, truth, scientific explanation, and scientific description. He would not have liked to be enrolled among the followers of a naïve and dogmatic scientism, but at the same time he would not have liked to be enrolled among the followers of a disparaging criticism of scientific practice. As a scientist and a believer, Duhem was also aware of the problematic relationships among science, philosophy and religion. The linguistic choice of labelling a scientific theory as a natural classification allowed him to get in touch with the tradition of nineteenth-century French philosophy of science, and allowed him to pursue a third way between dogmatism and scepticism. At the same time, the concept of natural classification was a scientifically and philosophically frail concept because it resisted any attempt at formalisation and even clear definition. Obviously, this frailty did not escape Duhem's sight, but for the time being he had no other way to point out the fact that scientific practice was a complex and pliable process, where many skills and many kinds of practices were at stake. More specifically, different kinds of rational practices (*esprit de finesse* and *esprit de géométrie* included) were involved in the building up of a scientific theory. Even experiments and empirical practices in a broad sense were embedded in a web of conjecture and interpretations. The history of science was the open theatre where the drama of scientific development was actually represented: general ideas, specific conjectures, logical lines of reasoning, mathematical structures, and different kinds of empirical activities interacted with each other in order to give birth to those provisional and dynamic entities, rich in meaning, that were labelled scientific theories. All was uncertain and revisable, and no happy end could be rationally forecast. This might be the cautious and consistent conclusion of Duhem's sophisticated historiography and epistemology. Nevertheless, in 1896 Duhem hesitated, and gave way to the temptation of a more comfortable finalism: his generalised mechanics or Energetics could become the suitable happy ending of the drama. In the end, Duhem's sophisticated historiography and epistemology tottered dangerously, and showed those uncertainties and imperfections that he had skilfully analysed in scientific theories and their histories.

36 This dogmatism was in opposition to what Duhem had repeatedly written: see, for instance, the theoretical pluralism he had expressed in Duhem 1892a, p. 270.

Conclusion

In the last chapter, I have interpreted Duhem's philosophy of science and his first papers on the history of science as the provisional accomplishment of a cultural process that had been ushered in by Cournot in the early 1860s. From the *Introduction* onwards, I have shown that a new cultural sensitivity emerged in the last decades of the nineteenth century: sophisticated histories and philosophies of sciences were among the most meaningful outcome of that new sensitivity. Afterwards, intellectual attitudes and themes that had emerged in those decades underwent cycles of disappearances and questionable re-appearances. The scholars who developed that cultural stream were sometimes forgotten and sometimes misinterpreted. The body of knowledge that had grown up and been accumulated before the turn of the twentieth century became a sort of buried memory: fragments of that body of knowledge periodically re-emerged, and gave birth to new intellectual streams which renewed the original inspiration. I venture to say that the intellectual commitment which inspired at least two generations from Cournot to Duhem has subsequently re-emerged whenever scholars have decided to undertake the narrow and uneven pathway between dogmatic scientism and scepticism.

The sophisticated philosophy of science that had been put forward by mathematicians, scientists, and in general scientifically-oriented scholars, did not become a hegemonic trend in the first decades of the twentieth century. Duhem's theoretical physics faded into the background together with some historiographical and epistemological remarks that had stemmed from his research programme. In the meantime, the interest in ancient science that had even attracted scientists of positivist leanings like Berthelot, also faded away. The early scientific contributions of ancient Greek *philosophers*, and mediaeval mathematics, natural philosophy and alchemy lost their appeal. Scientists focused on the new experimental and theoretical achievements: new particles and new rays, the competition between different electromagnetic theories, and the astonishing features of electromagnetic radiation. Microscopic particles and fields took centre stage: historical researches and meta-theoretical frameworks did not attract scientists any more.

In general, ideas and practices of the authors under scrutiny in the present book show meaningful affinities. Cournot, Boussinesq, and Duhem had put forward bold mathematisations of fresh research fields such as economic processes, physical and chemical instabilities, and thermodynamics of irreversible processes. Not only had Cournot, Bernard, Boussinesq, and Duhem put forward an abstract generalisation of their research fields, or the math-

 | DOI 10.1163/9789004315235_012

ematisation of specific issues stemming from those fields, but they had also attempted to cast light on the plurality of scientific methods and their histories. At the same time, their philosophical commitments resisted any attempt to frame them into a well-defined intellectual stream. They were in search of a third way between scepticism and dogmatism, but they were also suspicious of every rigid philosophical system. We certainly find in them some philosophical influences, and sometimes we can pinpoint explicit references to Leibniz and Pascal: in general we find traces of a philosophical tradition that had emerged together with modern science in the seventeenth century but had subsequently been overshadowed by more triumphant epistemologies and historiographies. For instance, Cournot and Boussinesq envisaged a more general kind of determinism where both deterministic and non-deterministic processes were submitted to the normative role of mathematics. In some way, freedom was the highest expression of determinism: freedom could not be pursued in a world where only chaos and chance occurred [Cournot 1875, pp. 113-7, 241, 243, and 249-52].

Even some unfortunate choices of words and the corresponding blurred meanings, such as Bernard's "*idée directrice*," or Boussinesq's "*principe directeur*," or Duhem's "*classification naturelle*" highlight an essential feature of their meta-theoretical researches: the impossibility of a definite legislation for scientific practice. Those ambiguous concepts might be interpreted as "the stuttering [balbutiements] of the early stages" of philosophy of science, and might be underestimated when compared with "the definiteness of theses, and their subtle organisation" that had been put forward from the 1920s onwards [Brenner 2003, p. 4]. Nevertheless I interpret those undefined concepts as the hallmark of a suitable and reliable meta-theoretical practice. The fact is that the philosophies of science of the main characters of this book did not stem from purely speculative interests but from their actual scientific practices. Some differences among them could be explained by the differences among the different sciences that they practiced, which ranged from mathematical physics to life sciences. They never claimed to be philosophers, but even today their meta-theoretical remarks appear definitely more convincing than more recent and apparently *refined* philosophies of sciences. In general, strict historiographical and epistemological frameworks prove to be unrealistic and ineffective.

To different degrees, our main characters were aware of the essential tension between rational and empirical practices, and some of them, notably Cournot and Duhem, explicitly acknowledged the existence and necessity of a third kind of intellectual practice that dealt with a complex network of conventions, hypotheses, and intuitive skills. Different labels were put forward for qualifying that third component: in any case, the existence of a conceptual, theoretical, or intuitive component was at stake in scientific practice. In brief, they were aware of the complexity and plurality of natural processes and at the same time they acknowledged the complexity and plurality of scientific practices.

Paradoxically enough, old-fashioned historical reconstructions seem more suitable for grasping the essential features of the intellectual stream that crossed the second half of the nineteenth century and involved some scholars from Cournot to Duhem. The first label the historian Benroubi associated with that intellectual stream, namely *Idealism*, seems unsuitable and potentially misleading, whereas the associated adjectives, *Critical* and *Epistemological*, seem more in tune with the actual commitments and theses of those scholars. Benrubi also stressed that they had sought "to emphasise the part played by the intellect... in the formation of exact science," and this might explain the choice of the word *Idealism*. When he wrote "Kant is unquestionably the chief pioneer of this movement," and "Kant dethrones Dogmatism and Scepticism at the same time," this is another clue that the label *Idealism* should be intended in a broad sense. In the end, Benrubi's last linguistic choice, namely "The Critical School in France" appears definitely more appropriate [Benrubi 1926, pp. 84-7].

The dissemination of Cournot, Tannery, Duhem, and Milhaud's meta-theoretical researches on the history and philosophy of science contributed to the professionalisation of French history and philosophy of science in the twentieth century, and the establishment of a more specific research tradition that is known as *historical epistemology*.[1] It is true that, around the mid-twentieth century, Koyré's "concept of scientific revolution," and Bachelard's

1 The emergence of this tradition has frequently been associated with Gaston Bachelard and Georges Canguilhem. We can certainly find in Bachelard and Canguilhem definite research programmes that developed the nineteenth-century heritage, but we also find nuances of scientism that Cournot, Tannery, Milhaud, and Duhem would not have endorsed. When Canguilhem agreed with Bachelard on the existence of "two kinds of history of science," namely the history of the out-of-date body of knowledge, and the history of legitimate science in the sense of contemporary standards, we find traces of an unhistorical undertone: the concept itself of *legitimate* science has its own history, and can be defined only historically [Canguilhem 1979, pp. 13 and 20-1]. At the same time, Canguilhem clarified the unhistorical concept of *precursor* in a way that neither Cournot nor Duhem had been able to do, even though Hélène Metzger had already criticized the concept in 1939 [Metzger 1939 (1987), pp. 79 and 83]. It seems to me that Canguilhem's fluctuating attitude towards history is consistent with his conception of the history of sciences as one of the most demanding tasks pertaining to "philosophical epistemology" [Canguilhem 1979, p. 23].

notion of "intellectual change," made philosophy of science undertake the pathway towards "a radical discontinuism, which was both historical and epistemological." Nevertheless, a sophisticated reconstruction of epistemological continuities and discontinuities in the history of science had already been undertaken by Cournot [Brenner 2003, pp. 103-4 and 107]. The philosophies of history that we find in Cournot and Duhem's writings are definitely more suitable for the comprehension of continuities and discontinuities in the history of science. Cournot, Naville, and Duhem had inquired into the superposition of cyclic and linear processes, and the persistence of structural continuities throughout scientific revolutions.

We find meaningful connections between Poincaré and Duhem's meta-theoretical commitments and twentieth-century French historical epistemology. Nevertheless we find meaningful connections even with logical empiricism, as has already been acknowledged and widely discussed by historians and philosophers. In fact, both Duhem and Poincaré were privileged points of reference for the *Vienna Circle*. The question that has spontaneously emerged is in some way rhetorical: how could the two scholars have inspired "both logical positivists and their opponents?" [Brenner 2003, pp. 67 and 153]. The answer is that neither Poincaré nor Duhem put forward philosophical systems: the scholars who developed a modern philosophy of science from Cournot to Duhem could offer many meaningful hints to the builder of different historiographical and epistemological systems. Their meta-theoretical researches can still help us understand what science aims to be, what scientific practice actually is, and how many different sciences and scientific practices have been put forward over time.

Afterword: Disappearances and Questionable Reappearances

Rediscovering Cournot

Our intellectual journey from Cournot to Duhem through thirty-five years of French cultural history has pointed out the emergence of a new awareness of the features and limits of scientific practice in the last decades of the nineteenth century. As a matter of fact, some protagonists of that intellectual stream did not enjoy great success. Cournot was definitely underestimated, and the same can be said of Tannery and Duhem. The former did not succeed in being appointed to the Chair of History of Science at the *Collège de France*, and the latter to a Chair of Theoretical Physics in Paris. However, they managed to attract the attention of some scientists and philosophers, and we find a periodical resurgence of interest in their research and specific meta-theoretical theses. Some brief biographical notes: Tannery died in 1904, Poincaré in 1912, Duhem in 1916, and Milhaud in 1918. Cournot and Bernard had already died in 1877 and 1878 respectively, and only Boussinesq managed to survive until 1929. Nobody handed over his own intellectual legacy to an identifiable school or at least to a small group of appreciative followers.

However, in the early twentieth century, here and there we find traces of an interest in Cournot's philosophy of science. In 1905 the journal *Revue de Métaphysique et de Morale* devoted a 250-page special issue to a critical appraisal of Cournot's scientific and philosophical legacy. The editors pointed out the value of a "generally ignored intellectual work" that was "still waiting for the appreciation it deserved." Henri Poincaré wrote the first paper on "*Cournot et les principes du calcul infinitésimal*," and Milhaud a "Note sur *la raison chez Cournot*." Other scholars intervened on Cournot's philosophy of history, and his studies in economics, statistics, and politics.

In his paper "*Le criticisme de Cournot*," the philosopher Dominique Parodi stressed the persistent value of Cournot's philosophy of science, at least with regard to his attitude and method: Cournot's best heritage could be found in the problems he discussed rather than "the solutions" he offered. His starting point appeared not so different from Kant's: the conditions of possibility for reliable knowledge, in particular scientific knowledge. Cournot's philosophy could be qualified as a philosophy of sciences in this Kantian sense. However, according to Parodi, Cournot had managed to go beyond Kant, because he could rely on contemporary developments in life sciences, from which he had

 | DOI 10.1163/9789004315235_013

drawn a conception of scientific law more pliable than the traditional concept of physical law. Definitely less ambitious than Kant's philosophical system, the whole of Cournot's philosophical researches had realised a convincing balance between "idealistic scepticism and illusory dogmatism." Cournot had highlighted "the value of science," and at the same time, the *relativity* and *probability* of that value. He had also managed to go beyond the positivist tradition since he had acknowledged the necessity of a third kind of *praxis* besides "strict deductions and definite experiments." That third component involved a marked "philosophical sense," rational wisdom, and critical skills [Parodi 1905, pp. 451-4 and 459].

Parodi found in Cournot an explicit awareness of the role of history in science. That role could be understood in two senses: the provisional and historical character of scientific enterprise, on the one hand, and an intrinsic historical component in the objects of scientific inquiry. The relevance of the second kind of historicity increased with the complexity of the systems under investigation: it could be discarded in inanimate bodies but could not be disregarded in living beings. Moreover, Cournot had acknowledged the intrinsic limits of scientific enterprise, namely "the impossibility of a formal demonstration for every statement of sufficient generality." For these reasons, Parodi looked upon Cournot as the intellectual avant-garde of his time: he had ventured to put forward a philosophy of science that was to be developed by Boutroux and Poincaré. Cournot's *critical realism* had prevented him from setting up a complete philosophical system. He had confined himself to "fragmented conclusions," where a clear and consistent line of reasoning stood beside partial and provisional theses. In this way, his intellectual legacy had exerted a mild but extraordinarily subtle influence. At the beginning of the twentieth century, that influence could be overtly acknowledged, and its durable effect properly appreciated [Parodi 1905, pp. 473-5 and 483-4].

In 1908, the philosopher Francois Mentré published a 600-page book on Cournot. The book, *Cournot et la Renaissance du Probabilisme*, was published in the series *Bibliothèque de Philosophie Expérimentale* that had hosted Duhem's *La théorie physique* two years earlier. Mentré reminded readers that Cournot's mathematical theory of wealth had been underestimated by contemporary economists but subsequently re-evaluated and further developed in the twentieth century. He also analysed Cournot's complex epistemology, and the complex relationship between rational and empirical practices: the more a physical law was general, the less it was affected by experience. Cournot had also managed to understand the pliability of scientific theories: sometimes the phenomena that were not in tune with a theory could however be interpreted in a way that did not contradict the theory itself [Mentré 1908, pp. IV-V].

In 1911 Milhaud published the paper "Cournot et le pragmatisme scientifique contemporain" in the Italian journal *Scientia*, which frequently hosted contributions to history and philosophy of science. He followed in Parodi's footsteps, and stressed the analogies between Cournot and Poincaré. More specifically, he claimed that Cournot's philosophy of science was more profound than Poincaré's, even though the former had began to "fight the naïve and positivistic conception of science" considerably earlier. Milhaud reminded readers that the scholars who had opposed naïve scientism had been charged with "scepticism and pragmatism," and he found that the charge deserved to be carefully discussed. The hallmark of Cournot's philosophy of science was the explicit acknowledgement of an essential feature of science: scientific practice could not be based only on "the domain of facts and demonstrable truths." Alongside experimental practice and the application of formal procedures there was a theoretical practice that dealt with "the domain of ideas and reason" in a wide sense. Facts and experiences had to be interpreted by a network of hypothesis and rational connections. In Milhaud's words, the passage from the empirical domain to the theoretical domain involved "some kind of indeterminism," namely a multiplicity of explicative theories. This was the natural consequence of Cournot's approach, and Milhaud claimed that the awareness of this consequence had already come to prominence in the *Essai* Cournot had published in 1851. However, in no way could that indeterminism be identified with scepticism [Milhaud 1911a, pp. 370-2].[1]

Milhaud's article reproduced many passages from Cournot's 1851 and 1861 books at length, and in the end, three interpretative theses emerged. First, Cournot had put forward in advance a philosophical and historiographical framework that had subsequently been developed "in different ways" by Mach, Duhem, and Poincaré. Second, Cournot had gone far beyond Milhaud's contemporary scholars. Third, in no way could Cournot be associated with "some kind of intellectual pragmatism." It was true that, in Cournot, no "logical or empirical demonstration" of the truth of a hypothesis could be given. Nevertheless, according to Milhaud, this indeterminism had not led Cournot to pragmatism. On the contrary, he had made "every effort to overcome pragmatism." He had looked for a concept of "objective reality" that did not collide with the impossibility of a formal or empirical demonstration. For the

1 Milhaud's separation between methodological indeterminism and scepticism seems really useful for the comprehension of Cournot's philosophy of science: indeterminism, or better theoretical pluralism, did not involve scepticism, which was a broader attitude, as naïve as its opposite attitude, namely scientism. Definitely less convincing is Milhaud association of scepticism with pragmatism, which may be considered as quite close to methodological pluralism.

appraisal of a scientific theory, he had had recourse to concepts like *order*, *harmony*, and *fruitfulness*. Although these concepts could not be formally defined, they could easily be grasped, and were endowed with unquestionable effectiveness. In the end, Milhaud found that Cournot had been more a realist than a pragmatist: he had been even "too much of a realist, too close to a scientific Pythagorism," even though never confident in whatever *absolute* entity [Milhaud 1911a, pp. 374-7 and 380].

It seems to me that the first part of Milhaud's appraisal is in tune with Cournot's philosophy of science. On the contrary, the subsequent representation of Cournot as a realist rather than a pragmatist seems untenable. More in general, one of the hallmarks of French philosophy of science from Cournot to Duhem was a balanced and original attitude towards scientific practice. That attitude distanced itself from naïve realism and idealism: if I were forced to make use of simplified philosophical labels, I might qualify Cournot as a pragmatist, but it seems to me that, in this context, every label is definitely unsuitable and intrinsically misleading.[2]

Besides the paper he had published in the special issue of the *Revue de Métaphysique et de Morale* in 1905, and the paper published in *Scientia* in 1911, Milhaud's three other papers appeared in 1911.[3] In the first, "Le développement de la pensée de Cournot," which was published in *La Revue du Mois*, Milhaud remarked that, at the time, "Cournot had become trendy." The same concept was restated in the second paper, "La définition de hasard de Cournot," which was published in the *Revue philosophique*: at the turn of the twentieth century, Cournot had "willingly been studied" by some scholars. In the third paper, "La Science & la Religion chez Cournot," an extract from the *Bulletin de la Société française de philosophie*, Milhaud stressed Cournot's systematic interest in science, and its commitment to "updating" contents and methods of "philosophical speculation" by means of a cross-fertilisation between science and philosophy [Milhaud 1911b, p. 7; Milhaud 1911c, p. 37; Milhaud 1911d, p. 110]. In the debate that followed, and which was briefly reported by the *Bulletin*, Mentré suggested that Cournot's religious attitude was akin to Pascal's, and this fact

2 Cournot had always rejected any kind of dogmatism. In 1905 the economist Henry L. Moore reminded readers that, even though Cournot was a Catholic, he had opposed "the prevalent role of the Church in education" because he thought that the weight of religion "would have damaged religion itself" [Moore 1905a, p. 542]. In the same year, in *The Quarterly Journal of Economics*, Moore stressed the wide scope of Cournot's scientific practice, which encompassed "pure mathematics, logic, philosophy, philosophy of history, the theory of statistics," and "mathematical economics" [Moore 1905b, p. 370].

3 The papers Milhaud published on Cournot between 1902 and 1911 were collated and republished in 1927, after the author's death.

could be associated with the fact that Catholics of neo-Thomist attitude had never laid claim to Cournot's affinity. Although Cournot had been "a believer and a Christian," more specifically a Catholic, he had never been a *clerical scholar*. Once more Parodi stressed "the modernity of Cournot's views," and the fact that those views were put forward when mechanistic and reductionist attitudes prevailed. The stress on Cournot's *modernity* was another way to point out the condition of relative intellectual isolation in which he had worked, and at the same time, the fruitfulness of his cogitations [Milhaud 1911d, pp. 111, 129, and 134-6].

Still in 1926 Benrubi stressed the influence of Cournot on "the present generation," and the importance of his meta-theoretical researches "for the contemporary *critique de la science*" [Benrubi 1926, p. 92].[4]

The process of disappearance and partial reappearance of themes and interests that we find in the twentieth century also involved determinism and reductionism. It seems that, in the first years of the new century, the debates that had attracted some scientists and philosophers in the last decades of the nineteenth century had lost any appeal. However some traces of those debates can be found, and deserve to be briefly reported.[5]

In 1914 the physical chemist Wilhelm Ostwald wrote down some remarks on Mayer's 1876 essay, and developed the concept of *Auslösung* in the context of the second principle of thermodynamics. The manuscript remained

4 When, in 1927, the English translation of Cournot's 1838 book *Recherches sur les principes mathématiques de la théorie des richesses* was re-edited together with an essay by the economist Irving Fisher, the mathematician Griffith Conrad Evans reviewed the book. He concluded his review by stressing "after ninety years, the eternal freshness of this brief classic" [Evans 1929, p. 271]. Some years later, a century after the publication of Cournot's *Recherches*, a journal of econometrics reminded readers of Cournot's life and scientific achievements. The author stressed that "in America and England" Cournot was known "as an economist" whereas in France he was "chiefly remembered as a philosopher." He also regretted that Cournot's innovative research in mathematical economics had brought him "no financial reward, little academic recognition, and scarcely any intellectual comradeship" [Nichol 1938, p. 193].

5 It is worth stressing that, after Bernard's death, the debate on *vitalism* had continued, and Berthelot and Louis Pasteur had been involved in a scientific argument where philosophical and theological issues were also at stake [Virtanen 1960, p. 12]. In December 1913, during the commemoration of the Bernard centennial at the *Collège de France*, Henri Bergson stressed the importance of Bernard's reflections on scientific practice, and associated Bernard's *Introduction* with Descartes' *Discours sur la méthode*. He also focused on Bernard's awareness of the essential tension between rational and empirical practices, and on the awareness of the intrinsic historicity of scientific enterprise [Bergson 1913, pp. 229-30 and 236].

unpublished until 1953, when the chemist and historian of science Alwin Mittasch published it with the sponsorship of the *German Association of Chemists* [Ostwald 1914 (1953), pp. 19 and 21-2]. In 1922 a new edition of Boussinesq's essay appeared, and the following year a review in the journal *Isis* also appeared. Boussinesq's main thesis was synthesised into the claim that physical laws and differential equations should not be identified with "absolute determinism," and should not be in contradiction with "free will and responsibility" [Guinet 1923, pp. 483-4]. Philosophers continued to debate on determinism, and Laplace and his powerful metaphor played an important role in it, but it seems that they had forgotten Boussinesq's research programme. Nevertheless a short passage can be found in a lecture the mathematician (with wide cultural interests) Karl Pearson gave on the early history of statistics presumably in the first term of the academic year 1925-6. He remarked that, however "remote from morality" the singular solutions of differential equations might appear, Saint-Venant and Boussinesq had looked upon them "as the great solutions of the problem of Freewill." He made also reference to a letter by Maxwell where the latter had qualified Boussinesq's researches as "epoch-making" [Pearson 1925 (1978), pp. viii, xiv, and 360].[6]

In 1937, in the book *Determinismus und Indeterminismus in der modernen Physik*, Ernst Cassirer addressed indeterminism in the context of classical and quantum physics. The first chapter was devoted to "Laplace's Mind" because he acknowledged that Laplace's figure of speech had played an important role in subsequent debates. According to Cassirer, Emil Du Bois-Reymond had been the first scholar to emphasise this metaphor: he had given it "prominence in the contexts of science and the theory of knowledge." He had put forward a sort of *ideal* or mythology of omniscience [Allwissenheit], where the past and future course of the world would be perfectly clear in all details. This mythology corresponded to "an empty concept," devoid of any empirical content, and it was useless even as "a methodological precept or guideline for our knowledge." In the book posthumously published in 1950, *The Problem of Knowledge – Philosophy, Science and History since Hegel*, which can be looked upon as the last volume of the series *Das Erkenntnisproblem in der Philosophie und Wissenschaft der neueren Zeit*, Cassirer commented on Du Bois-Reymond once more. In the talks the authoritative physiologist had addressed to the German Association of scientists and physicians in the 1870s and 1880s, he saw a mixture of "scepticism and dogmatism." In other words, the impossibility of dog-

6 Pearson probably made reference to the already mentioned letter Maxwell had sent to Galton. As far as I know, Ian Hacking was the first to mention Pearson's remarks [Hacking 1983, pp. 464-5].

matism had automatically become scepticism, because Du Bois-Reymond had identified the impossibility of a mechanical explanation with the impossibility of any explanation [Cassirer 1937, pp. 7-9, 11, 16, and 32; Cassirer 1950, p. 87].[7]

In reality, the opposition between dogmatism and scepticism had already found a sophisticated solution in some French philosophers of science from Cournot to Duhem. The provisional and probabilistic nature of scientific knowledge had already been interpreted as a hallmark of strength, creativity, and progress. Not only had Boussinesq's theses long disappeared from the cultural horizon but other sophisticated contributions to philosophy of science had apparently also disappeared or did not deserve to be mentioned any more.

Debating Duhem

In 1904 Abel Rey, a French scholar trained both in science and philosophy, published a long paper on Duhem's "scientific philosophy" in the philosophical journal *Revue de Métaphysique et de Morale*. He really managed to draw up a competent and insightful review of Duhem's theoretical physics and philosophy of science, and focused on some problematic or paradoxical meta-theoretical theses. In the first passage of the paper, when he introduced readers to Duhem's scientific and philosophical work, he mixed a respectful style with slightly ironic remarks: no field of physics and chemistry had been left untouched by the renowned physicist. Not only had Duhem "overturned traditional mechanics," but he had also offered a clear historical reconstruction of the scientific tradition he had claimed to have accomplished. Duhem's meta-theoretical design consisted in framing specific chemical and physical laws into a clear and logical structure, which was "as sophisticated as it was rigorous." Unfortunately experimental scientists had not appreciated a theoretical effort which could not lead to new, useful applications: chemists had probably thought that Duhem's theory was "a good piece of physics," and physicists that it was "interesting mathematics." With regard to meta-theoretical aspects,

7 In a manuscript written in 1932, and unpublished until 1990 (after the author's death in 1968), the French philosopher Alexandre Kojeve had already stressed the difference between *causality* and *legality*: he claimed that the latter deserved more attention in the debate on determinism. In its "hypothetical version," the metaphor that Laplace had put forward was in some sense a tautology, whereas the demand for "a universal application" of that principle corresponded to a stronger kind of determinism, or "the doctrine of causal determinism" [Kojeve 1990 (1932), pp. 27-8 and 44-5].

Rey found that the cumulative and "linear conception of experimental science," which had enthusiastically been promoted by Berthelot, had already been overcome by "Rankine, Helmholtz, Dubois-Reymond, Ostwald, Poincaré, G. Milhaud, etc." Apart from specific and important differences, all of them shared a common conclusion: science could "explain nothing," nor could it attain the causes of phenomena. Science could only offer "connections among phenomena," the outline of a formal description, and some previsions, at least "to a certain extent." No hypothesis could be looked upon as necessary, because "an infinity of equivalent ones" could be consistently devised. Moreover, theoretical and experimental practices could not be rigidly separated, because what was qualified as "the results of experience" involved complex mental processes rather than "passive observation" [Rey 1904, pp. 699-700 and 702-4.].[8]

According to Rey, in no way could Duhem's philosophy of science be associated with "Poincaré's ideas" or the conceptions of "Bergson's radical followers." The keystone of Duhem's original meta-theoretical commitment was the role of mechanics: traditional mechanics, or what might be labelled as the *mechanics of machinery*, was inadequate. A new mechanics was required, in order to describe electric, magnetic, and thermal processes, and even changes of physical state and chemical reactions. What *Peripatetic* tradition had labelled *degradation* and *creation* had to be included into the field of a wider and more powerful mechanics. In Duhem's perspective, generalised mechanics was a mathematical-physical language rather than a well-defined subject matter. The theoretical design could not stem from specific experimental data: it had to be set up rationally, or *more geometrico*, and had to be "compared to experience only afterwards," namely at the end of a complex process that linked wide-ranging hypotheses to specific events. Therefore Duhem's design was not so different from Descartes,' the founding father of that mechanical world-view that Duhem had so fiercely opposed. However, Rey specified that the structural analogy was not so strong as to associate Duhem with Descartes' "rational dogmatism" [Rey 1904, pp. 704, 707, 710-11, and 718-20].

According to Rey, the most questionable feature of Duhem's philosophy was the combination of the freedom to devise hypotheses with the existence of only *one* theoretical physics, namely "a system that was better than the others." Rey found that "the original compromise" was extremely frail, and substantially unsatisfactory. The differences between Duhem and nominalists were "most in the words than in the real foundations." In the end, Duhem's

8 For some details on Rey's philosophical and scientific training and interests see Delorme (DSB 1970-80), vol. 11, pp. 388-9.

qualitative conception of the universe, his distrust in a complete explication in mechanical terms, and his purely formal opposition to a radical scientific scepticism corresponded to "the scientific philosophy of a believer" [Rey 1904, pp. 731, 733-4, and 742-4.].[9]

Rey managed to pinpoint the specific features of Duhem's theoretical physics and philosophy of science, the specific differences between Duhem and his contemporaries, and the weakness of some meta-theoretical concepts and theses. More specifically he saw the frailty of the concept of *natural classification*, and the frailty of the more general commitment to balancing the reliability of scientific practice with its provisional and partial achievements. Nevertheless, I find that the association of Duhem with nominalism cannot be accepted, because Duhem really relied on the soundness and usefulness of scientific practice, and theoretical physics in particular. Duhem's philosophy of science was primarily addressed to explore the boundaries of that practice rather than to evaluate or underestimate it.

The implicit reference to Duhem's Catholic faith just in the last lines of Rey's paper triggered Duhem's reaction. He published a paper in the *Annales de philosophie chrétienne*, where he restated the clear separation between scientific practice and religious commitment, and attempted to clarify his concept of natural classification. In 1906 Rey briefly replied that the term *believer* made reference to the distrust in the possibility of a material self-explanation of the world, and it did not have to be intended as a specific reference to religious commitment. The debate was not important in itself, because Rey had simply made use of that adjective once, and at the end of a long paper, and because Duhem had already stated a clear separation between science and faith in the 1890s. The fact is that the ideological and political context of the French Third Republic at the turn of the twentieth century made personal religious commitment a very sensitive issue. In reality, the most problematic issue was the concept of natural classification, and on this specific subject Rey's reference to Duhem as a believer was probably misleading.[10] Natural classification was

9 Rey insisted on the actual metaphysical character of Duhem's meta-theoretical remarks in many passages of his paper. See *Ibidem*, p. 740: "Il prend donc parti, qu'il le veuille ou non pour une hypothèse métaphysique, tout comme le mécanisme."

10 On the timing of the debate between Duhem and Rey see Deltete 2008, pp. 627, 629, 634, and 636 (fn. 6). On the metaphysical meaning of Duhem's natural classification, see *Ibidem* p. 636: "Duhem sought to separate physics from metaphysics, but, I have argued, he also tried to bring them into contact. The key to this rapprochement was the concept of NC [natural classification], the idea that physical theory tends to a classification of physical and chemical phenomena that mirrors the ontological order of nature."

an intrinsically puzzling concept, because it was linked to the pursuit of an intellectual task that could never be accomplished. At the same time it was an authentic philosophical concept: it was equivalent to Pascal's *esprit de finesse*. I find that the philosophical intimacy between Duhem and Pascal is the key, but the fact that Pascal was also a believer might be misleading once more.[11]

In 1906 Duhem published the book *La théorie physique, son objet — sa structure*, where he collected and sometimes updated the content of the papers he had published in the 1890s. He stressed that his remarks had emerged from his "daily scientific practice." Once more he claimed that physical theories could not be "*explications*," but simply mathematical interpretations that stemmed from a few physical principles. At the same time, a physical theory was something more interesting than a mere combination of mathematical structures and empirical laws. There was also a conceptual structure where "common sense and mathematical logic" interacted with each other "in an inextricable way." The soundness of that conceptual network depended neither on empirical nor on formal procedures. It dealt with what Pascal had labelled "esprit de finesse": it was a meta-theoretical sensitivity that allowed scientists to overcome the essential tension between dogmatism and scepticism. The concept of natural classification made sense in the context of a long-lasting commitment to pursuing a demanding and slippery third way [Duhem 1906, pp. 1-2, 26, 36, 440-1, and 444]. Duhem frequently mentioned and explicitly quoted from Pascal, who represented his methodological landmark. He also focused on the complex relationship between experimental and theoretical practices. According to Duhem, a physical theory could be looked upon as a complex network of conceptual links between the domain of scattered facts and the domain of mathematical procedures. We find here one of the hallmarks of late nineteenth-century theoretical physics [Duhem 1906, pp. 217, 274, 303, and 328].

In 1911 Duhem published the two-volume *Traité d'énergétique ou de thermodynamique générale*, where he collected and updated most of his researches in theoretical physics. He stressed the importance of *Rational Mechanics* as the formal structure or the formal language for physics, even for the fields of physics outside mechanics. The language of rational mechanics had nothing to do with the specific mechanical models that had been used by some physicists in the context of thermodynamics or in the "*mechanical explanation of the Universe*." The generalisation of that language encompassed both

11 Stoffel has repeatedly stressed Duhem's familiarity with Pascal's texts, scientific texts included. There are convergent statements that Duhem knew Pascal's meaningful passages "by heart" [Stoffel 2007, pp. 280-5].

mechanics and thermodynamics without any recourse to microscopic models, and could be labelled "*Energetics*." Once again, on the track of his 1906 book, Duhem stressed the intrinsic tension between empirical and theoretical practice. When a theory was conceived, in the first stage it was not required to "take into account the facts of experience," but only to take care of its internal consistency. Only at the end of a complex process, the results of mathematical procedures had to be compared with experimental laws. Nevertheless, a theory could not be designed at random; it required a *justification*, and that justification had to be as historical as logical. The history of physics was a melting pot of experiences, hypotheses, mathematical tools, specific models, wide-ranging conceptual frameworks, and meta-theoretical options. History was the stage where the emergence, development, and fall of physical theories took place [Duhem 1911, tome I, pp. 1-5].

In the first years of the new century Duhem published a huge amount of historical research on ancient and mediaeval applied mathematics and natural philosophy. He could rely on Tannery's previous researches, which he frequently mentioned in the two volumes of *Les origines de la statique* (1905 and 1906), the three volumes of *Études sur Léonard de Vinci* (1906, 1909, and 1913), and the first volumes of *Le système du monde, histoire des doctrines cosmologiques de Platon à Copernic*, which were published between 1913 and 1917, the year after his death.

In the last years of his relatively brief life, Duhem was appreciated in the German intellectual environment. I have already remarked that *La théorie physique* was translated into German in 1908 by Friedrich Adler, and *L'évolution de la mécanique* in 1912 by Philipp Frank: Ernst Mach sympathetically undertook the *Introduction* to the German edition of *La théorie physique*.[12] After that, his name appeared in the early works of the scholars who gave birth to the *Vienna circle*, but his original interplay between philosophical and historical analysis slowly disappeared. Nevertheless we find some traces of historical interests and pluralistic epistemology in the first works of the Austrian economist and social scientist Otto Neurath, one of the founding fathers of the *Vienna circle*. In 1916, the year of Duhem's death, Neurath published the paper "Zur Klassification von Hypothesensystemen," where he focused on the history of physical theories, in particular in the field of optics. The author repeatedly mentioned Mach in the context of a wider tradition of research that involved other scholars from Whewell to Duhem, and their researches in the history of physics [Neurath 1983 (1916) p. 14].

12 Both Adler and Frank were qualified by Don Howard as *ardent* followers of Ernst Mach [Howard 1990, pp. 364-5].

Neurath did not mention explicitly the paper Duhem had published in 1894 on the history of optics, nor did he mention Duhem's 1906 book, but the French tradition of philosophy of science was on the stage when he cautiously stated that "one is involuntarily impelled to accord equal value to different systems of hypotheses." Equally cautious was his criticism of scholars who abruptly dismissed this point of view "as a new fashion that was introduced by Poincaré, Duhem and others." He remarked that scholars who had criticised Duhem and Poincaré had decided to "entirely overlook" history, because the competition between opposite conceptual models, and the fluctuating confidence in their actual explicative power, was continuously present in the history of science. It was a matter of fact that "several systems of hypotheses for the explanation of the same complex of facts" had been put forward over time. At the end of the paper, once more Neurath acknowledged that the "theory of systems of hypotheses had been greatly advanced by men like Mach, Duhem, Poincaré." However Duhem would probably not have appreciated what Neurath ironically qualified as the concluding remark of his considerations, namely "that the great physicist must necessarily be a bad philosopher" [Neurath 1983 (1916), pp. 28-9 and 31].

After his death, Duhem was sometimes celebrated in the French environment, but his influence slowly faded away. In December 1921, at the annual session of the *Académie des sciences*, the mathematician and mathematical physicist Émile Picard reported on Duhem's life and work. In the paper that was published in 1922 in the *Mémoires de l'Académie des Sciences de l'Institut de France*, Picard reminded readers that Duhem had been appreciated by mathematicians rather than physicists and chemists, although he would have liked to be considered, and really was, a theoretical physicist. He had been looked upon as "too much of a physicist by mathematicians, but also too much of a mathematician by physicists and chemists." According to Picard, although Duhem had sometimes been associated with pragmatists, and had been qualified as a Kantian during a conference of Catholic scholars held in Brussels, in reality he had always been in tune with Pascal, "whom he had continuously quoted." Picard also stressed that Duhem's overconfidence in some analogies between peripatetic physics and general thermodynamics was independent of any religious commitment. Duhem had managed to grasp the deep conceptual links between some "physical insights" stemming from ancient ages, and "certain views of contemporary science" [Picard 1922, pp. XCIX-CVI, CXVII-XX, and CXXXV-VIII].

The same year, in the neo-Thomist *Revue de philosophie*, Mentré reminded readers that Duhem had published many important researches on the history and philosophy of science in the journal, chapter after chapter of *La théorie physique* included. He praised Duhem for having always been insensitive to "personal ambitions and honours," and he appreciated the cross-fertilization among science, philosophy, and history in Duhem's researches. At the same time, he qualified Duhem's philosophy of science as "disappointing and ambiguous." After some puzzling and inconclusive remarks, he clarified the reason for that disappointment: although quite sophisticated and essentially correct, "his religious philosophy" had merely been *defensive*. In other words, Duhem's philosophy of science did not allow religion and philosophy to offer a general framework for scientific practice. Mentré did not appreciate the intellectual harmony between Duhem and Pascal, and claimed that Pascal's position was *dangerous*, because of its rigid separation between science and faith, and its underestimation of the necessary interplay between "scientific and theological reasons" [Mentré 1922, pp. 450, 454, 459-60, and 464].[13]

Mentré's paper is quite interesting because it shows the personal contiguity between Duhem and the Catholic environment, and at the same time the philosophical distance. It is not strange that neo-Thomist scholars considered Duhem and Pascal dangerous: they had been both scientists and philosophers, and their twofold competence had led them to perceive clearly the complex and plural links among scientific practice, philosophical commitment, and religious faith. That complexity and plurality had prevented them from putting forward a reassuring convergence of science, philosophy and religion.

However critical, Mentré showed a clear comprehension of Duhem's historiographical and epistemological theses, and a clear awareness of the sensitive issues at stake. In the following years, that awareness faded away, and even the detailed reconstruction of Duhem's physical theories published by the physicist Octave Manville in 1928, and the sympathetic biography published by the mathematician Pierre Humbert in 1932, overlooked the most problematic aspects of Duhem's historiography and epistemology. However Humbert remarked that Duhem had lived in a context of "extreme scientism," which was one the hallmarks of late nineteenth-century French cultural environments. He looked upon Duhem as one of the leaders of a new, more sophisticated,

13 The Catholic journal had been founded in 1900, and Duhem had written extensively in it. On Duhem's collaboration with *La revue de philosophie*, see Hilbert 2000, pp. 112-19. In November 1922 Mentré gave three lectures on Duhem's history and philosophy of science at the department of philosophy of the *Institut Catholique de Paris*. The first, "La vie, l'homme, le savant" was delivered on November 8th, the second "L'histoire de la science et de la philosophie. Comment il a renouvelé l'étude de la scolastique" was delivered on November 15th, and the third, "Le philosophe et le croyant. Des rapports de la science et de la foi selon Duhem," on November 22nd [*Ibidem*, p. 541].

approach to science: what Duhem, Poincaré, Boutroux, and Le Roy had in common, in spite of their specific conceptions, was the awareness of the complexity of scientific practice, and the awareness of its intrinsic limitations, in opposition to a "blind faith" in science [Humbert 1932, pp. 17, 62, 69, and 73].[14]

However, in the first decades of the twentieth century, that awareness survived to a certain extent, and some traces can be found in the texts of certain scientists and philosophers. Since the 1910s, we find Federigo Enriques' inquiry into the historical and theoretical dimensions of science, more specifically "the implicit philosophy" or "the philosophical core" of scientific theories besides "a historical criticism of scientific concepts." The contribution of Emile Meyerson, Lucien Lévy-Bruhl, Léon Brunschvicg, and Hélène Metzger, and "the institutional lineage" that had led from Milhaud to Gaston Bachelard through Rey have been widely studied. In the 1930s, Albert Lautman explored "the hidden history" of mathematics, namely the history of the philosophical core of mathematical theories [Brenner 2003, pp. 1, 2, 4-5, and 7-8; Chimisso 2008, pp. 1-2, and 5-6; Castellana 2014, pp. 314; Castellana 2015, pp. 57-62, and 66].

It seems to me that Metzger's views deserve to be briefly outlined. In her researches, she made reference to a history of science that cast light on the evolution of mental representations. She hoped that the alliance between the history of science and philosophy of science would have contributed to a better comprehension of "the theory of scientific knowledge" [Metzger 1929 (1987), p. 106].[15] The intrinsic historicity of scientific practice could easily be noticed because scientific interests and perspectives had frequently changed over time. According to Metzger, the historian should read the primary sources of distant ages as the contemporary readers of those sources would have read them. Historical objectivity did not correspond to a detached approach but a

14 Humbert was a Catholic mathematician who spent most of his career at Montpellier, and did some researches on the history of science. He dared to overturn the received view on Duhem as a right-winger, and claimed that he had been a democrat. See *Ibidem*, p. 126, footnote 1: "ses préférences secrètes le poussaient du côté des démocrates, chez qui il comptait beaucoup d'amis." He also reported Duhem's talk in favour of university training for girls [*Ibidem*, pp. 133-4]. He could rely on some information received from his father, the mathematician George Humbert, who had personally known Duhem since his youth [Stoffel 2007, p. 283].

15 See *Ibidem* p. 106: "Espérons que désormais entre les historiens et les philosophes qui veulent contribuer au perfectionnement de la théorie de la connaissance scientifique, il s'établira une collaboration de plus en plus intime et féconde." Among the scholars who set up that tradition, she mentioned Comte, Cournot, and Renouvier [Metzger 1929 (1987), p. 104].

sort of identification with the author. That identification would have allowed the researcher to get in touch with "the deep motivations" that had given rise to the texts under investigation. A critical and philological analysis was also required in order to attain the final target, namely "the global comprehension" of those texts. She focused on tacit knowledge, namely the beliefs that had received "the unanimous and spontaneous agreement of scientists" at all times. Explicit and implicit beliefs had played an important role even in empirical practice: on the track of Duhem, whom she explicitly mentioned, she claimed that scientific objects had been "created by theory," and could be considered "material outcomes of theory" [Metzger 1933 (1987), pp. 9, 11-13, 16, and 18-9; Metzger 1937a (1987), pp. 59 and 65-7].[16]

Metzger rightly pointed out that the concept of *precursor* stemmed from a naïve philosophy of history. The so-called precursors were created by historians who had arbitrarily modernized some theses put forward at a certain historical stage. Those theses were re-interpreted and reduced to other theses that more authoritative scholars were to put forward afterwards. In brief, the concept of precursor was the consequence of a deliberate misunderstanding [Metzger 1939 (1987), pp. 79 and 83]. History was not a mere collection of events, in the same way as science was not a mere collection of facts. Both historical and scientific practices required a subject and an interpretation: historians of science should explicitly deploy their intellectual attitudes and the guidelines of their historiographies [Metzger 1934 (1987), p. 148; Metzger 1937a (1987), pp. 58-9]. She could not be in tune with the new trend of logical empiricism: in the *Vienna circle* she saw a dramatic linguistic turn that narrowed the practice and scope of philosophy of science. She feared that neo-positivism or logical empiricism could lead to "the most barbarous scholasticism." Broadly speaking, Duhem was considered by Metzger as a positivist, but when she addressed specific issues, she pointed how much Duhem had detached himself from the positivist tradition. More specifically, she pointed out Duhem's stress on pragmatic or intuitive components in scientific practice (elements that could not be logically proved nor refuted), and Duhem's problematic search for a *natural classification* [Metzger 1936 (1987), pp. 55-6; Metzger 1937b (1987), p. 156].[17]

16 In different texts Metzger insisted on the necessity of a *global* reconstruction of every cultural context. She urged historians to reconstruct thoughts and feelings [les états d'âme] of the ancient scholars, their doubts and disappointments included [Metzger 1936 (1987), p. 45]. For a scientific biography of Hélène Metzger, see Freudenthal (ed.) 1990.

17 On the track of Canguilhem, Sophie Roux remarked that Duhem's concept of precursor required "a logical time", or better "some kind of logical space" besides "the historical

More in general, Duhem has persistently puzzled historians and philosophers until recent years. If we read the considerable amount of secondary literature published over a number of decades, a wide range of contradictory appraisals emerges. In 1941 Armand Lowinger qualified Duhem's epistemology as "methodological positivism." In 1979, Harry W. Paul remarked that Duhem had be looked upon as a Thomist, but his views were "savagely contested by the hard-line Thomists," who would have appreciated a more "aggressive philosophy needed for modern Catholicism." In 1985, Roberto Maiocchi found that Duhem was isolated because of his "intermediate position between neo-Thomism and modernism," and in 1987, Jaki labelled Duhem as a naive neo-Thomist: in his words, "Duhem's Thomism was that of a passionately independent amateur."[18] In 1989 Bas van Fraassen qualified Duhem as "an empiricist hero"; in 2002 Jean-Francois Stoffel described Duhem as a phenomenalist; in 2011, Paul Needham credited Duhem with "moderate realism" [Lowinger 1941 (1967), p. 19; Paul 1979, pp. 3 and 159; Maiocchi 1985, p. 13; Jaki 1987, p. XI; van Fraassen 1989, p. 353, fn. 2; Stoffel 2002, pp. 17, 24, 27, 47, and 367; Needham 2011, p. 7].[19]

The Invention of the Duhem-Quine Thesis

From the 1940s onwards, some references to Duhem became superficial and misleading. In 1941, in the Prefatory Note to his book *The Methodology of Pierre Duhem*, the American scholar Armand Lowinger remarked that "extended comments on Duhem's work" existed only in French, apart from Mach's

---

time". Canguilhem had seen a close connection between Duhem and logical empiricism but Roux found it "historically false". Unlike Adler, Frank, and Neurath, the most influential neo-positivists Hans Reichenbach, Moritz Schlick, and Rudolf Carnap had never studied Duhem [Roux 2016, pp. 19-21].

18 I find that Duhem was neither a naïve nor a sophisticated neo-Thomist. As Robert Deltete recently remarked, Duhem "tried to distance himself" from Thomists, and discouraged "fellows Catholics from using the results of science to promote Christian apologetics." He undertook a two-fold task: "to cut off *both* any science-based attacks on religion *and* all possibility of a science-based natural theology" [Deltete 2011, pp. 19-21]

19 Even more astonishing are the appraisals given on Duhem's political leanings, many decades ago. In 1932 the mathematician Pierre Humbert claimed that Duhem was a democrat, and in 1967 the scientist Donald G. Miller, who was sympathetic to Duhem's scientific enterprise, qualified him as a "man of the right, royalist, anti-Semitic, and extremist in religion" [Humbert 1932, pp. 126, fn. 1, and pp. 133-4; Miller 1967, pp. 463 and 468].

Foreword to the German translation of *La Théorie physique* and a paper published in *Isis* in 1936. Lowinger's book contained oversimplified statements on Duhem's methodology: it seems that the echo of late nineteenth-century debates on theoretical physics and philosophy of science were substantially forgotten. During and immediately after the Second World War, scientific practice in physics focused on huge research projects in nuclear physics. Apart from ethical issues stemming from the close links between scientific research and new weaponry, meta-theoretical commitments intrinsically linked to this practice lost their cogency, and were systematically disregarded. Lowinger stressed that issues such as "abstract methodological philosophy," or "the general history of science," or "the history of the particular theory in question" had no relevance for professional scientific practice. It was definitely true. Ultimately, Duhem's conceptions were looked upon as too sophisticated: Lowinger found that in actual scientific practice, scientists were "enabled to rehabilitate induction and crucial experiment as integral parts of scientific methodology" [Lowinger 1967 (1941), pp. 19, 165, and 170].

In France, epistemology and history of science continued to interact with each other, but Cournot, Poincaré, Duhem, and Milhaud's criticism gave way to more traditional accounts of scientific progress. Around the mid-twentieth century, Gaston Bachelard claimed that the success of recent physical theories had overcome "the shyness of philosophy." The physical entities of contemporary microphysics could be looked upon as "things in themselves": paradoxically enough, those entities beyond any direct experience were considered the best implementation of a sound realism. According to Bachelard, microphysics called for a new philosophical synthesis, where "the negative notion of the thing in itself" conflated with "the positive notion of the noumenon." He was interested in a history of science lit up by a specific finalism, namely a history that had to start from the certainties of present-day science in order to track down "the progressive development of truth." It was "a recurrent history" in the sense of an edifying history of progress throughout subsequent discontinuous achievements. In this perspective, the history of science appeared as "the most irreversible history," in the sense of a history of the permanent defeats of irrationalism. Bachelard's rationalism leant towards a fascinating and pedagogical *story* of physics rather than a strict history, or at least a history that gave prominence to the finality of reason and truth. He specified that his "epistemological recurrence" could not be interpreted in the sense of Duhem's underdetermination or theoretical plurality: he found that, at every time, some theoretical options had to be excluded as definitely irrational [Bachelard 1951, pp. 15-6, 26-7, and 47].

However, in the 1950s, some themes that had been put forward in the late nineteenth-century philosophy of science re-emerged in an unexpected way. Although Duhem's books and papers had become faraway objects of erudite researches in philosophy, a new interest in some of his meta-theoretical theses emerged in the context of analytical philosophy, wherein philosophy of science had to be understood as the logic of science. It is well known that in 1951 Willard van Orman Quine published a paper where he strongly criticised both the dichotomy analytic/synthetic and reductionism. He claimed that modern empiricism had been conditioned "in large part by two dogmas." One of them was the belief "in some fundamental cleavage between truths which are *analytic*," namely statements independent of matters of fact, and "truths which are *synthetic*, or grounded in facts." He interpreted the other dogma, namely reductionism, as the belief that each meaningful statement was equivalent to "some logical construct upon terms which refer to immediate experience." He found that both beliefs were ill founded: he rather leant towards "a blurring of the supposed boundary between speculative metaphysics and natural science." Quine's naturalism assumed a sort of continuity among logic, scientific research, and philosophical practice: in the end, he found that an inescapable effect followed: "a shift toward pragmatism" [Quine 1951, p. 20].

According to Quine, the distinction between analytic and synthetic statements was nothing other than "an unempirical dogma of empiricists," namely a metaphysical belief. The second dogma, namely the dogma of reductionism, was closely linked to the confidence in the existence of "a cleavage between the analytic and the synthetic." Although he acknowledged that the most radical reductionism had "long since ceased to figure in Carnap's philosophy," he maintained that the dogma of reductionism in a milder form had continued to influence empiricists. That milder version implied that "each statement, taken in isolation from its fellows, can admit of confirmation or infirmation at all." On the contrary, he claimed that statements about the external world answer to the tribunal of experience "not individually but only as a corporate body" [Quine 1951, pp. 34 and 38].

The meta-theoretical thesis that no isolated scientific statement could undergo corroboration or refutation in itself but only as a component of a wider body of knowledge, was looked upon by scholars as a holistic thesis and was subsequently labelled as *Duhem's thesis*. Nevertheless, in his paper Quine neither quoted from Duhem nor mentioned him. In some sense, with regard to Duhem's actual views, Quine's paper represented a borderline between two ages. It came after the disappearance of Duhem's intellectual framework, and before the emergence of a new interest in the dynamical structure of actual scientific practice, which had been one of Duhem's hallmarks. In the Preface to his 1963 collection of essays under the title *From the logical point of view*, Quine specified that his "critique of analyticity" had stemmed from informal talks with Carnap, Alonzo Church, Nelson Goodman, Alfred Tarski, and Morton White, talks which had taken place from 1939 onwards. This means that neither before nor after the Second World War was Quine aware of Duhem's epistemological theses or was he interested in taking them into account explicitly [Quine 1963, p. viii].[20]

However the label *Duhem's thesis* survived, and subsequent debates made reference to it even though Duhem's actual views were substantially overlooked. In 1960 the philosopher of science Adolf Grünbaum put forward a refutation of what he called "Duhemian argument." This expression and the other he made use of in the first pages of his paper, namely "Duhem's contention," "Duhemian fashion," and "Duhem's thesis," did not make reference to Duhem's specific papers or books [Grünbaum 1976 (1960) p. 119].[21] From the outset we see that the language, the lines of reasoning, and the insistence on logical aspects, specifically logical calculus, are far from Duhem's language and views, where extra-empirical and extra-logical features of scientific theories were at stake. In brief, the core of Duhem's meta-theoretical reflections vanished in a net of logical deductions that were formally *Duhemian* but non-Duhemian as to the intellectual context. Historical sensitivity had disappeared as well: the

20 In 1990 Don Howard reported that in a private communication, Quine had acknowledged that Carl Hempel and Philipp Frank had pointed out "the kinship" of his views with Duhem's [Howard 1990, p. 376, fn. 2]. See also Howard 1990, p. 363: "Duhem's influence on Neurath was significant, direct, and generously acknowledged. His influence on Quine was equally significant, though indirect (Neurath was the principal intermediary), with Quine himself having been unaware of the parallel between his and Duhem's views until it was pointed out by others. And it was primarily through Quine's writings that Duhem's ideas have retained what currency they have in contemporary debates in the philosophy of science."

21 Only the 1914 English translation of the second edition of Duhem's *La théorie physique* was mentioned in the first footnote, even though the book did not appear in the short bibliography. Although Quine was mentioned in the second line of the paper, he did not appear in the bibliography either. On the contrary, Einstein's *Autobiography* was mentioned, because Grünbaum claimed that "Duhem's argument" had been "*articulated* and endorsed by Einstein" a decade before [Grünbaum 1976 (1960) p. 119]. Grünbaum also made use of the expression "Duhemian thesis" and the substantives *Duhemism* and *Duhemian*. The adjective Duhemian appeared in the expressions "Duhemian schema," "Duhemian view," and "Duhemian ambiguity." For the influence of Duhem on Einstein, see Howard 1990, pp. 364-73: I cannot agree with Howard on the label "Duhem's holistic variety of conventionalism" [Howard 1990, p. 363].

emergence and hegemony of neo-positivism had divorced the philosophy of science from the history of science. In the end, the re-emergence of what was called Duhem's philosophy of science was in part a misinterpretation, and in part an over-simplification.[22]

It seems to me that, some years later, the philosopher of science Laurens Laudan managed to grasp the gulf between Duhem and Grünbaum's intellectual frameworks. In the short paper he published in 1965 on Grünbaum's approach to Duhem's meta-theoretical remarks, he claimed that the former had "misconstrued Duhem's views on falsiafibility," and that "the logical blunder" he had discussed could not be ascribed to Duhem. He stressed the conceptual gap between Duhem and Grünbaum, and reminded readers that over time some cultural processes had transformed "Duhem's conventionalism into the doctrine which Grünbaum attacks." More specifically, he pointed out the necessity of undertaking "a careful analysis of ... Duhem's account of crucial experiments." According to Laudan, that analysis would have shown how far Grünbaum's argument had missed the mark. Laudan did not put forward a detailed historical reconstruction but managed to highlight the context and the main issue at stake in the late nineteenth-century philosophy of science. He mentioned the naïve realism of most scientists in the late nineteenth century, and the reaction triggered by that naïve realism. Duhem belonged to the community of scientists who had not accepted a simplified representation of scientific theories: in that historical and cultural context, Duhem had put forward a philosophy of science that opposed such simplified realism [Laudan 1965, p. 295].[23]

The 1960s represented a crucial turning point in the philosophy of science: we find both the persistence of a logical point of view on scientific theories, in accordance with the neo-positivistic tradition, and the re-emergence of historical and critical interests. As is well known, Norwood Russell Hanson's 1958 *Patterns of Discovery*, and Thomas Kuhn's 1962 *The structure of Scientific Revolutions* widened the scope of philosophy of science by opening research to historical and sociological issues. Laudan's contribution might be placed at the crossroads of the two attitudes.

22 In the brief abstract, Grünbaum had claimed that he was to offer "a refutation of P. Duhem's thesis that the *falsifiability* of an isolated empirical hypothesis H as an *explanans*" was *unavoidably inconclusive*. In reality he probably refuted an abstract *Duhemian* fellow who made sense only in the context of Grünbaum's philosophical framework.

23 He made reference to the English translation of the second French edition (1914) of Duhem's *La théorie physique; son objet et sa structure*.

In 1957, in the Preface to his *From the Closed World to the Infinite Universe*, Alexandre Koyré had already stressed the existence of deep revolutions in the history of scientific and philosophical thought: in the sixteenth and seventeenth centuries, one of those revolutions had changed "the very framework and patterns of our thinking." As he had already done in his *Galilean Studies*, he was to explore "the structural patterns of the old and the new world-views" [Koyré 1957, p. 6]. In 1958 Hanson regretted that philosophers had regarded "as paradigm of physical inquiry not unsettled, dynamic, research sciences" but "finished systems." In other words, he criticised the tendency to focus on the logical structure of theories rather than their actual, uneven development, and on actual scientific practice. He also regretted that philosophers had contented themselves with "an artificial account" of scientific practice even when the founding fathers of modern science were involved. In the wake of Tannery and Duhem, he was interested in the persistent features of scientific explanations over time, beyond the misleading separation classical-modern. The insistence on structural features and patterns of discovery led him beyond logical analyses and experimental tests. This demanding task, which involved the analysis of tacit and elusive elements of scientific practice, could lead to a "profitable philosophical discussion of any science" provided that the philosopher was endowed with "a thorough familiarity with its history and its present state" [Hanson 1958, pp. 1-3]. Actually, Tannery and Duhem's meta-theoretical aims seemed still alive.

In general, in the early 1960s some historiographical and epistemological theses and remarks, which had emerged in the late nineteenth century and which I have analysed in this book, were debated once more, but the memory of the historical context was rarely taken into account. In 1968, Thomas Kuhn claimed that the history of science was a new research field "still emerging from a long and varied prehistory." Only in the twentieth century had the history of science become something different from "a chronology of accumulating positive achievements in a technical specialty defined by hindsight." At the same time Kuhn acknowledged that in Whewell, Mach, and Duhem's writings, "philosophical concerns" had become "a primary motive for creative activity in the history of science." Beside this philosophical influence, he also found in Duhem the roots of another attitude that had positively influenced the early stages of contemporary history of science: the re-evaluation of mediaeval applied mathematics and natural philosophy. Moreover he found in "the universally venerated Paul Tannery" a third intellectual path that had influenced the subsequent generation of historians from George Sarton onwards: the replacement of the histories of special sciences with "general histories of science," and the awareness that "the division of knowledge embodied in contemporary science curricula" did not fit with the past bodies of knowledge [Kuhn 1968 (1977), pp. 105-9].

However, from the 1970s onwards some historians, historians of science, and philosophers of science began to draw the attention of scholars to that context.[24] Duhem's intellectual enterprise continued to be explored as well. In 1990 the journal *Synthese* devoted a special issue to a selection of papers presented at the conference "Pierre Duhem: Historian and Philosopher of Science," which had been held at Virginia Polytechnic Institute and State University in 1989. The editors acknowledged that a recent "change of climate in the history and philosophy of science," namely "the decline of logic-based philosophy of science," allowed scholars to search for a "real contact with Duhem" [Ariew and Barker 1990, pp. 179-81].

The historian of science Robert Westman stressed that "the spirit of Duhem" was still alive, and Brenner pointed out two important issues: first, methodology and history of science were intimately connected in Duhem's philosophy, and second, Duhem's specific kind of holism was really a novelty. Other important issues were emphasised by Maiocchi: Duhem's epistemological views had already been put forward before the alleged *crisis* that had affected physics at the end of the nineteenth century. More specifically, Duhem's 1906 epistemological theses had already been fully expressed in the papers he had published in the 1890s. Moreover, the theme of crisis was "totally absent in Duhem." The generalised mechanics Duhem had devised in the same years did not reject classical mechanics, but "enlarged and generalised it," in accordance with Rankine's *Energetism*. Nevertheless, it seems to me that when Maiocchi qualified Duhem as "radically anti-inductivist" and anti-empiricist, or he found in Duhem "a realistic and cognitive vision of scientific enterprise," he overstretched Duhem's intrinsically pliable philosophy of science [Westman 1990, p. 261; Brenner 1990, pp. 326 and 334; Maiocchi 1990, pp. 385-6, 388, and 395]. Two remarks deserve to be made: first, Duhem's philosophy was really pliable because it rested upon his actual scientific practice; second, according to the most widespread meaning of realism, Duhem cannot be qualified as a realist. His *Pascalian view* had led him into a sort of no-man's-land for the philosophy of science.

---

24 With regard to historical reconstructions of the cultural environment of French-speaking countries in the late nineteenth century, I can mention Paul 1968, Paul 1972a, Paul 1972b, Redondi 1978, Paul 1979, Maiocchi 1985, Brouzeng 1987, Brenner 1990, Martin 1991, Deltete 1999, Stoffel 2002, Needham 2002, Stoffel 2008, and Brenner, Needham, Stump, and Deltete 2011. With regard to the logical-philosophical debate see Ariew 1984, and van Fraassen 1989. More detailed references will be given in the following pages.

This was the historiographical and philosophical problem on which the historian and philosopher of biology Richard M. Burian attempted to cast light. He pointed out "the discrepancies between the standard current versions" of labels such as *realism* and *conventionalism*, and the actual views of scholars who had been living more than a century ago. I venture to add a more radical remark: when we are dealing with original thinkers, standard philosophical labels conceal more than they reveal. Philosophical labels can puzzle and mislead anyone who is really interested in the comprehension of a sophisticated and articulated point of view. The philosopher Andrew Lugg also highlighted Duhem's commitment to describing actual scientific practice rather than defining "the logical status of theoretical entities." Duhem's "sensitivity to history" was "an effective antidote both to dogmatic claims ... and to sceptical counterclaims," and had led him to undertake a demanding third way. The philosopher Ernan McMullin looked upon Duhem as a tightrope walker who dodged scientific realism, Poincaré's conventionalism, and the inductivism that can be found "in the physics textbooks of his day." This sounds right, but, in the end, McMullin agreed ("with some reservations") with Lugg that Duhem "was not an instrumentalist," and disagreed that "he was a realist." It seems that the war of labels can never be stopped [Burian 1990, p. 401 and 403; Lugg 1990, pp. 409 and 412-3; McMullin 1990, pp. 421-4].

Once more I must stress that any label whatsoever seems unsuitable for Duhem's complex and pliable philosophy of science, which was in tune with the complexity and pliability of actual scientific practice. More in general, I find misleading words and concepts such as *realism, instrumentalism, conventionalism, ...* when referred to the critical practice of some scholars from Cournot to Duhem. They were aware that scientific practice had frequently led scientists to undertake theoretical pathways where neither the alleged certainty of facts nor the alleged firmness of logic and mathematics could help find the best direction. Even worse, scientific principles such as the principle of inertia conflicted with both the mere observation of facts, and the rational expectation of definite, unambiguous definitions.

However, the skirmish among labels continued. In 1991 Needham defended Duhem against the charge of instrumentalism: he found that Duhem could be qualified as "a moderate realist" with an "antireductionist, holistic attitude." More specifically, he found that Duhem was not extraneous to the concept of scientific truth, and the concept of natural classification was in tune with such an attitude. In 2000 he examined similarities and differences in Duhem and Quine's general philosophical views, and re-stated his opposition to an instrumentalist interpretation of Duhem. At the same time he highlighted the similarity between Duhem and Quine's holistic views, and smoothed the differences [Needham 1991, pp. 96, 103, and 108; Needham 2000, pp. 123 and 130]. Needham's approach is convincing from the logical point of view. The fact is that Quine had actually started from a logical point of view, namely a logical analysis of scientific practice, whereas Duhem had been interested in grasping the essential feature of actual scientific practice, non-logical components included. According to Duhem, logic was an essential component in the structure of a scientific theory, but the actual development of science over time, and the actual processes of devising or updating scientific theories, also called for skills and practices which stood outside the domain of logic.

Further Debates on Reductionism and Determinism

After the Second World War, Karl Popper remarked that, differently from quantum physics, which was "generally admitted to be indeterministic," classical physics had usually been "taken to be deterministic," but he found that the opposition was misleading. According to Popper, "classical mechanics was not deterministic, and was forced to "admit the existence of unpredictable events." A mythological or "theological view of science" had led to the attribution of omniscience to science. What was labelled "Laplace's determinism" was nothing else but "a misinterpretation" [Popper 1950a, pp. 117, 122, and 128; Popper 1950b, pp. 173 and 193]. In 1953, Mittasch published Ostwald's 1914 essay together with a new layout of Mayer's 1876 paper: it does not seem that the booklet reached a wide audience.[25] Mittasch had already published a book in 1940, *Julius Robert Mayers Kausalbegriff*, and two papers in 1942, in the volume that celebrated the centenary of "the discovery of the energy principle." In the book he had credited Mayer with "having begun to transform the essentially limited *mechanical* physics of his time" into a wide-ranging "physics of energy." He mentioned Saint Venant's "travail décrochant" and Boussinesq's "principe directeur": Boussinesq had associated the domain of "inanimate entities" with general solutions of differential equations, whereas "the domain of living beings" had to be ruled by singular solutions [Mittasch 1940, pp. 55 and 126; Mittasch 1953b, p. 7].[26]

25 Only three years later a short review appeared, where the physicist and historian of science Hans Schimank qualified Mittasch's booklet as "excellent for a seminar on the history of science or medicine" [Schimank 1956, p. 190].

26 In the first paper he published in the centenary volume, he focused on the "meaning of Mayer's *Auslösung* for chemistry," and interpreted that concept as "a noteworthy comple-

In brief, only scattered references to the debate on determinism in the last decades of the nineteenth century can be found in the twentieth century. The debate on reductionism underwent a similar fate: in the central decades of the twentieth century a renewed *Comtian* attitude prevailed. After the Second World War, scientific practice experienced a deep transformation: in huge laboratories, hundreds of researchers pursued long-term research projects that required a huge amount of money. The widespread success of scientific practice during and after the war led to two intertwined consequences: fast scientific progress became an undisputed matter of fact, and researches on past theories or historical-critical reconstructions became definitely inappropriate. Systematic interests in the history and philosophy of science would have brought scientists into disrepute.

With regard to the main issues raised by the scholars who had given birth to French philosophy of science, namely the different aspects of determinism, the questionable soundness of a strict reductionism, the existence of metaphysical elements in scientific theories, the problematic link between empirical and theoretical statements, and the necessity of extra-empirical and extra-logical components in scientific practice, it seems that a widespread drift towards oversimplification took possession of many scientists and philosophers from the 1930s to 1950s. The influence of neo-positivism or logical empiricism was sill strong after the Second World War. Philosophers of science narrowed their field of research and confined themselves to a logical analysis of scientific theories. In this perspective, many problems that French scientists and philosophers had debated, and many features of actual scientific practice, were wittingly disregarded.

In 1956, in a paper on reductionism in a philosophical journal, the philosopher Paul Oppenheim and the mathematician and philosopher John Kemeny remarked that there had been "a good deal of controversy" as to just what the observational and theoretical terms were. Nevertheless they found that, in the context of their attempts to give a formal definition of reductionism, they could only suppose "that a separation has been made." It was irrelevant for them "how the two are distinguished" [Kemeny and Oppenheim 1956, p. 8]. In 1967 the young philosopher of science Kenneth Schaffner attempted to put forward a complex formal definition of scientific reduction after having discussed "Four Reduction Paradigms." He found that a reduction was actually possible, or rather, it was "a scientific fact" [Schaffner 1967, pp. 138-40, and 145-6].

ment" to the principle of energy," however "merely formal, essentially elementary, and non-mathematical" it might be [Mittasch 1942a, p. 281].

It was only after the pluralist and relativist shift in the context of philosophy of science, and after the re-emergence of interest in the history of science, that a more sophisticated approach to reductionism was put forward. A new language and new concepts emerged, and expressions such as normal science, scientific revolution, incommensurability, and research programme entered the cultural toolbox of educated people. In the 1960s and 1970s, the new trend in philosophy of science also managed to impose its language and its intellectual agenda on social sciences.

In 1974 the philosopher of science Frederick Suppe published the proceedings of an "important and pivotal conference" held in Chicago in 1969, the *Illinois Symposium on the Structure of Scientific Theories*, and extensively commented on the issues under investigation in a long introductory essay of almost 250 pages. In the resulting book, *The Structure of Scientific Theories*, he put forward a historical-critical reconstruction of the so-called *Received View on Theories*, namely the logical positivism or logical empiricism that had exerted "near total dominance" over the philosophy of science for over thirty years from the 1920s to the 1950s. According to the received view, scientific theories were looked upon as "axiomatic calculi which are given a partial observational interpretation by means of correspondence rules." In the 1950s, and more extensively in the 1960s, the received view was criticized and "alternative philosophies of science" were advanced by Hanson, Kuhn, and then Feyerabend. According to Suppe, "the logical or structural analysis" of scientific theories was replaced by "a general account of the nature of scientific theorizing": the new approach paid attention to processes through which theories emerged, developed, and changed over time, and also focused on "the various *Weltanschauungen*" through which science approached and interpreted Nature and scientific practice.[27] In 1977, in the second edition of the volume, he added an *Afterword* of almost 120 pages wherein he stressed the traumatic

27 In 2000, Suppe synthetically listed the main critical assessments of the positivistic received view put forward in the 1960s:

(1) its observational-theoretical distinction was untenable;

(2) correspondence rules were a heterogeneous confusion of meaning relationships, experimental designs, measurement, and causal relationships, some of which are not properly parts of theories;

(3) the notion of partial interpretation associated with more liberal correspondence rules was incoherent;

(4) theories are not axiomatic systems;

(5) symbolic logic is an inappropriate formalism;

(6) theories are not linguistic entities and thus theories are individuated incy [Suppe 2000, pp. S103].

character of that philosophical shift. That "acute state of disarray" had produced a "widespread confusion and disagreement among philosophers as to what the main problems in philosophy of science were," and a new direction of research had emerged from "the then disordered house of philosophy of science" [Suppe 1977, pp. 3-4, 233, 617 and 729].

In 1988, the physicist and philosopher of science Fritz Rohrlich examined the historical process wherein a scientific theory became "the most fundamental theory of the day," namely the most satisfactory theory that had replaced a less satisfactory theory, and that would have been replaced by an even more satisfactory theory in an indefinite future. He claimed that although "the *continuous* change" from one theory to a more general or fundamental one could be mathematically conceived, the interpretation was *discontinuous* in the transition. The discontinuity was due to "the *cognitive emergence* of qualitatively new and different descriptions." His "*pluralistic epistemology of science*" was based on two meta-theoretical assumptions: "the surviving coarser theory" was not "unnecessary or useless," and even a scientific revolution did not "call for the replacement of the old theory within its domain" [Rohrlich 1988, pp. 296-7, 307, and 309].

However the conceptual stream that had crossed French cultural environment from Cournot to Duhem seemed permanently forgotten: the new trend appeared as a fresh creation rather than the re-emergence of previous research programmes in the philosophy of science. In reality, Cournot had been almost forgotten but Duhem had not, even though sometimes he had been misunderstood rather than interpreted. With regard to Boussinesq, in the 1970s the historian of science Mary Jo Nye reminded readers of his existence and his research programme, and framed his meta-theoretical design into a conceptual stream that had crossed French-speaking countries between the 1870s and the 1890s. She also acknowledged the influence "the mathematician A.A. Cournot and physiologist Claude Bernard" had had on him [Nye 1976, pp. 274, 276, and 289-90; Nye 1979, pp. 107, 110, and 117-9].[28] In 1983 Ian Hacking endorsed, at

He found that "the most significant objections, in order," were (6) and (2), whereas (4) was irrelevant and (3) false. I find that the most significant issues are (4), (6), and (1) in order.

28 She pointed out the existence of a generation of talented scientists and mathematicians who were sensitive to historical and philosophical issues: among them she mentioned Paul Tannery, Pierre Duhem, and Henri Poincaré. She listed Joseph Boussinesq in the number of prominent Catholic members of the *Brussels Scientific Society*, which had been founded in 1875, and in 1877 had given birth to the scientific journal *Revue des questions scientifiques*.

least in part, Cassirer's thesis on Du Bois-Reymond's role in the emergence of "Laplace's Mind" as a modern mythology of determinism. However, Hacking thought that "Laplace's celebrated determinist dictum" was not only a metaphor. He found that Laplace, like Kant and Hume, was "a determinist about external bodies but not about the mind." He also found in Saint-Venant, his former pupil Boussinesq, and Maxwell a common attitude towards the query of determinism, and the emergence of "a completely new idea," which he qualified as "completely crazy" [Hacking 1983, pp. 455-60 and 462-5].

In the meantime, from the 1980s onwards, a new analytic trend became hegemonic in the philosophy of science. That new attitude could be looked upon as a new implementation of the previous logical empiricism: it overlooked the tradition of the discipline, and the complex interaction between historiography and epistemology, but focused on the formal structures of theories and the consistency of mathematical models.[29] A new generation of philosophers skilfully focused on the logical foundations of science, and lost interest in the complex history of science and philosophy. In 1986, John Earman, in his introductory but authoritative treatise on determinism, depicted determinism as "a perennial topic of philosophical discussion" even though classical physics had been looked upon by philosophers as "a largely deterministic affair." On the contrary, he found that "classical worlds prove to be an unfriendly environment for any form of Laplacian determinism." In general, he claimed that ontology and epistemology should be carefully disentangled: the ontological problem of determinism should not be confused with its epistemological counterpart, namely the possibility of prediction. In particular, he found that the determinism-free will controversy had "all of the earmarks of a dead problem": to study the history of those debates was "an exercise in frustration" [Earman 1986, pp. 1-2, 7-8, 24, 34-5, 41-2, 235, and 243].

In 1987 the mathematician and historian of science Giorgio Israel, after having qualified Laplace's research programme as an "ideal Newtonian programme," also stressed the difference between determinism and predictability, which involved the difference between ontological and epistemological levels. A physical system could be ruled by definite mathematical procedures but accurate predictions might be problematic [Israel 1987, pp. 84, 86, and 93-4]. In some way, here we are at the core of Boussinesq's intellectual pathway:

29 In 1979 the philosopher of science Wolfgang Stegmüller pointed out the emergence of two different trends in the philosophy of science, namely "the *formal language approach*" and a less formal approach wherein real physical theories could be "axiomatized in a precise way *without recourse to formal languages.*" In both cases, some kind of axiomatisation was at stake: only analytic and axiomatic approaches deserved consideration [Stegmüller 1979, p. 4].

mathematical-mechanical determinism can give rise to unpredictable processes. The following year, the mathematician and historian of mathematics Michael Deakin explored Maxwell, Saint-Venant and Boussinesq's approaches to what he called an "apparent paradox." In his short paper, "Nineteenth Century Anticipations of Modern Theory of Dynamical Systems," in accordance with his *whig* perspective, Deakin claimed that Maxwell was "distinctly modern," while Saint-Venant and Boussinesq were definitely not [Deakin 1988, pp. 183 and 186-9].

At the end of the twentieth century, a century after the developments in the history and philosophy of science discussed in the present book, after Cournot, Tannery, and Duhem's researches, we find naïve and simplified accounts of old scientific theories. In the last decades of the nineteenth century, Cournot, Duhem, and Milhaud had pointed out the complexity of history, the complexity of scientific method, and the complex interaction between historiographical frameworks and epistemological assumptions. Duhem had systematically explored the borderline between historiography and epistemology, but histories with the benefit of hindsight continued to flourish even a century later. In reality, the decline of the logically-based philosophy of science as supposed by Ariew and Barker in 1990 was more apparent than real. From the 1980s onwards we find two concurrent trends: on the one hand a historically and critically oriented philosophy of science, and on the other a logically oriented philosophy of science. These two different approaches also affected the debate on determinism, but the second trend took the lead.

In 1991 Israel undertook a historical research on determinism and its "translation in mathematical language," namely "the existence and uniqueness theorem for a system of ordinary differential equations." He appositely remarked that the first explicit statement of determinism did not belong to the history of mathematical physics "but to the history of medical science." It had been the French physiologist Claude Bernard "to first introduce this term in the scientific language, at least the French one." He pointed out Laplace's influence on Bernard, but stressed that Bernard's determinism was devoid of "any metaphysical interpretation" and any materialistic hypothesis. He found that Boussinesq had managed to rightly point out the problem but his solution was "technically weak." Neither Cournot nor Saint-Venant had managed to do better: they had put forward "metaphysical hypotheses" rather than scientific theories [Israel 1991, pp. 305, 307, 313, 331-3, 339, 344-5, and 352].[30]

30 Israel acknowledged that the French philosopher and historian of science Georges Canguilhem had been the first to stress the link between the debate on determinism and Claude Bernard's scientific work. See, for instance, Canguilhem 1998, pp. 64-5.

In the meantime, chaotic systems and chaos theory had become a relevant research topic. In 1995 the philosopher James W. Garson pointed out how attractive chaos theory was "for resolving the problem of freedom," since it could blend "underlying physical determinism with an emerging unpredictability." Chaos theory offered some kind of "*unpredictable* determinism," which was different from the indeterminism that had emerged from twentieth-century microphysics. Freedom involved "rational control," and indeterminism could not be looked upon as a good foundation for rational control. What chaos theory could teach us was that "non-determinism does not entail randomness," and therefore the "dichotomy between rules and randomness" could be overcome [Garson 1995, pp. 366-9 and 371-2].[31] In 1996 the philosopher John Dupré also addressed the vexed question of the supposed conflict between determinism and human freedom. Not only did he find that "indeterminism makes the conception of freedom of the will even less tenable," but also having recourse to chaotic systems was meaningless. According to Dupré, the solution rested on "a complete reversal of the traditional non-compatibilist approach," where the natural world consisted of "a network of causal connections," whereas human beings lay "outside or partially outside this web." In reality humans were "sources of causal order," and provided "oases of order and predictability" [Dupré 1996, pp. 385-7, 392-4, and 398-400].[32]

31 In 1991 the philosopher Jesse Hobbs had come back to the difference between predictability and determinism. In classical physics, unstable equilibriums were confined to isolated points or singularities, whereas in chaotic systems, unstable equilibriums could "no longer be isolated or kept under control": it could be extended to "every point in phase space." In the end, chaos theory really challenged determinism "as a serious theory or world view" [Hobbs 1991, p. 142, 160, 162, and 164].

32 In 1977, Suppe had found that the most promising perspective in philosophy of science could rely on a new "metaphysical and epistemological realism" that focused on "rationality in the growth of scientific knowledge." At the end of the twentieth century, he came back to the historical reconstruction of "the chaos evidenced in 1969" and the main critical assessments addressed to the positivistic received view. Differently from 1974, when he had assumed that philosophy of science was "little more than an analysis of theories and their roles in the scientific enterprise," he acknowledged that the actual scientific landscape was notably more complex, and the claim on what was really important in philosophy of science depended on the actual development of science over time. More specifically, he pointed out that much of theoretical practice had become *atheoretical* and "computationally intensive": some scientific practices dealt with "modeling data," and devising models had become "the main vehicle of scientific knowledge." In the end, he found that "semantic analysis of theories" would have led to "escalating understanding of theories, models, and how science *really* works" [Suppe 1977, pp. 618 and 730; Suppe 2000, pp. S103, S109, and S114].

With regard to logically oriented philosophy of science, in 2003 John Norton focused on causation and determinism in science, and on what he looked upon as the received view on the subject. He claimed that a principle of causality was only consistent with specific and simplified implementations of scientific theories. He discussed a simple but interesting case: a point-like mass at rest upon the top of a rotationally symmetric surface [Norton 2003, pp. 1-5, 8-9, 13-14, and 19-21].[33] The mass upon the dome was looked upon as a recent issue, a problem without any history. In subsequent *philosophical* literature, it was addressed as "Norton's dome." Boussinesq and the debate that involved philosophers, scientists, and mathematicians in the second half of the nineteenth century were definitely out of reach.

In 2008, in the context of a debate hosted by the journal *Philosophy of Science*, Norton went back to his dome and the corresponding indeterminism. He focused on the dynamical aspect of what should happen at the top of the dome whereas David Malament focused on the geometrical aspect.[34] The philosopher Alexandre Korolev also devoted his paper to geometrical and physical technicalities: more specifically he focused on "the initiating cause that sets the mass in motion," and the dome curvature. Earman took part in the debate on the so-called Norton dome, and focused on the problematic link between mathematics and physics. Classical mechanics was consistent with more exotic systems, for instance "an infinite system of particles that interact via elastic collisions." In the end, 130 years after Boussinesq's essay, a renowned philosopher of science acknowledged that determinism in the context of physical sciences still managed "to spring surprise on us" [Norton 2008, pp. 789-90 and 792-3; Malament 2008, pp. 799, 808, and 815); Korolev 2008, pp. 945 and 949-51; Earman 2008, pp. 817-9 and 828].

The philosopher Mark Wilson showed a more historical sensitivity, although he did not make specific reference to the existing literature. He claimed that the supposed indeterminism stemmed from a sort of incompleteness of the different approaches to mechanics, an incompleteness which he variously labelled "missing physics," "foundational gaps," "*descriptive holes*," and "strange descriptive gaps" [Wilson 2009, pp. 174-7 and 181-2]. More recently van Strien discussed Boussinesq's research programme in connection with the debates on Norton's dome around 2008, and acknowledged that Boussinesq and Bertrand's approaches to "indeterministic systems similar to the Norton

33 Norton mentioned Russell's similar approach, in particular the statement that "The law of causality, ..., is a relic of a bygone age" [Norton 2003, p. 3; Russell 1917, p. 132].

34 The dome surface was everywhere infinitely differentiable apart from the apex, where it was only once differentiable.

dome" were not in tune with recent ones [van Strien 2014a, pp. 167 and 170]. As Hélène Metzger taught us more than eighty years ago, history of science requires a painstaking and passionate inquiry into intellectual contexts and ways of thinking that might lead us far from our commitments and meanings.

At the end of this *Afterword*, it is worth mentioning that at the turn of the twentieth-first century, the economist James W. Friedman remembered Cournot to his professional community for his "clear and sophisticated treatment of market demand, monopoly, competitive markets, and above all, oligopoly" [Friedman J.W. 2000, p. 31]. In 2007, Thierry Martin and Jean-Philippe Touffut reminded readers of Cournot's mathematical theories of economics: unfortunately, there had been "a large temporal gap between their application and their original publication." When Cournot's *Recherches* had been published in 1838, the book had gone "almost completely unnoticed": he had definitely been "ahead of his time," and he had found "almost no interlocutors." Only after the first translation into English in 1898 had the book been taken seriously into account, and "in the early 1940s, Cournot's contributions to economic theory" had been "widely recognized and discussed." The authors stressed that Cournot's approach had "initiated the mathematical modelling of social phenomena," and his subsequent epistemology could be looked upon as a sophisticated outcome of his scientific researches. They found a meaningful connection between the application of probability theory to social phenomena, and Cournot's probabilistic epistemology [Martin and Touffut 2007, pp. 1-2 and 4].

The cycles of disappearances and questionable reappearances of meaningful issues that were at stake in the last decades of the nineteenth century shows us how fruitful buried memories can be, and how interesting the scientific and philosophical landscape of those decades really was.

References

Achistein, P. and Kargon, R. (eds.) (1987), *Thomson's Baltimore lectures and modern theoretical physics: historical and philosophical perspectives*, Cambridge, Massachusetts: MIT Press.

Alunni, C. and André Y. (eds.) (2015), *Federigo Enriques o le armonie nascoste della cultura europea*, Pisa: Scuola Normale Superiore.

Apmann, R.P. (1964), A case history in theory and experiment: Fluid flow in bends, *Isis*, 55(4), 427-434.

Ariew, R. (1984), The Duhem Thesis, *The British Journal for the Philosophy of Science*, 35, 4, 313-325.

Ariew, R. and Barker, P. (1990), Introduction, *Synthese*, 83, 179-182.

Audierne, R. (1905), La classification des connaissances dans Comte e dans Cournot, *Revue de Métaphysique et de Morale*, 13 (3), (Mai 1905), 509-519.

Bachelard, G. (1951), *L'activité rationaliste de la physique contemporaine*, Paris: Presses Universitaires de France.

Barbin, E. and Cléro J.P. (eds.) (2014), *Les mathématiques et l'expérience*, Paris: Hermann.

Barnes, J. (1982), *The Presocratic Philosophers*, London: Routledge.

Barraclough, G. (1964), *An Introduction to CONTEMPORARY HISTORY*, London: Watts & co.

Becker, P. and Wetzell, R.F. (eds.) (2006), *The History of Criminology in International Perspective*, Cambridge/New York/Melbourne: Cambridge University Press.

Benrubi, I. (1926), *Contemporary Thought of France*, London: Williams and Norgate.

Benrubi, I. (1933), *Les sources et les courantes de la philosophie contemporaine en France*, Paris: Alcan.

Bergson, H. (1889), *Essai sur les données immédiates de la conscience*, Paris: Alcan.

Bernard, C. (1864), *Des organismes. De l'anatomie générale et de son histoire*; in Bernard, C. (1866), pp. 3-25.

Bernard, C. (1865), *Introduction à l'étude de la médicine expérimentale*, Paris: Baillière.

Bernard, C. (1866), *Leçons sur les propriétés des tissus vivants*, Paris: Baillière.

Bernard, C. (1867), *Rapport sur le progrès et la marche de la physiologie générale en France*, Paris: Hachette.

Bernard, C. (1869), *Discours de M. Claude Bernard prononcé à sa réception à l'Académie Française le 27 mai 1869*, Paris: Didier.

Bernard, C. (1872), *De la physiologie générale*, Paris: Hachette.

Bernard, C. (1878-9), *Leçons sur les phénomènes de la vie communs aux animaux et aux végétaux*, 2 vols., Paris: Baillière.

Berthelot, M. (1860), *Traité de chimie organique fondée sur la synthèse*, Paris: Mallet-Bachelier.

Berthelot, M. (1885), *Les origines de l'alchimie*, Paris: Geoges Steinheil.

Berthelot, M. (1886), *Science et philosophie*, Paris: Calmann Lévy.

Berthelot, M. (1893a), *La chimie au moyen âge. Tome II. L'alchimie Syriaque*, Paris: Imprimerie Nationale.

Berthelot, M. (1893b), *La chimie au moyen âge. Tome III. L'alchimie Arabe*, Paris: Imprimerie Nationale.

Bertrand, J. (1878), Conciliation du véritable déterminisme mécanique… par J. Boussinesq,… , *Journal des Savants*, Septembre 1878, 517-523.

Bhakthavatsalam, S. (2015), The rationale behind Pierre Duhem's natural classification, *Studies in History and Philosophy of Science*, 51, 11-21.

Biographie Nationale publiée par l'Académie Royale de Belgique (1969), tome trente-cinquième, Supplément, tome VII (fascicule 1er), Bruxelles: Établissements Émile Bruylant.

Boëns, H. (1878), La physiologie et la psychologie ou le corps et l'âme, *La philosophie positive*, XX (janvier à juin 1878), 343-360.

Boëns, H. (1879), *La science et la philosophie ou nouvelle classification des sciences*, Paris: Baillière.

Boltzmann, L. (1877), Über die Beziehung zwischen dem zweiten Hauptsatze der mechanischen Wärmetheorie und der Wahrscheinlichkeitsrechnung respektive den Sätzen über das Wärmegleichgewicht; in Boltzmann, L. (1909), II Band, pp. 164-223.

Boltzmann, L. (1892), On the methods of theoretical physics; in Boltzmann, L. (1974), pp. 5-12.

Boltzmann, L. (1896), Zur Energetik, *Annalen der Physik*, 58, 595-598.

Boltzmann, L. (1897), Vorlesungen über die Principen der Mechanik, vol. 1, Leipzig: Barth; partially translated into English in Boltzmann, L. (1974), pp. 223-254.

Boltzmann, L. (1899), On the Development of the Methods of theoretical Physics in recent Times; in Boltzmann, L. (1974), pp. 77-100.

Boltzmann, L. (1909), *Wissenschaftlichen Abhandlungen*, Leipzig: Barth.

Boltzmann, L. (1974), *Theoretical Physics and Philosophical Problems*, (McGuinness B. ed.), Dordrecht, Holland/ Boston, U.S.A.: Reidel Publishing Company.

Bordoni, S. (2008), *Crossing the boundaries between matter and energy*, Pavia: Università degli Studi di Pavia – La Goliardica Pavese.

Bordoni, S. (2012a), Unearthing a Buried Memory: Duhem's Third Way to Thermodynamics. Part 1, *Centaurus*, 54 (2), 124-147.

Bordoni, S. (2012b), Unearthing a Buried Memory: Duhem's Third Way to Thermodynamics. Part 2, *Centaurus*, 54 (3), 232-249.

Bordoni, S. (2012c), *Taming Complexity. Duhem's third pathway to Thermodynamics*, Urbino: Editrice Montefeltro.

Bordoni, S. (2013a), Routes towards an Abstract Thermodynamics in the late nineteenth century, *European Physical Journal – H*, 38, 617-660.

Bordoni S. (2013b), Looking for a Rational Thermodynamics in the late XIX century, *Preprint 443*, Berlin: Max-Planck-Institut für Wissenschaftsgeschichte (Max Planck Institute for the History of Science).

Bordoni, S. (2014a), J.J. Thomson and Duhem's Lagrangian Approach to Thermodynamics, *Entropy*, 16, 5876-5890.

Bordoni, S. (2014b), La science de la thermodynamique dans la deuxième moitié du dix-neuvième siècle, *Revue des Questions Scientifiques*, 185 (4), 399-420.

Bordoni, S. (2015a), Pierre Duhem le physicien et la recherche d'une thermodynamique rationnelle, *Revue des questions scientifiques*, 186 (4), 549-572.

Bordoni, S. (2015b), On the borderline between Science and Philosophy: a debate on determinism in France around 1880, *Studies in History and Philosophy of Science*, 49, 27-35.

Boussinesq, J. (1868), Mémoire sur l'influence des frottements dans les mouvements réguliers des fluides, *Journal des mathématiques pures et appliquées*, Series 2, 13, 377-424.

Boussinesq, J. (1877a), Essai sur la théorie des eaux courantes, *Mémoires présentés par divers savants à l'Académie royale des sciences*, 23, 1-680.

Boussinesq, J. (1877b), Sur la conciliation de la liberté morale avec le déterminisme scientifique, *Comptes Rendus de l'Académie des Sciences*, LXXXIV, 362-4.

Boussinesq, J. (1878a), *Conciliation du véritable déterminisme mécanique avec l'existence de la vie et de la liberté morale*, Paris: Gauthier-Villars.

Boussinesq, J. (1878b), Conciliation du véritable déterminisme mécanique avec l'existence de la vie et de la liberté morale, *Extrait du compte-rendu de l'Académie des Sciences morales et politiques*, (rédigé par M. Ch. Vergé), tome IX (mai 1878), 696-757.

Boussinesq, J. (1879a,) Conciliation du véritable déterminisme mécanique avec l'existence de la vie et de la liberté morale", *Mémoires de la société des sciences, de l'agriculture et des arts de Lille*, 6 (4), 1-257.

Boussinesq, J. (1879b), Le déterminisme et la liberté, *Revue Philosophique de la France et de l'Etranger*, 7, 58-66.

Boussinesq, J. (1879c), *Étude sur divers points de la philosophie des sciences*, Paris: Gauthier-Villars.

Boutroux, E. (1874), *De la contingence des lois de la nature*, Paris: Baillière.

Boutroux, E. (1895), *De l'idée de loi naturelle dans la science et la philosophie contemporaines*, Paris: Oudin/Alcan.

Boutroux, P. (1938), L'oeuvre de Paul Tannery; in Boutroux P,. and Sarton, G. (1938), pp. 691-705.

Boutroux, P. and Sarton, G. (1938), L'oeuvre de Paul Tannery, *Osiris*, 4, 690-705.

Bowler, P.J. and Morus, I.R. (2005), *Making Modern Science*, Chicago and London: The University of Chicago Press.

Braunstein, J.F. (2012), Historical Epistemology, Old and New; in *Epistemology and History. From Bachelard and Canguilhem to Today's History of Science, Conference, Preprint 434*, Max-Planck-Institut für Wissenschaftsgeschichte (Max Planck Institute for the History of Science), pp. 33-40.

Brenner, A. (1990a), *Duhem. Science, réalité et apparence*, Paris: Vrin.

Brenner, A. (1990b), Holism a century ago: the elaboration of Duhem's thesis, *Synthese*, 83, 325-335.

Brenner, A. (1996), La nature des hypothèses physiques selon Poincaré, à la lumière de la controverse avec Duhem; in Greffe, J.L., Heinzmann, G. and Lorenz K. (eds.) (1996), pp. 389-396.

Brenner, A. (2003), *Les origines françaises de la philosophie des sciences*, Paris: Presses Universitaires de France.

Brenner, A. (2015), *Les textes fondateurs de l'épistémologie Française*, Paris: Hermann.

Brenner, A. (2016), L'épistémologie historique d'Abel Rey, *Revue de Métaphysique et de Morale*, 2, 159-175.

Brenner, A. and Gayon, J. (2009a), *French Studies in the Philosophy of Science*, Boston Studies in the Philosophy of Science 276, Berlin, New York: Springer.

Brenner, A. and Gayon, J. (2009b), Introduction; in Brenner, A. and Gayon, J. (2009a), pp. 1-22.

Brenner, A., Needham, P., Stump, D.J. and Deltete, R. (2011), New Perspectives on Pierre Duhem's *The aim and structure of physical theory*, *Metascience*, 20, 1-25.

Brock, W.H. (2009), J. R. Partington (1886-1965): Physical Chemistry in Deed and Word, *Bulletin for the History of Chemistry*, 34 (1), 11-20.

Brouzeng, P. (1981), *L'oeuvre scientifique de Pierre Duhem et sa contribution au développement de la thermodynamique des phénomènes irréversibles*, 2 vols., Thèse (pour obtenir le grade de Docteur d'Etat ès Sciences), Bordeaux: Université de Bordeaux I.

Brouzeng, P. (1987), *Duhem 1861-1916 – Science et Providence*, Paris: Belin.

Brunschvicg, L. (1922), La philosophie d'Émile Boutroux, *Revue de Métaphysique et de Morale*, 29 (3), 261-283.

Buchwald, J.Z. (1985a), *From Maxwell to Microphysics*, Chicago: University of Chicago Press.

Buchwald, J.Z. (1985b), Oliver Heaviside, Maxwell's Apostle and Maxwellian Apostate, *Centaurus*, 28, 288-330.

Buchwald, J.Z. (1985c), Modifying the continuum: methods of Maxwellian electrodynamics; in Harman, P.M. (ed.) (1985), *Wrangler and Physicists*, Manchester: Manchester University Press, pp. 225-41.

Buchwald, J.Z. (1994), *The Creation of Scientific Effects – Heinrich Hertz and Electric Waves*, Chicago and London: The University of Chicago Press.

Burian, R.M. (1990), Maiocchi on Duhem, Howard on Duhem and Einstein: historiographical comments, *Synthese*, 83, 401-408.

Canguilhem, G. (1943), Essai sur quelques problèmes concernant le normal et le pathologique; in Canguilhem, G. (1966), pp. 7-167.

Canguilhem, G. (1958), La philosophie biologique d'Auguste Comte et son influence en France au XIXe siècle, *Bulletin de la Société française de Philosophie*; in Canguilhem, G. (1979), pp. 61-74.

Canguilhem, G. (1965), Claude Bernard et Bichat, *XI International Conference of History of Sciences, Warsaw-Krakow*, August 28, 1965; in Canguilhem, G. (1979), pp. 156-62.

Canguilhem, G. (1966), *Le Normal et le Pathologique*, Paris: Presses Universitaire de France.

Canguilhem, G. (1977), *Idéologie et Rationalité*, Paris: Vrin.

Canguilhem, G. (1979), *Études d'histoire et de philosophie des sciences*, Paris: Vrin.

Canguilhem, G. (1998), *Le normal et le pathologique*, Paris: Presses Universitaires de France.

Cassirer, E. (1937), *Determinismus und Indeterminismus in der modernen Physik*, Göteborg: Wettergren & Kerbers.

Cassirer, E. (1950), *The problem of Knowledge. Philosophy, Science, and History since Hegel*, (ed. by Woglom, W.H. and Hendel, C.W.), New Haven: Yale University Press.

Castellana, M. (2014), Les mathématiques et l'expérience selon Albert Lautman; in Barbin, E. and Cléro J.P. (eds.) (2014), pp. 311-338.

Castellana, M. (2015), Su alcune armonie nascoste in Federigo Enriques: continuità/discontinuità; in Alunni C. and André Y. (eds.) (2015), pp. 53-80.

Catana, L. (2011), Tannery and Duhem on the Concept of a System in the History of Philosophy and History of Science, *Intellectual History Review*, 21 (4), 515-531.

Chimisso, C. (2008), *Writing the History of the Mind. Philosophy and Science in France, 1900 to 1960s*, Aldershot UK, Burlington USA: Ashgate.

Clevelin, M. (2006), Duhem et Tannery, lecteurs de Galilée, *Galileiana*, III, 3-17.

Coleman, W. (1985), The Cognitive Basis of a Discipline: Claude Bernard on Physiology, *Isis*, 76, (1), 49-70.

Comte, A. (1830), *Cours de philosophie positive*, Tome premier, Paris: Rouen Frères.

Cournot, A.A. (1838), *Recherches sur les principes mathématiques de la théorie des richesses*, Paris: Hachette.

Cournot, A.A. (1841), *Traité élémentaire de la théorie des fonctions et du calcul infinitésimal*, tome second, Paris: Hachette.

Cournot, A.A. (1843), *Exposition de la théorie des chances et des probabilités*, Paris: Hachette.

Cournot, A.A. (1851), *Essai sur les fondements de nos connaissances et sur les caractères de la critique philosophique*, Paris: Hachette.

Cournot, A.A. (1857), *Traité élémentaire de la théorie des fonctions et du calcul infinitésimal*, Hachette, Paris.

Cournot, A.A. (1861), *Traité de l'enchainement des idées fondamentales dans les sciences et dans l'histoire*, 2 tomes, Paris: Hachette.

Cournot, A.A. (1872), *Considérations sur la marche des idées et des événements dans les temps modernes*, Paris: Hachette.

Cournot, A.A. (1875), *Matérialisme, vitalisme, rationalisme. Études sur l'emploi des données de la science en philosophie*, Paris: Hachette.

Crocco, G. (2016), Poincaré et le problème de l'esprit, *Revue de Métaphysique et de Morale*, 2, 209-224.

Darrigol, O. (1993), The Electrodynamic Revolution in Germany as documented by Early German Expositions of « Maxwell's Theory », *Archive for History of Exact Sciences*, 45, 189-280.

Darrigol, O. (2002), Between Hydrodynamics and Elasticity Theory: The First Five Births of the Navier-Stokes Equation, *Archive for History of Exact Sciences*, 56, 95-150.

Darrigol, O. (2009), *Worlds of Flow: A history of hydrodynamics from the Bernoullis to Prandtl*, Oxford, New York: Oxford University Press.

Darwin, C.R. and Wallace, A.R. (1858), On the Tendency of Species to form Varieties; and on the Perpetuation of Varieties and Species by Natural Means of Selection, *Journal of the Proceedings of the Linnean Society. Zoology*, 3 (20 August), 45-62.

Daston, L. (2009), Comments on Schickore and Sturm: Where do Epistemological Problems Come From?; in Sturm, T. and Feest, U. (eds.) (2009), pp. 35-7.

Deakin, M.A.B. (1988), Nineteenth Century Anticipations of Modern Theory of Dynamical Systems, *Archive for History of Exact Sciences*, XXXIX (2), 183-194.

Delboeuf, J. (1877), Les mathématiques et le transformisme, *La Revue Scientifique de la France et de l'étranger, Revue des cours scientifiques* (2e série), 29, 669-79.

Delboeuf, J. (1882a), Déterminisme et liberté. La liberté démontrée par la mécanique, *Revue Philosophique de la France et de l'Étranger*, 13, 453-480 and 608-638.

Delboeuf, J. (1882b), Déterminisme et liberté. La liberté démontrée par la mécanique, *Revue Philosophique de la France et de l'Étranger*, 14, 156-189.

Delorme, S. (1970-80), *Rey, Abel*; in Gillispie, C.C. (ed.), (1970-80), vol. 11, pp. 388-9.

Deltete, R. (1999), Helm and Boltzmann: Energetics at the Lübeck Naturforscherversammlung, *Synthese*, 119, 45-68.

Deltete, R. (2008), Man of science, man of faith: Pierre Duhem's « Physique de croyant », *Zygon*, 43 (3), 627-637.

Domet de Vorges, E. (1893), Les hypothèses physiques sont-elles des explications métaphysiques?, *Annales de philosophie chrétienne*, 127, 137-51.

Doncel, M.G. (1991), On the Process of Hertz's Conversion to Hertzian Waves, *Archive for History of Exact Sciences*, 43 (1), 1-27.

Du Bois-Reymond, E. (1872), Über die Grenzen des Naturerkenntnis; in Du Bois-Reymond, E. (1912), I Band, pp. 441-473.

Du Bois-Reymond, E. (1880), Die sieben Welträtsel; in Du Bois-Reymond, E. (1912), II Band, pp. 65-98.

Du Bois-Reymond, E. (1912), *Reden*, 2 Bänder, Leipzig: Veit & Comp.

Duhamel, J.M.C. (1847), *Cours d'Analyse de l'Ecole Polytechnique*, Paris: Bachelier.

Duhamel, J.M.C. (1853), *Cours de Mécanique*, Paris: Mallet-Bachelier.

Duhem, P. (1886), *Le potentiel thermodynamique et ses applications à la mécanique chimique et à la théorie des phénomènes électriques*, Paris: Hermann.

Duhem, P. (1891), Sur les équations générales de la Thermodynamique, *Annales Scientifiques de l'Ecole Normale Supérieure*, 3e série, tome 8, 231-266.

Duhem, P. (1892a), Commentaire aux principes de la Thermodynamique – Première partie, *Journal de Mathématiques pures et appliquées*, 4e série, tome 8, 269-330.

Duhem, P. (1892b), "Quelques réflexions au sujet des théories physiques", *Revue des questions scientifiques*, 31, 139-177.

Duhem, P. (1892c), Notation atomique et hypothèses atomistiques, *Revue des questions scientifiques*, 31, 391-454.

Duhem, P. (1893a), Commentaire aux principes de la Thermodynamique – Deuxième partie, *Journal de Mathématiques pures et appliquées*, 4e série, tome 9, 293-473.

Duhem, P. (1893b), *Introduction à la mécanique chimique et à la théorie des phénomènes électriques*, Gand: Hoste.

Duhem, P. (1893c), Une nouvelle théorie du monde inorganique, *Revue des questions scientifiques*, 33, 90-133.

Duhem, P. (1893d), Physique et métaphysique, *Revue des questions scientifiques*, 34, 55-83.

Duhem, P. (1893e), L'école anglaise et les théories physiques, à propos d'un livre de W. Thomson, *Revue des questions scientifiques*, 34, pp. 345-78; in Duhem, P. (1987), pp. 113-46.

Duhem, P. (1894a), Commentaire aux principes de la Thermodynamique – Troisième partie, *Journal de Mathématiques pures et appliquées*, 4e série, tome 10, 207-285.

Duhem, P. (1894b), Sur les déformations permanentes et l'hystérésis, *Mémoires couronnées et Mémoires des savants étrangères, publiés par l'Académie royale des sciences, des lettres et des beaux-arts de Belgique*, tome LIV; in Duhem, P. (1896c), pp. 3-61.

Duhem, P. (1894c), Quelques réflexions au sujet de la physique expérimentale, *Revue des questions scientifiques*, 36, 147-197.

Duhem, P. (1894d), Les théories de l'optique, *Revue des deux mondes*, 123, 94-125.

Duhem, P. (1896a), *Théorie thermodynamique de la viscosité, du frottement et des faux équilibres chimiques*, Paris: Hermann.

Duhem, P. (1896b), L'évolution des théories physiques du XVIIe siècle jusqu'à nos jours, *Revue des questions scientifiques*, 40, 463-499.

Duhem, P. (1896c), *Sur les déformations permanentes et l'hystérésis*, Bruxelles: Hayez.

Duhem, P. (1905), Paul Tannery, *Revue de philosophie*, 6, 216-230.

Duhem, P. (1906), *La théorie physique. Son objet et sa structure*, Paris: Chevalier & Rivière.

Duhem, P. (1906a), *Les origines de la statique*. Tome II, Paris: Hermann.

Duhem, P. (1906b), *Études sur Léonard de Vinci: ceux qu'il a lu et ceux qui l'ont lu*. Première Série, Paris: Hermann.

Duhem, P. (1908), *Essai sur la notion de théorie physique de Platon à Galilée*, Paris: Hermann.

Duhem, P. (1909), *Études sur Léonard de Vinci: ceux qu'il a lu et ceux qui l'ont lu*. Seconde Série, Paris: Hermann.

Duhem, P. (1911), *Traité d'énergétique ou de thermodynamique générale*, 2 vols., Paris: Gauthier-Villars.

Duhem, P. (1913a), *Études sur Léonard de Vinci: ceux qu'il a lu et ceux qui l'ont lu*. Troisième Série. Les précurseurs parisiens de Galilée, Paris: Hermann.

Duhem, P. (1913b), *Le système du monde. Histoire des doctrines cosmologiques de Platon à Copernic*. Tome I. La cosmologie hellénique, Bordeaux.

Duhem, P. (1978), *Ziel und Struktur der physikalischen Theorien*, Hamburg: Felix Meiner.

Duhem, P. (1987), *Prémices philosophiques*, (présenté avec une introduction en anglais par S.L. Jaki), Leiden, New York, København, Köln: Brill.

Duhem, P. (2002), *Mixture and Chemical Combination – And Related Essays* (Boston Studies in the Philosophy of Science), Edited and translated with an Introduction, by Paul Needham, Dordrecht, Boston, London: Kluwer.

Dupré, J. (1996), The Solution to the Problem of the Freedom of the Will, *Philosophical Perspectives*, 10, 385-402.

Earman, J. (1986), *A Primer on Determinism*, Dordrecht, Boston, Lancaster, Tokio: Reidel.

Earman, J. (2008), How Determinism Can Fail in Classical Physics and How Quantum Physics Can (Sometimes) Provide a Cure, *Philosophy of Science*, 75 (5), 817-829.

Eastwood, D.M. (1936), *The Revival of Pascal – A study of his relation to French modern thought*, Oxford: Clarendon Press.

Egger, V. (1877), La physiologie cérébrale et la psychologie, *Revue des deux mondes*, XXIV (1er novembre 1877), 193-211.

Elkana, Y. (1974), *The Discovery of the Conservation of Energy*, London: Hutchinson Educational.

Evans, G.C. (1929), Cournot on Mathematical Economics, *Bulletin of the American Mathematical Society*, 35, 269-71.

Fano, V. (2013), *I paradossi di Zenone*, Roma: Carocci.

Faure, F. (1905), Les idées de Cournot sur la statistique, *Revue de Métaphysique et de Morale*, 13 (3), (Mai 1905), 395-411.

Fouillé, A. (1872), *La liberté et le déterminisme*, Paris: Librairie Philosophique de Ladrange.

Fouillée, A. (1882), Les nouveaux expédients en faveur du libre arbitre. I. Expédients logiques et mécaniques, *Revue Philosophique de la France et de l'Étranger*, 14, 585-617.

Fouillée, A. (1883), Notes et discussions. Le libre arbitre et le temps, *Revue Philosophique de la France et de l'Étranger*, 15, 86-88.

Fraassen (van), B. (1980), *The Scientific Image*, Oxford: Clarendon Press.

Fraassen (van), B. (1989), *Law and Symmetry*, Oxford and New York: Oxford University Press.

Fraassen (van), B. (2008), *Scientific Representation*, Oxford: Clarendon Press.

Freudenthal, G. (ed.) 1990, *Études sur / Studies on Hélène Metzger*, Leiden: Brill.

Friedel, C. (1884), La théorie atomique; in Stallo, J.B. (1884), pp. VII-XII.

Friedman, J. W. (2000), The Legacy of Augustin Cournot, *Cahiers D'économie Politique – Papers in Political Economy*, 37, 31-46.

Galison, P. (2003), *Einstein's Clocks, Poincaré's Maps – Empires of Time*, New York and London: Norton & Company.

Garson, J.W. (1995), Chaos and free will, *Philosophical Psychology*, 8 (4), 365-374.

Giannetto, E. (1995), Physical Theories and Theoretical Physics; in Rossi, A. (ed.) (1995), *Atti del XIII Congresso Nazionale di Storia della Fisica*, Lecce: Conte, pp. 163-167.

Gibbon, E. (1854), *The History of the Decline and Fall of the Roman Empire*, vol. III, Boston: Phillips, Sampson, and Company.

Gilain, C. (1994), Ordinary differential equations; in Grattan-Guinness, I. (1994), vol. 1, pp. 440-451.

Gillispie, C.C. (ed.), (1970-80), *Dictionary of Scientific Biography*, New York: Charles Schreibner's Sons.

Giusti Doran, B. (1975), Origins and Consolidation of Field Theory in Nineteenth-Century Britain: From the Mechanical to Electromagnetic View of Nature, *Historical Studies in the Physical Sciences*, VI, 133-260.

Grattan-Guinness, I. (1990), *Convolutions in French Mathematics, 1800-1840*, 3 vols., Basel, Berlin, Boston: Birkhäuser.

Grattan-Guinness, I. (ed.) (1994), *Companion Encyclopedia of the History and Philosophy of the Mathematical Sciences*, 2 vols., London and New York: Routledge.

Greffe, J.L., Heinzmann, G. and Lorenz, K. (eds.) (1996), *Science et philosophie. Science and Philosophy. Wissenschaft und Philosophie*, Paris: Blanchard.

Grünbaum, A. (1960), The Duhemian Argument, *Philosophy of Science*, 27 (1), 75-87; in Harding, S. (ed.) (1976), pp. 116-31.

Guckenheimer, J. (1984), The Lorenz Equations: Bifurcations, Chaos, and Strange Attractors, *The American Mathematical Monthly*, 91 (5), 325-326.

Guinet, L. (1923), Conciliation du véritable déterminisme mécanique avec l'existence de la vie et de la liberté morale – Review, *Isis*, 5 (2), 483-484.

Gusdorf, G. (1966), *Les sciences humaines et la pensée occidentale I – De l'histoire des sciences à l'histoire de la pensée*, Paris: Payot.

Gusdorf, G. (1969), *Les sciences humaines et la pensée occidentale III, La révolution galiléenne*, Paris: Payot.

Guzzardi, L. (2001), Trasformazioni e inneschi. Occasionalismo nel XIX secolo, *Rivista di estetica*, 18 (3), 142-155.

Hacking, I. (1983a), Nineteenth Century Cracks in the Concept of Determinism, *Journal for the History of Ideas*, 44 (3), 455-475.

Hacking, I. (1983b), *Representing and Intervening*, Cambridge and New York: Cambridge University Press.

Hacking, I. (1990), *The Taming of Chance*, Cambridge, UK: Cambridge University Press.

Hacking, I. (2002), *Historical Ontology*, Cambridge, Massachusetts: Harvard University Press.

Hankins, T.L. (1985), *Science and the Enlightenment*, Cambridge, New York, Melbourne: Cambridge University Press.

Hanson, R.N. (1958), *Patterns of Discovery*, London and New York: Cambridge University Press.

Harding, S.G. (ed.) (1976), *Can Theories be Refuted?*, Dordrecht Holland, Boston U.S.A.: Reidel.

Harman, P.M. (ed.) (1990), *The scientific letters and papers of James Clerk Maxwell*, volume I, Cambridge: Cambridge University Press.

Harman, P.M. (ed.) (1995), *The scientific letters and papers of James Clerk Maxwell*, volume II, Cambridge: Cambridge University Press.

Harman, P.M. (1998), *The Natural Philosophy of James Clerk Maxwell*, Cambridge: Cambridge University Press.

Harman, P.M. (ed.) (2002), *The scientific letters and papers of James Clerk Maxwell*, volume III, Cambridge: Cambridge University Press.

Heaviside, O. (1893), *Electromagnetic theory*, London: The Electrician Printing and Publishing Company.

Helm, G. (1898), *Die Energetik nach ihrer geschichtlichen Entwicklung*, Leipzig: Veit & Comp; in Helm, G. (1981), *Die Energetik....*, New York: Arno Press.

Helm, G. (2000), The Historical Development of Energetics (Translated, and with an Introductory Essay by R.J. Deltete), Kluwer Academic Publishers. Dordrecht/ Boston/ London.

Helmholtz, H. (1847), *Über die Erhaltung der Kraft*; in Helmholtz, H. (1889), *Über die Erhaltung der Kraft*, Ostwald's Klassiker der exakten Wissenschaften, Leipzig: Engelmann.

Hertz, H. (1892), *Untersuchungen über die Ausbreitung der elektrischen Kraft*, Leipzig: Barth; English translation Hertz, H. (1962), *Electric Waves*, New York: Dover.

Hilbert, M. (2000), *Pierre Duhem and neo-Thomist Interpretations of Physical Sciences*, PhD thesis, Institute for the History and Philosophy of Science and Technology, Toronto: University of Toronto.

Hobbs, J. (1991), Chaos and Indeterminism, *Canadian Journal of Philosophy*, 21 (2), 141-164.

Hoffmann, D., Bevilacqua F. and Stuewer R.H. (eds.) (1996), *The Emergence of Modern Physics*, Proceedings of a Conference held at Berlin in 1995, Pavia: Università degli Studi di Pavia.

Holmyard, E.J. (1922a), A Critical Examination of Berthelot's Work upon Arabic Chemistry, *Isis*, 6 (4), 479-499.

Holmyard, E.J. (1922b), Arabic Chemistry, *Nature*, CX, 2765 (Oct 28, 1922), 573-574.

Howard, D. (1990), Einstein and Duhem, *Synthese*, 83, 363-384.

Humbert, P. (1932), *Pierre Duhem*, Paris: Bloud et Gay.

Israel, G. (1987), L'interminabile crisi del meccanicismo, *Rivista di Storia della Scienza*, IV (1), 73-99.

Israel, G. (1991), Il determinismo e la teoria delle equazioni differenziali ordinarie, *Physis*, XXVIII, 305-358.

Ivanova, M. (2015), Conventionalism about what? Where Pierre Duhem and Poincaré part ways, *Studies in History and Philosophy of Science*, 54, 80-89.

Jaki, S.L. (1984), *Uneasy Genius: the Life and Work of Pierre Duhem*, The Hague: Martinus Nijhoff Publishers.

James, W. (1884), The Dilemma of Determinism; in James, W. (1897), pp. 145-83.

James, W. (1897), *The Will to Believe and other Essays*, New York, London, and Bombay: Longmans Green and Co.

James & James (1992), *Mathematics Dictionary* (5th ed.), New York: Chapman & Hall.

Janet, P. (1878), Rapport de M. Paul Janet à l'Académie des Sciences Morales et politiques sur un mémoire de M. Boussinesq; in Boussinesq, J. (1878a), pp. 3-23.

Jenkins, E.W. (2014), E. J. Holmyard and the Historical Approach to Science Teaching; in Matthews M.R. (ed.) (2014), pp. 2383-408.

Kant, I. (1787), *Kritik der reinen Vernunft*; in Kant, I. (1853), *Kritik der reinen Vernunft*, Leipzig: Leopold Voss.

Kemeny, J.G. and Oppenheim P. (1956), On Reduction, *Philosophical Studies*, VII (1-2), 6-19.

Kingsley, P. (1994), From Pythagoras to the Turba philosophorum: Egypt and Pythagorean Tradition, *Journal of the Warburg and Courtauld Institutes*, 57, 1-13.

Knight, D. (2008), Series Editor's Introduction; in Chimisso, C. (2008), p. vii.

Kojeve, A. 1990 [1932], *L'idée du déterminisme dans la physique classique et dans la physique moderne*, Paris: Librairie Générale Française.

Korolev, A. (2008), Asymptotic Reasoning, and Time Irreversibility in Classical Physics, *Philosophy of Science*, 75 (5), 943-956.

Koyré, A. (1957), Preface; in Koyré, A. (2008), pp. 6-7.

Koyré, A. (2008), *From the Closed World to the Infinite Universe*, Radford, USA: Wilder Publications.

Kragh, H. (1996), The New Rays and the Failed Anti-Materialistic Revolution; in Hoffmann, D., Bevilacqua, F. and Stuewer, R.H. (eds.) (1996), pp 61-77.

Kragh, H. (2002), The Vortex Atom: A Victorian Theory of Everything, *Centaurus* 44 (1-2), 32-114.

Kragh, H. (2008a), *Entropic Creation – Religious context of Thermodynamics and Cosmology*, Farnham, England: Ashgate.

Kragh, H. (2008b), Pierre Duhem, Entropy, and Christian Faith, *Physics in Perspective*, 10, 379-395.

Kuhn, T.S. (1968), The History of Science, *International Encyclopedia of the Social Sciences*, 14, Crowell Collier and Macmillan, pp. 74-83; in Kuhn, T.S. (1977), pp. 105-126.

Kuhn, T.S. (1976), Mathematical vs. Experimental Traditions in the Development of Physical Science, *Journal of Interdisciplinary History*, VII (I), 1-31.

Kuhn, T.S. (1977), *The Essential Tension*, Chicago and London: The University of Chicago Press.

Kuhn, T.S. (1996), *The Structure of Scientific Revolutions*, 3rd edition, Chicago and London: The University of Chicago Press.

Lacki, J. (2007), Les Principes de la Mécanique de Heinrich Hertz: une prélude à l'axiomatique; in Pont, J.C., Freland, L., Padovani, F. and Slavinskaia, L. (eds.) (2007), pp. 241-62.

Lami, E.O. (ed.) 1891, *Dictionnaire encyclopédique et biographique de l'Industrie et des Arts industriels, Supplément 1*, Paris: Librairie des Dictionnaires.

Laplace, P.S. (1772), Mémoire sur les solutions particulières des équations différentielles et sur les inégalités séculaires des planètes"; in Laplace, P.S. (1891), 8, pp. 325-66.

Laplace, P.S. (1799), *Traité de mécanique céleste*, tome 1, Paris: Crapelet-Duprat.

Laplace, P.S. (1814), *Essai philosophique sur les probabilités*, 2nd edition, Paris: Courcier.

Laplace, P.S. (1825), *Essai philosophique sur les probabilités*, 5th edition, Paris: Bacelier.

Laplace, P.S. (1891), *Oeuvres Complètes*, 8, Paris: Gauthier-Villars.

Laudan, L. (1965), Grünbaum on the 'Duhemian Argument', *Philosophy of Science*, 32, (3-4), 295-299.

Lechalas, G. (1893a), Quelques réflexions soumises à M. Vicaire, *Annales de philosophie chrétienne*, XXVIII, 278-282.

Lechalas, G. (1893b), M. Duhem est-il positiviste?, *Annales de philosophie chrétienne*, 127, 312-314.

Lesch, J.E. (1984), *Science and medicine in France: the emergence of experimental physiology, 1790-1855*, Cambridge – Massachusetts, and London: Harvard University Press.

Locher, F. (2007), L'enseignement de l'histoire des sciences en France sous la Troisième République, *Histoire de l'éducation*, 114, 217-219.

Lodge, O. (1883), The ether and its functions, *Nature*, 27, 304-306 and 328-330.

Lodge, O. (1885), On the Identity of Energy: in Connection with Mr. Poynting's Paper on the Transfer of Energy in an Electromagnetic Field; and on the two Fundamental Forms of Energy, *Philosophical Magazine* 19, 482-487.

Lowinger, A. (1941), *The Methodology of Pierre Duhem;* reprinted in Lowinger A. 1961, *The Methodology of Pierre Duhem*, New York: AMS Press.

Lugg, A. (1990), Pierre Duhem's conception of natural classification, *Synthese*, 83, 409-420.

Luys, J. (1876), *Le cerveau et ses fonctions*, Paris: Baillière.

Mach E. (1883), *Die Mechanik in ihrer Entwickelung Historisch-kritisch dargestellt*, Leipzig: Brockhaus.

Mach, E. (1908), Vorwort zur deutschen Ausgabe; in Duhem, P. (1978), pp. III-V.

Magendie, F. (1836), Première leçon; in Magendie, F. (1837), pp. 1-17.

Magendie, F. (1837), *Leçons sur les phénomènes physiques de la vie*, tome II, Paris: Angé.

Maiocchi, R. (1985), *Chimica e filosofia – Scienze, epistemologia, storia e religione nell'opera di Pierre Duhem*, Firenze: La Nuova Italia.

Maiocchi, R. (1990), Pierre Duhem's The aim and structure of physical theory: a book against conventionalism, *Synthese*, 83, 385-400.

Malament, D.B. (2008), Norton's Slippery Slope, *Philosophy of Science*, 75 (5), 799-816.

Martin, R.N.D. (1991), *Pierre Duhem – Philosophy and History in the Work of a Believing Physicist*, La Salle, Illinois: Open Court.

Martin, T. and Touffut, J.P. (2007), Introduction; in Touffut J.P. (ed.) (2007), pp. 1-7.

Matthews, M.R. (ed.) (2014), *International Handbook of Research in History, Philosophy and Science Teaching*, Netherlands: Springer.

Max-Planck-Institut für Wissenschaftsgeschichte, (2012), *Epistemology and History. From Bachelard and Canguilhem to Today's History of Science*, Conference, Preprint 434, Berlin: MPIWG.

Maxwell, J.C. (1862), From a letter to Lewis Campbell; in Harman, P.M. (ed.) (1990), pp. 711-712.

Maxwell, J.C. (1873), Does the progress of physical science tend to give any advantage to the opinion of necessity (or determinism) over that of the contingency of events and the freedom of the will?; in Harman, P.M. (ed.) (1995), pp. 814-823.

Maxwell, J.C. (1878), Paradoxical Philosophy, *Nature*, XIX, (December 19, 1878), 141-143; in Niven, W.D. (ed.) 1890, pp. 756-62.

Maxwell, J.C. (1879), Letter to Francis Galton; in Harman, P.M. (ed.) (2002), pp. 756-758.

Maxwell, J.C. (1890), *The Scientific Papers of James Clerk Maxwell*, (edited by W.D. Niven) 2 vols, Cambridge: University Press.

Mayer, J.R. (1842), Bemerkungen über die Kräfte der unbelebten Natur, *Annalen der Chemie und Pharmacie*, XLII, 233-240; in Mayer, J.R. (1867), pp. 1-12.

Mayer, J.R. (1844), Mayer an Griesinger 20. Juli 1844; in Preyer, W.T. (1889), pp. 96-103.

Mayer, J.R. (1867), *Die Mechanik der Wärme*, Stuttgart: Cotta.

Mayer, J.R. (1876), Ueber Auslösung, *Staatsanzeiger für Württemberg*, Besondere Beilage (22 März 1876); in Mittasch, A. (ed.) (1953a), pp. 9-18.

Mazliak, L. (2013), Poincaré and Probability, *Lettera Matematica International Edition*, 1 (1), 33-39.

McCormmach, R. and Jungnickel, C. (1986), *Intellectual Mastery of Nature*, 2 vols., Chicago and London: The University of Chicago Press.

Mcmullin, E. (1990), Comment: Duhem's middle way, *Synthese*, 83, 421-430.

Mendelsohn, E. (1965), Physical Models and Physiological Concepts: Explanation in Nineteenth-Century Biology, *The British Journal for the History of Science*, 2 (3), 201-219.

Mentré, F. (1905), Les raciness historiques du probabilisme rationnel de Cournot, *Revue de Métaphysique et de Morale*, 13 (3), (Mai 1905), 485-508.

Mentré, F. (1922), Pierre Duhem, le théoricien, *Revue de philosophie*, 29, 444-473 and 608-627.

Mentré, F. 1908, *Cournot et la Renaissance du Probabilisme*, Paris: Rivière.

Merz, J.T. (1912), *A History of European Thought in the Nineteenth Century*, vol. II, Edinburgh and London: Blackwood and Sons.

Metzger, H. (1929), La philosophie d'Émile Meyerson et l'histoire des science, *Archeion*, 11, XXXII-XLII; in Metzger, H. (1987), pp. 95-106.

Metzger, H. (1930), La philosophie de Lucien Lévy-Bruhl et l'histoire des science, *Archeion*, 12, pp. 15-24; in Metzger, H. (1987), pp. 113-24.

Metzger, H. (1933), L'historien des sciences doit-il se faire le contemporain des savants dont il parle ?, *Archeion*, 15, 34-44; in Metzger, H. (1987), pp. 9-21.

Metzger, H. (1934), Compte rendu de "Federigo Enriques, Signification de l'histoire de la pensée scientifique", *Archeion*, 16, 390-91; in Metzger, H. (1987), pp. 146-50.

Metzger, H. (1935), Tribunal de l'histoire et théorie de la connaissance scientifique, *Archeion*, 17, pp. 1-14; in Metzger, H. (1987), pp. 23-39.

Metzger, H. (1936), L'*a priori* dans la doctrine scientifique et l'histoire des sciences, *Archeion*, 18, 29-42; in Metzger, H. (1987), pp. 41-56.

Metzger, H. (1937a), La méthode philosophique dans l'histoire des sciences, *Archeion*, 19, 204-216; in Metzger, H. (1987), pp. 57-73.

Metzger, H. (1937b), Pierre Duhem, la théorie physique et l'histoire des science, *Archeion*, 19, 135-139; in Metzger, H. (1987), pp. 151-156.

Metzger, H. (1939), Le rôle des précurseurs dans l'évolution de la science, *Thales 1937-9*, 4, 199-209; in Metzger, H. (1987), pp. 75-91.

Metzger, H. (1987), *La méthode philosophique en histoire des sciences. Textes 1914-1939 (réunis par Gad Freudenthal)*, Paris: Fayard.

Milhaud, G. (1893), *Leçons sur les origines de la science grecque*, Paris: Alcan.

Milhaud, G. (1894), *Essai sur les conditions et les limites de la certitude logique*, Paris: Alcan.

Milhaud, G. (1906), Paul Tannery, *La Revue des Idées*, (15 janvier 1906), 1-14.

Milhaud, G. (1911a), Cournot et le pragmatisme scientifique contemporain, *Scientia*, 10, 370-380.

Milhaud, G. (1911b), Le développement de la pensée de Cournot, *La Revue du Mois*, (Oct. 1911); in Milhaud, G. (1927), pp. 7-34.

Milhaud, G. (1911c), La définition de hasard de Cournot, *Revue philosophique*, (Août 1911); in Milhaud, G. (1927), pp. 37-65.

Milhaud, G. (1911d), La Science & la Religion chez Cournot, (Extrait du *Bulletin de la Société française de philosophie*, Séance du 4 Mai 1911); in Milhaud, G. (1927), pp. 109-136.

Milhaud, G. (1927), *Etudes sur Cournot*, Paris: Vrin.

Miller, D.G. (1967), Pierre Duhem, un oublié, *Revue des Questions Scientifiques*, 138 (Octobre 1967), 445-470; French translation of Miller, D.G. (1966), Ignored Intellect: Pierre Duhem, *Physics Today*, 19, 47-53.

Mittasch, A. (1940), *Julius Robert Mayers Kausalbegriff*, Berlin: Julius Springer.

Mittasch, A. (1942a), J. R. Mayers Begriff der Auslösung in seiner Bedeutung für die Chemie; in Pietsch, E. and Schimank H. (eds.) (1942), pp. 281-294.

Mittasch, A. (1942b), Robert Mayer Anschauengen über das Leib-Seele-Verhältnis; in Pietsch, E. and Schimank, H. (eds.) (1942), pp. 329-354.

Mittasch, A. (ed.) (1953a), *Julius Robert Mayer über Auslösung von Wilhelm Ostwald*, Weinheim: Verlag Chemie.

Mittasch, A. (1953b), Vorwort des Herausgebers; in Mittasch, A. (ed.) (1953a), pp. 7-8.

Mittasch, A. (1953c), Nachwort; in Mittasch, A. (ed.) (1953a), pp. 48-49.

Moore, H.L. (1905a), Antoine-Augustin Cournot, *Revue de Métaphysique et de Morale*, 13 (3), (Mai 1905), pp. 521-43.

Moore, H.L. (1905b), The Personality of Antoine Augustin Cournot, *The Quarterly Journal of Economics*, 19 (3), 370-99.

Morus, I.R. (2005), *When Physics Became King*, Chicago and London: The University of Chicago Press.

Mucchielli, L. (2006), Criminology, Hygienism, and Eugencs in France, 1870-1914; in Becker, P. and Wetzell, R.F. (eds.) (2006), pp. 207-229.

Mueller, T. (2015), The Boussinesq Debate: Reversibility, Instability, and Free Will, *Science in Context*, 28 (4), 613-635.

Naville, E. (1879), La physique et la morale, *Revue Philosophique de la France et de l'Étranger*, 7, 265-286.

Naville, E. (1883), *La physique moderne. Études historiques et philosophiques*, Paris: Baillière.

Naville, E. (1890a), *La physique moderne. Études historiques et philosophiques*, 2nd edition, Paris: Félix Alcan.

Naville, E. (1890b), *Le libre arbitre. Études philosophiques*, Paris: Fischbacher.

Needham, P. (1991), Duhem and Cartwright on the truth of laws, *Synthese*, 89 (1), 89-109.

Needham, P. (2000), Duhem and Quine, *Dialectica*, 54 (2), 109-132.

Needham, P. (2002), Introduction; in Duhem, P. (2002), pp. ix-xxx.

Needham, P. (2011), Duhem's moderate realism; in Brenner, A., Needham, P., Stump, D.J., and Deltete, R. (2011), pp. 7-12.

Neurath, O. (1916), Zur Klassification von Hypothesensystemen, *Jahrb. D. phil. Gesell. A. d. Univ. Wien. GpmS*, pp. 85-101; in Neurath, O. (1983), pp. 13-31.

Neurath, O. (1983), *Philosophical Papers 1913-1946*, (Cohen, R.S. and Neurath, M. eds.), Dordrecht, Boston, and Lancaster: Reidel.

Nichol, A.J. (1938), Tragedies in the life of Cournot, *Econometrica*, 6 (3), 193-197.

Niven, W.D. (ed.) (1890), *The scientific papers of James Clerk Maxwell*, volume II, Cambridge: The University Press.

Norton, J. (2003), Causation as Folk Science, *Philosophers' Imprint*, 3 (4), www.philosophersimprint.org/003004/, pp. 1-22 (accessed December 2015).

Norton, J. (2008), An Unexpectedly Simple Failure of Determinism, *Philosophy of Science*, 75 (5), 786-798.

Nye, M.J. (1976), The Moral Freedom of Man and the Determinism of Nature: The Catholic Synthesis of Science and History in the Revue des questions scientifiques, *The British Journal for the Philosophy of Science*, 9 (3), 274-292.

Nye, M.J. (1979), The Boutroux Circle and Poincaré's Conventionalism, *Journal for the History of Ideas*, 40 (1), 107-120.

Olson, R.G. (2008), *Science and Scientism in Nineteenth-Century Europe*, Urbana and Chicago: University of Illinois Press.

Oppenheim, J. (1985), *The Other World: Spiritualism and Psychical Research in England, 1850-1914*, Cambridge, London and New York: Cambridge University Press.

Ostwald, W. (1896), Zur Energetik, *Annalen der Physik und Chemie*, 58, 154-167.

Ostwald, W. (1914), *Julius Robert Mayer über Auslösung* (manuscript); in Mittasch, A. (ed.) (1953a), pp. 19-47.

Parodi, D. 1905, Le criticisme de Cournot, *Revue de Métaphysique et de Morale*, 13 (3), (Mai 1905), pp. 451-484.

Partington, J.R. (1923), The Identity of Geber, *Nature*, 111 (2781), 219-220.

Pascal, B. (1651), *Fragment d'un traité du vide*; in Pascal, B. (1872), pp. 158-162.

Pascal, B. (1872), *Œuvres complètes de Blaise Pascal*, tome troisième, Paris: Hachette.

Pascal, B. (1897), *Pensées (Texte établi par* Léon Brunschvicg), Hachette, Paris; in Pascal, B. (1976), *Pensées*, Paris: Garnier-Flammarion.

Pascal, B. (1951), *Pensées* (Introduction et notes de Louis Lafuma), 3 vols., Paris: Editions du Luxembourg.

Paul, H.W. (1968), The Debate Over the Bankruptcy of Science in 1895, *French Historical Studies*, 5 (3), 299-327.

Paul, H.W. (1972a), The crucifix and the crucible catholic scientists in the third republic, *The Catholic Historical Review*, 58 (2), 195-219.

Paul, H.W. (1972b), The Issue of Decline in Nineteenth-Century French Science, *French Historical Studies*, 7 (3), 416-450.

Paul, H.W. 1979, *The Edge of Contingency. French Catholic Reaction to Scientific Change from Darwin to Duhem*, Gainesville: University Presses of Florida.

Pearson, E.S. (ed.) (1925), *The History of Statistics in the 17th and 18th centuries.... Lectures by Karl Pearson given at University College London during the academic session 1921-1933* (manuscript); in Pearson, E.S. (1978), *The History of Statistics in the 17th and 18th centuries*, New York: Macmillan.

Picard, É. (1922), La vie et l'œuvre de Pierre Duhem", *Mémoires de l'Académie des sciences de l'Institut de France*, 57, CIII-CXLII.

Pietsch, E. and Schimank, H. (eds.) (1942), *Robert Mayer und das Energieprinzip 1842-1942*, Berlin: Verein Deutscher Ingenieure.

Poincaré, H. (1889), *Leçons sur la Théorie Mathématique de la lumière, Cours de Physique Mathématique*, Paris: Carré.

Poincaré, H. (1890), *Électricité et Optique, Cours de Physique Mathématique*, Paris: Carré.

Poincaré, H. (1891a), Les géométries non Euclidiennes, *Revue générale des sciences pures et appliquées*, 11 (23), 769-774.

Poincaré, H. (1891b), Le problème des trois corps, *Revue générale des sciences pures et appliquées*, II (1), 1-5.

Poincaré, H. (1892), *Thermodynamique, Cours de Physique Mathématique*, Paris: Carré.

Poincaré, H. (1893), Le mécanisme et l'expérience, *Revue de Métaphysique et de Morale*, 1 (6), 534-537.

Poisson, S.D. (1833), *Traité de Mécanique*, Paris: Bachelier.

Pont J-C., Freland, L., Padovani F. and Slavinskaia, L. (eds.) (2007), *Pour comprendre le XIXe. Histoire et philosophie des sciences à la fin du siècle*, Firenze: Olschki.

Popper, K.R. (1950a), Indeterminism in Quantum Physics and Classical Physics. Part I, *The British Journal for the Philosophy of Science*, 1 (2), 117-133.

Popper, K.R. (1950b), Indeterminism in Quantum Physics and Classical Physics. Part II, *The British Journal for the Philosophy of Science*, 1 (3), 173-195.

Preyer, W.T. (1889), *Robert von Mayer über die Erhaltung der Energie. Briefe an Wilhelm Griesinger nebst dessen Antwortschreiben aus den Jahren 1842-1845*, Berlin: Gebrüder Paetel.

Purrington, R.D. (1997), *Physics in the Nineteenth Century*, New Brunswick, New Jersey, and London: Rutgers University Press.

Quine, W.V.O. (1951), Main Trends in Recent Philosophy: Two Dogmas of Empiricism, *The Philosophical Review*, 60, 20-43.

Quine, W.V.O. (1963), *From a logical point of view*, New York and Evanston: Harper and Row.

Rankine, W.J.M. (1855), "Outlines of the Science of Energetics"; in Rankine, W.J.M. (1881), *Miscellaneous Scientific Papers*, London: Charles Griffin and Company, pp. 209-228.

Redondi, P. (1978), *Epistemologia e storia della scienza – Le svolte teoriche da Duhem a Bachelard*, Milano: Feltrinelli.

Renan, E. (1890), *L'avenir de la science. Pensées de 1848*, Paris: Calmann Lévy.

Renn, J. (2006), *Auf den Schultern von Riesen und Zwergen*, Weinheim: Wiley-VCH.

Renn, J. (2009), Comments on Friedman and Barker: The Dynamics of Scientific research, in Sturm T. and Feest U. (eds.) (2009), pp. 141-4.

Renouvier, C. (1878), Des notions de matière et de force dans les sciences de la nature III. L'unité des forces physiques, *La critique philosophique*, 37, 161-170.

Renouvier, C. (1883a), Les nouvelles chicanes contre la possibilité du libre arbitre, *La critique philosophique*, 50, 371-383.

Renouvier, C. (1883b), Les objections de M. Fouillée contre la conciliation du libre arbitre avec les lois du mouvement, *La critique philosophique*, 51, 389-400.

Renouvier, C. (1885), *Esquisse d'une classification systématique des doctrines philosophiques*, Paris: Au Bureau de la Critique Philosophique.

Rey, A. (1904), La philosophie scientifique de M. Duhem, *Revue de Métaphysique et de Morale*, 12 (4), 699-744.

Rey, A. (1906), La physique de M. Duhem, *Annales de philosophie chrétienne, CLI, 4e série, I* (5), 535-537.

Rey, A. (1907), *La Théorie de la physique chez les physiciens contemporaines*, Paris: Alcan.

Rheinberger, H.J. (2010), *On Historicizing Epistemology*, California: Stanford University Press.

Rohrlich, F. (1988), Pluralistic Ontology and Theory Reduction in the Physical Sciences, *The British Journal for the Philosophy of Science*, 39, 295-312.

Ross, S. (1962), Scientist: the story of a word, *Annals of Science*, 18, 65-86.

Roux, S. (2016), Lire la théorie physique aujourd'hui; in Duhem, P. (2016), *La théorie physique. Son objet, sa structure*, Lyon: ENS Éditions, pp. 6-31.

Russell, B. (1918), On the Notion of Cause; in *Mysticism and Logic and Other Essays*, London: Longmans, pp. 180-208.

Russo, L. (2013), *La rivoluzione dimenticata*, Milano: Feltrinelli.

Ryding, K.C. (1994), Islamic Alchemy According to Al-Khwarizmi, *Ambix*, 41 (3), 121-134.

Saint-Venant, A.B. (1877), Accord des lois de la mécanique avec la liberté de l'homme dans son action sur la matière, *Comptes Rendus de l'Académie des Sciences*, LXXXIV, 419-423.

Sarton, G. (1938), L'oeuvre de Paul Tannery; in Boutroux, P. and Sarton, G. (1938), pp. 690-691.

Sarton, G. (1948), *The Life of Science*, New York: Schuman.

Schaffner, K.F. (1967), Approaches to Reduction, *Philosophy of Science*, 34 (2), 137-147.

Schimank, H. (1956), Julius Robert Mayer über Auslösung bei Wilhelm Ostwald: Alwin Mittasch, *Sud Hoffs Archiv für Geschichte der Medizin und der Naturwissenschaften*, 40 (2), 189-190.

Schöttler, P. (2012), Szientismus, Zur Geschichte eines schwierigen Begriffs, *NTM Zeitschrift für Geschichte der Wissenschaften, Technik und Medizin*, 20 (4), 245-269.

Shea, W. (2003), *Designing Experiments and Games of Chance*, Sagamore Beach, USA: Science History Publications.

Smith, C. and Wise, M.N. (1989), *Energy and Empire. A biographical study of Lord Kelvin*, Cambridge, New York, and Sidney: Cambridge University Press.

Soury, J. (1878), *Jésus et les Évangiles*, Paris: Charpentier.

Soury, J. (1881), *Théories naturalistes du monde et de la vie dans l'antiquité*, Paris: Charpentier.

Soury, J. (1882), *Philosophie naturelle*, Paris: Charpentier.

Stallo, J.B. (1882), *The concepts and theories of modern physics*, New York: Appleton & C.

Stallo, J.B. (1884), *La matière et la physique moderne* (avec une préface par C. Friedel), Paris: Alcan.

Stegmüller, W. (1979), *The Structuralist View of Theories*, Berlin, Heidelberg, and New York: Springer-Verlag.

Stewart, B. (1873), *The Conservation of Energy being an elementary treatise on energy and its laws*, London: King & Co.

Stoffel, J.F. (1996a), *Pierre Duhem et ses doctorants*, Louvain-la-Neuve: Centre interfacultaire d'étude en histoire des sciences.

Stoffel, J.F. (1996b), De la nécessité de l'histoire des sciences pour les études de philosophie des sciences, *Revue Philosophique de Louvain*, 94 (3), 415-427.

Stoffel, J.F. (2002), *Le phénoménalisme problématique de Pierre Duhem*, Bruxelles: Académie Royale de Belgique.

Stoffel, J.F. (2007), Pierre Duhem: dans le sillage di Blaise Pascal, *Revista Portuguese de Filosofia*, 63, 275-307.

Stoffel, J.F. (2008), Pierre Duhem avait-il « quelque théologien derrière lui » lors de l'élaboration de son articulation de la physique et de la métaphysique ? Le cas di Maurice Blondel, *Recherches philosophiques*, 89-116.

Strien (van), M. (2014b), On the origins and foundations of Laplacian determinism, *Journal for General Philosophy of Science*, 45, 167-185.

Strien (van), M. (2014c), Vital Instability: Life and Free Will in Physics and Physiology, 1860-1880, *Annals of Science*, 1-20.

Strien (van), M. (2014a), The Norton Dome and the Nineteenth Century Foundations of Determinism, *Studies in History and Philosophy of Science*, 45, 24-31.

Stump, D. (1989), Henri Poincaré's philosophy of science, *Studies in History and Philosophy of Science*, Part A, 20 (3), 335-363.

Sturm, T. and Feest, U. (2009), What (Good) is Historical Epistemology? Introductory Remarks; in Sturm, T. and Feest, U. (eds.) (2009), pp. 3-4.

Sturm, T. and Feest, U. (eds.) 2009, *What (Good) is Historical Epistemology?*, Preprint 386, Berlin: Max-Planck-Institut für Wissenschaftsgeschichte.

Suppe, F. (1977), *The Structure of Scientific Theories*, Urbana and Chicago: University of Illinois Press:.

Suppe, F. (2000), Understanding Scientific Theories: An Assessment of Developments, 1969-1998, *Philosophy of Science*, 67, (Proceedings), S102-S115.

Tannery, P. (1883), Notes et discussions. Le libre arbitre et le temps, *Revue Philosophique de la France et de l'Étranger*, 15, 83-85.

Tannery, P. (1887a), *Pour l'histoire de la science hellène*, Paris: Alcan.

Tannery, P. (1887b), *La Géométrie Greque*, Paris: Gauthier-Villars.

Tannery, P. (1893), *Recherches sur l'histoire de l'astronomie ancienne*, Paris: Gauthier-Villars.

Tannery, P. (1905), Auguste Comte et l'histoire des sciences, *Revue générale des sciences pures et appliquées*, XVI (9), 410-417.

Thomson, J.J. (1936), *Recollections and Reflections*, London: G. Bell and Sons.

Thomson, W. (1860), Royal Institution Friday Evening Lacture; in Thomson, W. (1872), *Reprint of papers on electrostatics and magnetism*, London: MacMillan, pp. 208-224

Touffut, J-P. (ed.) (2007), *Augustin Cournot: Modelling Economics*, The Cournot Centre series, Cheltenham: Edward Elgar Publishing.

Ugaglia, M. (2004), *Modelli idrostatici del moto da Aristotele a Galileo*, Roma: Lateran University Press.

Vacherot, E. (1858), *La métaphysique et la science ou principes de métaphysique positive*, tome 1, Paris: Chamerot.

Vicaire, E. (1893), De la valeur objective des hypothèses physiques, *Revue des questions scientifiques*, 33, 451-510.

Virtanen, R. (1960), *Claude Bernard and his place in the history of ideas*. Lincoln: University of Nebraska Press.

Westman, R.S. (1990), The Duhemian historiographical project, *Synthese*, 83, 261-272.

Whewell W. (1847a), *The Philosophy of Inductive Sciences founded upon their History*, volume the first, London: Parker.

Whewell, W. (1847b), *The Philosophy of Inductive Sciences founded upon their History*, volume the second, London: Parker.

Wilson, M. (2009), Determinism and the Mystery of the Missing Physics, *The British Journal for the Philosophy of Science*, 60, 173-193.

Wolloch, N. (2007), "Facts, or Conjectures": Antoine-Yves Goguet's Historiography, *Journal of the History of Ideas*, 68 (3), 429-449.

Index to Names

KEY
Philoponus, J. 27 means that the scholar is mentioned in the main text only (page 27)
Philoponus, J. 27f11 means that the scholar is mentioned only in the footnote 11 of page 27
Philoponus, J. 27(+f11) means that the scholar is mentioned in the main text and in the footnote 11
Philoponus, J. 27-30 means that the scholar is frequently mentioned in the page range
Philoponus, J. ***27-30*** (bold + italics) means that the scholar is the main subject in the page range

www.ingramcontent.com/pod-product-compliance
Lightning Source LLC
LaVergne TN
LVHW012052011026
849162LV00018B/184
9789004315228